The Mammals
OF NORTH DAKOTA

The Mammals
OF NORTH DAKOTA
SECOND EDITION

by Robert W. Seabloom

With contributions by
John W. Hoganson and William F. Jensen

North Dakota State University Press
Fargo, ND

North Dakota State University Press
NDSU Dept. 2360, P.O. Box 6050
Fargo, ND 58108-6050
www.ndsupress.org

THE MAMMALS OF NORTH DAKOTA, Second Edition
By Robert W. Seabloom
With contributions by John W. Hoganson and William F. Jensen

First Printing

The publication of *The Mammals of North Dakota*, Second Edition, is made possible by the generous support of donors to the NDSU Press Fund and the NDSU Press Endowed Fund, and other contributors to NDSU Press.

David Bertolini, Director
Suzzanne Kelley, Publisher
Zachary Vietz, Graduate Assistant in Publishing

Deb Tanner, Cover and Interior Design

International Standard Book Number: 9781946163158
Library of Congress Control Number: 2020931221

Printed in the United States of America

Publisher's Cataloging-In-Publication Data
(Prepared by The Donohue Group, Inc.)

Names: Seabloom, Robert W., 1932- author. | Hoganson, John W., contributor. | Jensen, William F. (William Fredrick), 1956- contributor.
Title: The mammals of North Dakota / by Robert W. Seabloom ; with contributions by John W. Hoganson and William F. Jensen.
Description: Second edition. | Fargo, ND : North Dakota State University Press, [2020] | Includes bibliographical references and index.
Identifiers: ISBN 9781946163158
Subjects: LCSH: Mammals--North Dakota. | Mammals--North Dakota--Identification.
Classification: LCC QL719.N9 S43 2020 | DDC 599.09784--dc23

Contents

Preface to the Second Edition
Acknowledgments

Introduction . . . 1

The Class Mammalia . . . 3

Early Mammalogy in North Dakota . . . 7

Mammalian Paleofaunas of North Dakota *John W. Hoganson* . . . 10
- Mammals during the Age of Reptiles . . . 11
- Mammals of the North Dakota Everglades . . . 11
- Mammals of the North Dakota Scrubland and Savanna . . . 16
- Mammals of the Great Ice Age . . . 19
- North Dakota's Existing Mammalian Fauna . . . 22

The Landscape and Flora of North Dakota: "An Acquired Taste" *William F. Jensen* . . . 23
- Topography . . . 24
- Climate . . . 26
- Landscape and Habitat Components . . . 27
  - Prairie . . . 28
  - Riparian and Upland Forests . . . 32
  - Wetlands . . . 37
  - Badlands . . . 40

Mammalian Biogeography of North Dakota . . . 42

Checklist of the Mammals of North Dakota . . . 46

Order Didelphimorphia—Opossums . . . 52
- Family Didelphidae–New World Opossums . . . 53

Order Eulipotyphla (Soricomorpha)—Shrews and Moles . . . 57
- Family Soricidae–Shrews . . . 58

Order Chiroptera—Bats . . . 73
- Family Vespertilionidae–Common Bats . . . 75

Order Carnivora—Carnivores 107
Family Felidae–Cats 108
Family Canidae–Dogs 124
Family Ursidae–Bears 146
Family Mustelidae–Weasels and Allies 151
Family Mephitidae–Skunks 193
Family Procyonidae–Raccoons 202
Order Rodentia—Rodents 207
Family Sciuridae–Squirrels 208
Family Castoridae–Beavers 256
Family Geomyidae–Pocket Gophers 263
Family Heteromyidae–Pocket Mice and Kangaroo Rats 271
Family Dipodidae–Jumping Mice 282
Family Cricetidae–New World Rats and Mice 288
Family Muridae–Old World Rats and Mice 330
Family Erethizontidae–New World Porcupines 338
Order Lagomorpha—Rabbits, Hares, and Pikas 343
Family Leporidae–Hares and Rabbits 344
Order Artiodactyla—Even-toed Ungulates 360
Family Cervidae–Deer 362
Family Antilocapridae–Pronghorns 386
Family Bovidae–Bovids 391
Species Potentially Occurring in North Dakota 403
Investigating Mammals 406
Keys to the Mammals of North Dakota 409
Glossary 437
References 443
Index to Common and Scientific Names 465

Preface to the Second Edition

When the first edition of *The Mammals of North Dakota* was published in 2011, I had no idea there might be a second. *The Mammals of North Dakota* was the only comprehensive reference on the state's mammals at the time, bringing up-to-date the eighty-five-year span from Vernon Bailey's (1926) "Biological Survey of North Dakota: I. Physiography and Life Zones, II." I believed that 2011 knowledge of species diversity, classification, nomenclature, and distributions would certainly stand for more than eight years. How wrong I was! Since 2011, there have been significant shifts in distributions for twenty-four species (that is more than one-fourth of the state's mammals), and three more species have been added to increase our checklist to eighty-eight. The water shrew (number eighty-eight) was added just before the draft went to the editor. Developments in molecular genetics have resulted in some reclassification and changes in scientific names.

What are the causes of all of these changes? Climate change? Shifts in land use? Habitat changes? Or have there just been more people out there looking? Such questions may take considerably more time to be answered, but I have to wonder about species such as opossums and spotted skunks, for many years confined to extreme southeastern North Dakota but now continuing to spread over much of the eastern half of the state. Species known only by one or two specimens since the turn of the twentieth century have turned up in new locations. Bats, once thought to be migratory out of the state have now been documented hibernating in the badlands. Exciting recent and current research provide new insights into regional mammalian ecology. Where it all ends up, I do not know, but this second edition is an attempt at keeping things current for now.

I am again indebted to those friends and colleagues who provided comments and current data for the second edition. Dr. Erik Fritzell and Dr. Bill Jensen (North Dakota Game and Fish Department) both gave the draft a thorough read and provided many helpful comments. Bill also provided revisions of his habitat section and updates on moose, elk, and deer. North Dakota Game and Fish Department biologists Pat Isaacson, Sandra Johnson, Stephanie Tucker, and Brett Wiedmann all provided comments and updates on non-game, furbearers, and bighorn sheep. And, I cannot ignore all of the personal, phoned, and written comments that have come in over the past eight years which have helped in encouraging me to keep plugging along. Finally, many thanks to North Dakota State University Press editor in chief Suzzanne Kelley and her staff for their patience and professionalism in bringing this edition to life.

—Bob Seabloom

Acknowledgments

This project began in 2003, when Dr. Marsha Sovada of the Northern Prairie Wildlife Research Center approached me about developing a comprehensive series of accounts for all of the mammal species of North Dakota that would be used by natural resource agencies in the state. This effort was significant, for there had been no such work since Vernon Bailey's (1926) "A Biological Survey of North Dakota: I. Physiography and Life Zones; II.The Mammals." Upon completion of the project, Marsha—along with the interested agencies—encouraged me to incorporate the accounts into a book on the mammals of the state.

Many thanks to Marsha for getting the project going and for her encouragement throughout. Financial support was provided by the North Dakota Game and Fish Department and U.S. Forest Service, Dakota Prairie Grasslands. Steve Dyke from the Game and Fish Department and Dan Svingen from the Forest Service were especially instrumental, and the Great Plains Natural Science Society agreed to receive and disburse funds for editing and image costs.

North Dakota Game and Fish Department biologists Craig Bihrle, Dorothy Fecske, Chris Grondahl, Bill Jensen, and Sandy Johnson were especially helpful in providing data, reviewing big game accounts, and providing image assistance. Dr. Jeremy Guinn and his students at Sitting Bull College, Fort Yates, reviewed the Native American names of mammals listed in Vernon Bailey's original work and provided additional Lakota names of species. Reviews of various sections were provided by Dan Svingen, Erin Gillam, John Hoganson, Bill Jensen, Cori Lausen, and Mike Oehler.

Co-authors John Hoganson (North Dakota Geological Survey) and Bill Jensen (North Dakota Game and Fish Department) significantly enriched this work by developing chapters on the fossil fauna and major habitats of the state. Good friend and colleague Dr. Erik Fritzell undertook the formidable task of editing the final product, for which I will be always grateful.

I am not sure if this project could have been completed without the office support provided by the University of North Dakota Biology Department. Special thanks go to Morris Pung and Jonathan McClure for their computer expertise. I certainly cannot fail to acknowledge all of my former students who, over thirty-five years, trapped, netted, measured, tagged, and radio-tracked everything from shrews to elk, often under nearly impossible conditions.

Images of mammals were provided by the American Society of Mammalogists Mammal Images Library, Bat Conservation International, California Academy of Sciences Manzanita Project, Craig Bihrle, Dan Brockman, Bob Gress, Chris Grondahl, Deborah Hanson, George Rohde, Robert Seabloom, Dave Strand, and Harold Umber. Specific images are acknowledged in each species account.

Finally, many thanks to my wife, Nikki, for her patience, advice, and putting up with my eccentricities and obsession in moving this project along over the past five years. Maybe there will now be a little more time for camping, birding, and fishing.

—Bob Seabloom

Introduction

North Dakota is in a unique location in North America for those interested in natural history. Its setting in the geographic center of the continent has resulted in climatic extremes, and a fascinating repertoire of adaptations for survival by its plants and animals. Three North American biomes—the eastern deciduous forests to the east, the boreal forests to the north, and the temperate grasslands dominating much of the state—have all contributed to the diversity of its fauna.

It is my intent in this book to fill a significant gap in the published information on the natural history of North Dakota: the mammals. Other aspects of the state's flora and fauna have been covered in the fairly recent past, including books on the plants, fish, amphibians and reptiles, and breeding birds. Not since Vernon Bailey's (1926) "A Biological Survey of North Dakota," however, has there been a single comprehensive work on the mammals. This does not imply that such an important component of North Dakota natural history has gone unstudied. Indeed, there have been hundreds of reports, theses, dissertations, and published articles on mammals by students, wildlife biologists, faculty, and other serious naturalists. Furthermore, web sites maintained by governmental agencies and private conservation organizations have provided additional valuable data on the mammals of the region. It remained for these works to be assembled into a single reference on the mammals of the state, a task which I have undertaken here. I have made a concerted attempt to emphasize research done within North Dakota, and readily accept responsibility for any reports that may have been missed. Regional accounts from neighboring states have also been emphasized, as well as species accounts such as "Mammalian Species" published by the American Society of Mammalogists.

Each species account begins with one or more current common names, known names used by regional Native American tribes, and the scientific name. The importance of the scientific name, consisting of genus and species, cannot be overemphasized, for common names can vary greatly through time and by region. For example, additional common names for the muskrat include "musquash," "marsh hare," and "Chesapeake terrapin," depending on the region of North America. However, the scientific name, *Ondatra zibethicus*, is consistently used by scientists throughout the world. The authority for scientific names used in this work is Wilson and Reeder's (2005) "Mammal species of the world," a publication of the National Museum of Natural History. In some cases, there have been more recent revisions of scientific names, and these are included as well. Following the nomenclature, each account has sections on morphologic description of the species, its North American and statewide distribution, ecology and behavior, reproduction, its status including assessment of relative abundance and any threatened or special concern legal categories, management strategies, and selected references.

North Dakota countywide distribution maps are also included with the species accounts. These maps are based on museum specimens, reports in the scientific literature, harvest records of game species, and confirmed sightings in recent years. It is important to note that these maps are based on current knowledge, and that significant areas of the state have not been systematically surveyed, especially with regard to the less common species. Darkly shaded counties are those with confirmed records of a given species, while lighter shades reflect probable occurrence or older records. In the case of big game, the lighter shading also reflects secondary vs. primary range of a species. Hopefully, these maps will stimulate interest and new efforts to update our knowledge of species distributions in the state.

This reference is intended to be used by students, professional biologists, and serious naturalists. Technical terms are used of necessity (how else can one deal with the "anterior condyloid foramen" in lay terms), but they are defined or clarified in the text, the glossary, or by diagrams. I also hope that this book will stimulate further inquiry on the mammals of North Dakota; there are many gaps in our knowledge, especially of the less common species. Other important sections include accounts on the ecology of the state and mammalian prehistory, the fossil record in North Dakota.

The Class Mammalia

The answer to the question, "What is a mammal?" seems obvious, but when we look at the group as a whole, it becomes elusive. Two external features, the presence of hair and mammary glands, are the most apparent and hold for all living species. Hair is unique to mammals; no other animals have it. Most mammals are covered by an insulating coat of hair, but some, for example whales and people, are mostly naked or sparsely covered. The principal hair types include vibrissae, guard hairs, and underfur. Vibrissae are long, stiff hairs with a sensory function. "Whiskers" of dogs and cats are examples. Guard hairs are coarse overlying hairs serving a protective function. They may be further specialized, such as the defensive spines of a porcupine. The underfur consists of short, fine hairs and provides much of the insulation for most terrestrial mammals.

All mammals have mammary glands and suckle their young, an adaptation unique in the animal kingdom. Mammary glands are derived from sweat or sebaceous glands and secrete milk, a stable form of nutrition for the developing young, thus enhancing juvenile survival.

Mammals are warm-blooded, an adaptation undoubtedly critical to their early survival and allowing them to live in a wide variety of environments. Their high metabolic rates provide an internal heat source (endothermy), and heat loss is regulated by complex morphologic, physiologic, and behavioral mechanisms (homeothermy). Although a critical adaptation to the development and success of mammals, it is not unique, for the birds share essentially the same mechanisms for temperature regulation. However, mammalian warm-bloodedness facilitated the development of an advanced central nervous system and complex behaviors seen in few other animals.

Mammals typically bear living young, but we must qualify this statement because of the egg-laying platypus and echidnas of Australia reminiscent of their reptilian ancestors. Other unique, but not obvious, mammalian features include details of their jaw articulation and middle ear structure, which have provided important clues separating early mammal fossils from the reptiles. More detailed discussion of mammalian characteristics may be found in general mammalogy textbooks, such as Feldhammer et al. (2007).

The Class Mammalia is not particularly large, with a little over 5,000 species compared, for example, to 9,600 species of birds, 40,000 species of fish, and millions of species of insects. Slightly over 400 species of mammals occur in North America and surrounding oceans. However, the adaptive and structural diversity exhibited by the mammals is impressive. Consider the range of size from a shrew weighing 3 grams (g) to a blue whale weighing 160,000 kilograms (kg)! Mammals are capable of living in nearly all environments. They are found from the high arctic to the tropics and all habitats in between. Some have conquered the air and exhibit true flight. Others are fully aquatic or marine, coming up only to breathe. Still others burrow underground, run on the surface, or climb trees. Desert species can get along with little or no free water, while others live in rain forests. Some mammals survive severe winters by going into hibernation; other species remain active throughout the year. Much of this diversity can be seen and appreciated here in the heart of North America. The following table outlines our current understanding of the classification of modern and extinct mammals.

CLASS MAMMALIA

Subclass Prototheria

INFRACLASS EOTHERIA

Order Triconodonta *(extinct)*: Small, primitive predators from North America and Europe

Order Docodonta *(extinct)*

INFRACLASS ORNITHODELPHIA

Order Monotremata: Primitive egg-laying mammals of the Australian region (Platypus and echidna)

INFRACLASS ALLOTHERIA *(extinct)*

Order Multituberculata: First mammalian herbivores; occurred in North America, Europe, and Asia

Subclass Theria

INFRACLASS TRITUBERCULATA *(extinct)*

Order Symmetrodonta: Among the oldest known mammals, probably herbaceous; from North America and Europe

Order Pantotheria: Small; predators from North America, Europe, and Africa; thought to be ancestral to modern placental mammals

INFRACLASS METATHERIA

Order Didelphimorphia (Opossums): Except for the North American opossum, occur in the Southern Hemisphere

Order Paucituberculata (Caenolestids, shrew-like marsupials): South America

Order Microbiotheria (Monito del monte): South America

Order Dasyuromorphia (Numbats, marsupial "mice," native "cats," Tasmanian devil): Australian region

Order Peramelemorphia (Bandicoots): Australian region

Order Diprotodontia (Koalas, wombats, cuscus, possums, gliders, kangaroos, wallabies): Australian region

Order Notoryctemorphia (Marsupial "moles"): Australian region

CLASS MAMMALIA (continued)

INFRACLASS EUTHERIA

Order Afrosoricida (Golden moles, tenrecs): Africa, Madagascar

Order Macroscelidea (Elephant shrews): Africa

Order Dermoptera (Gliding lemurs): Southeast Asia

Order Chiroptera (Bats): Worldwide, except Antarctica

Order Scandentia (Tree shrews): East Asia

Order Primates (Primates): Asia, Africa, North and South America, except for man which is worldwide

Order Cingulata (Armadillos): North and South America

Order Pilosa (Anteaters, Sloths): North and South America

Order Erinaceomorpha (Hedgehogs): Eurasia, Africa

Order Eulipotyphla (Soricomorpha) (Shrews, moles, solenodons): Worldwide

Order Pholidota (Scaly anteaters): Africa, Southeast Asia

Order Carnivora (Carnivores): Worldwide

Order Cetacea (Whales, porpoises, dolphins): Oceans worldwide

Order Lagomorpha (Rabbits, hares, pikas): Worldwide, except Antarctica and most islands

Order Rodentia (Rodents): Worldwide

Order Tubulidentata (Aardvarks): Africa

Order Proboscidea (Elephants): Africa, Asia

Order Hyracoidea (Hyraxes): Africa, southwest Asia

Order Sirenia (Dugongs, manatees): Tropical seas and rivers

Order Perissodactyla (Odd-toed ungulates, including horses, rhinos, tapirs): Africa, Asia, North and South America

Order Artiodactyla (Even-toed ungulates, including pigs, antelope, deer, cattle, etc.): Worldwide, except for Australian region

Early Mammalogy in North Dakota

As with other areas of biology, modern mammalogy involves a blend of conventional natural history, technology, mathematics, and statistics. Today, we can track animal movements with satellites and GPS systems, take remote measurements of their physiology, and use DNA sequencing to clarify relationships of species and subspecies. However, we, too, often forget the importance of natural curiosity, an awareness of our surroundings, and the utility of a good pair of eyes, some binoculars, and a notebook. Add some traps and a few skinning tools, and we may have the beginnings of a field mammalogist.

Native Americans had acquired an intimate knowledge of our native plants and animals through the centuries, and their very survival depended on that knowledge. They passed on much to the first European explorers, but have been rarely recognized for their insights. The various tribes were well-acquainted with many species of mammals and had given them names, yet white naturalists have been credited with their "discoveries." One of the goals of this book has been to include, so far as is known, the names given by various tribes to the mammals of North Dakota.

White settlement spurred the desire for knowledge of "unsettled" North America in the 18th and 19th centuries, resulting in the explorations of many now-famous naturalists. The first whites to record their observations of the animals of the region were fur traders such as Alexander Henry the Younger, who established trading posts in the Red River Valley between 1800 and 1808 (Coues 1897). Henry's Red River journals provide detailed records of furbearers taken during his eight years in North Dakota, and relative numbers of big-game animals as well. Immense herds of elk and bison were noted. In 1800, he recorded a continuous stream of drowned bison floating down the Red River for weeks following spring break-up. White-tailed deer were numerous in what is now Minnesota, but not in the Red River Valley. Black bears were common, but not grizzlies except in the Devils Lake area. The Lewis and Clark expedition spent the winter of 1804-1805 encamped north of Bismarck and returned through the state in 1806. Twenty-seven mammals were recorded, including bison, bighorn sheep, pronghorn (the first technical description), elk, prairie dogs, and white-tailed jackrabbits (Reid and Gannon 1927). Later 19th century surveys included those of Maximilian in 1833-1834 along the Missouri River, John James Audubon in 1843 along the Missouri River to Fort Buford, the Stevens expedition of 1853 seeking a railroad route from the Mississippi River to the Pacific Ocean, Elliott Coues in 1873 for the Northern Boundary Survey, and J. A. Allen in 1873 for the Northern Pacific Railroad Expedition from Fort Rice to the Yellowstone River (Bailey 1926). In addition to recording occurrence of species, attempts were also made to estimate numbers of game animals, especially bison. For example, the Stevens expedition encountered a huge herd of bison near Lake Jessie (Griggs County) and estimated it as 200,000 to 500,000 head. While likely an exaggeration, it does illustrate the incredible numbers occurring at that time.

The Bureau of Biological Survey, forerunner of the U.S. Fish and Wildlife Service, had its origins around 1885, and in 1893 began compiling all of the earlier work done in North Dakota. A systematic collecting effort was also initiated, which culminated in Vernon Bailey's (1926) "A Biological Survey of North Dakota," which continued for many years as the only comprehensive reference on the mammals of the state. One of Bailey's trappers, Henry V. Williams, continued for many years as a taxidermist in Grafton, and his collection has provided a major nucleus for the University of North Dakota Vertebrate Museum.

The 20th century saw the continuing work of the Bureau of Biological Survey and its successor, the U.S. Fish and Wildlife Service. The North Dakota Game and Fish Department began its annual surveys of game species and furbearers. The Northern Prairie Wildlife Research Center (originally operated by the U.S. Fish and Wildlife Service and more recently by the U.S. Geological Survey) was established in the 1960s, and has conducted wide-ranging studies on mammalian nest predators from Franklin's ground squirrels to raccoons, red foxes, and coyotes. The University of North Dakota's Institute for Ecological Studies conducted a wide variety of biological surveys, primarily associated with energy and water development in the state. The Regional Environmental Assessment Program (REAP), established by the Legislature in the 1970s, was responsible for wildlife surveys by the North Dakota universities in the coal lands of southwestern North Dakota. These studies clarified the status of a number of species which had gone unrecorded for over 75 years. As modern-day naturalists, it is well to give due credit to those who earlier traversed North Dakota by keel boat, canoe, horseback and buckboard (as well as their more modern counterparts) for their contributions to our knowledge of the state's diverse mammalian fauna.

Mammalian Paleofaunas of North Dakota

John W. Hoganson

Mammals evolved from mammal-like reptiles. General consensus is that this occurred during the latest Triassic about 210 million years ago, although the transition from reptile to mammal is not easy to define in the fossil record. Mammals, therefore, appeared during the beginning of the Mesozoic Era not long after the first dinosaurs. During the Mesozoic, also referred to as the Age of Reptiles, mammals were small, mostly the size of shrews, mice, and rats. The most common mammals that lived during the Mesozoic were multituberculates, although metatherians (marsupials) and eutherians (placentals) were also present. It is believed by many paleontologists that mammals did not attain larger sizes and play a more significant biological role during the Mesozoic because they were out-competed by dinosaurs.

After the demise of the last of the dinosaurs and about three-fourths of the rest of the world's biota at the Cretaceous-Tertiary (K-T) boundary about 65 million years ago, mammals radiated quickly. This extinction event, one of Earth's most devastating, marked the close of the Mesozoic Era and was the beginning of the Cenozoic Era, or the Age of Mammals. Several groups of multituberculates, metatherians, and eutherians survived the K-T boundary extinction. The beginning of the Cenozoic was a time of biotic renewal when mammals underwent a dramatic radiation. By 10 million to 12 million years after the K-T boundary extinction, all living mammalian orders had evolved.

Mammals during the Age of Reptiles

The oldest mammalian fossils in North Dakota are found in the Cretaceous Hell Creek Formation (about 68 million to 65 million years ago) (Figure 1). These fossils, mostly teeth and jaws, have been recovered from sites south of Mandan in Morton County and near Marmarth in Slope and Bowman counties. At that time a huge delta, similar to the Mississippi Delta, existed in western North Dakota. A shallow ocean persisted in eastern North Dakota; its western shoreline was in the vicinity of present-day Bismarck. This deltaic coastal plain was forested and the climate was warm-temperate to subtropical, similar to areas in the southeastern United States today. The lush forested delta provided habitats for many species of exotic plants and animals, all of which are now extinct. The dominant life forms inhabiting the delta were dinosaurs. Fossils of 14 species of dinosaurs have been recovered from the Hell Creek Formation in North Dakota. This delta community also included many species of plants, bivalves, gastropods, fish, salamanders, lizards, turtles, crocodiles, pterosaurs, champsosaurs (crocodile-like animals), birds, and mammals. Mammals were inconspicuous members of the Hell Creek community, and their fossils are scarce. Only 14 species of mammals representing eight genera have been recovered from the Hell Creek Formation in North Dakota. This fauna included six species of multituberceltes (e.g., *Meniscoessus*), six species of metatherians (e.g., *Didelphodon* and *Alphadon*), and two species of eutherians (e.g., *Cimolestes*). About the size of an opossum, *Didelphodon* was the largest of these animals (Figure 2).

Mammals of the North Dakota Everglades

Nonavian dinosaurs and at least two-thirds of mammalian lineages that existed during the Cretaceous were wiped out during the K-T boundary extinction. Most metatherians became extinct at that time. The Paleocene (first epoch of the Tertiary Period from about 65 million to 55 million years ago) was a time of recovery when the world began rebounding from this biological catastrophy. During the Paleocene, most of western North Dakota was a forested swampland, similar to today's

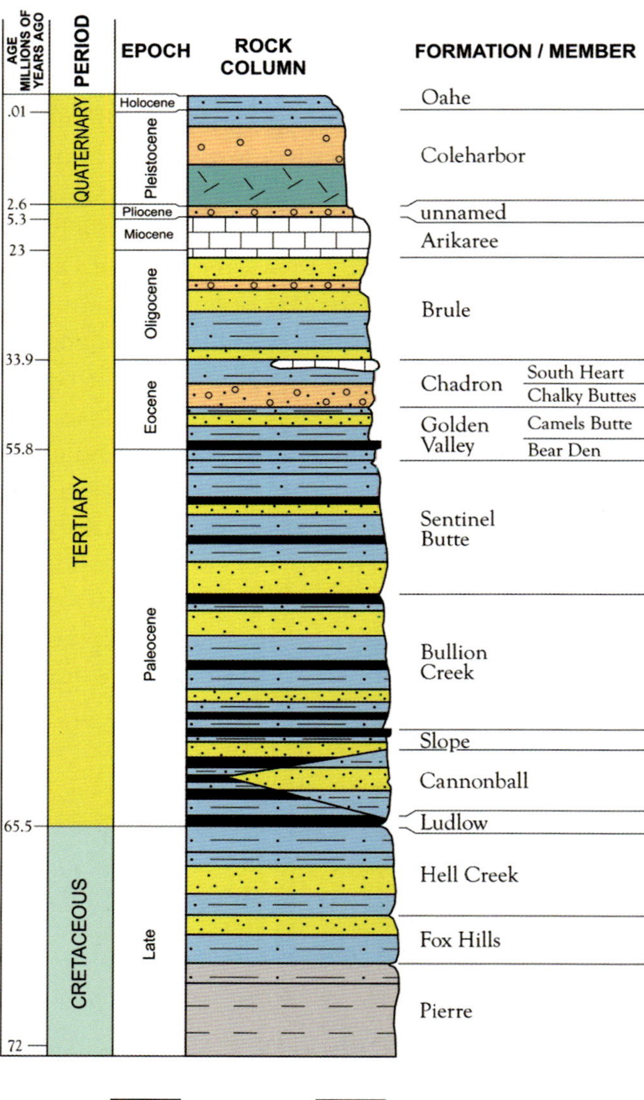

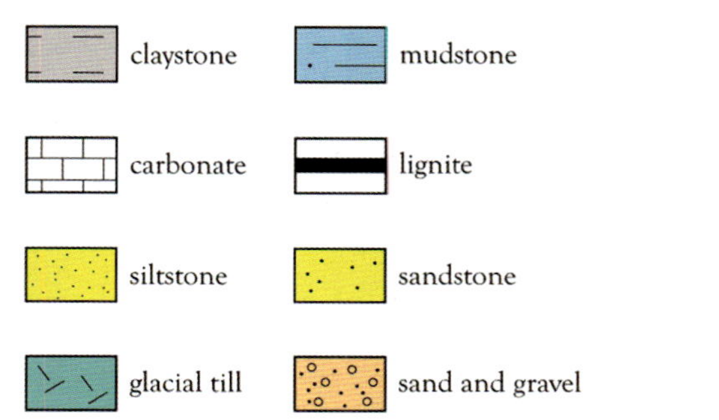

Figure 1. The stratigraphic column of North Dakota lists the names and ages of the rock formations that occur in North Dakota.

forested swamps in areas of Florida. During much of that time, eastern North Dakota was covered by the last ocean to inundate North Dakota, the Cannonball Sea. The climate was warm-temperate to subtropical. Ferns, cycads, figs, bald cypresses, magnolias, ginkgos, sycamores, dawn redwoods, palms and many other exotic plants flourished in the hot, steamy, swampy lowlands of western North Dakota. The largest predators that inhabited western North Dakota at that time were crocodiles; some grew to lengths of 4.5 meters (m). Mollusks, insects, fish, salamanders, lizards, turtles, alligators, champsosaurs, snakes, frogs, birds, and mammals also thrived in North Dakota's Paleocene swamplands.

The evolutionary radiation of mammals during the Paleocene is unprecedented. The Late Cretaceous mammalian community dominated

Figure 2. The Cretaceous age metatherian, *Didelphodon* being captured by a meat-eating dinosaur. Painting by and courtesy of Michael W. Skrepnick.

by metatherians and the now extinct multituberculates was replaced in the Paleocene by a diverse mammalian community dominated by eutherians. The most rapid phase of radiation was during the first few million years after the K-T boundary extinction. In North America, there were 84 genera of mammals by the middle Paleocene, and 100 by the late Paleocene, compared to 18 genera in the Late Cretaceous. Most living orders of eutherian mammals first appeared in the Paleocene. Fossils of over 40 species of mammals, including multituberculates, metatherians, carnivores, insectivores, "primates," condylarths, and pantodonts, have been collected from several Paleocene rock formations at numerous fossil sites in western North Dakota. The largest mammal that inhabited North Dakota during the Paleocene was the pantodont *Titanoides*. *Titanoides* was bear-like in appearance and possessed huge canine tusks and stout limbs with clawed digits adapted for browsing and digging roots. One of the more interesting mammals that lived in North Dakota at that time was the lemur-like *Plesiadapis* (Figure 3).

Figure 3. The Paleocene age primate-like *Plesiadapis*. Image from a mural at the North Dakota Heritage Center, Bismarck. Painting by Geoff Elson, Split Rock Studios.

Plesiadapis was somewhat larger than a squirrel, about 60 centimeters (cm) long, and had a long tail, agile limbs with claws, and eyes positioned on the sides of its head. It was well-adapted for climbing trees with its long, clawed fingers and toes. *Plesiadapis* was a primitive relative of primates and probably occupied an ecological niche similar to that of squirrels today. The largest mammalian predators during the Paleocene were only the size of small dogs.

The greenhouse conditions of the Paleocene continued into the Eocene, about 55 million years ago. The animal community that inhabited North Dakota was diverse and included reptiles, fish, amphibians, birds, and mammals. Insectivores, rodents, the early horse "*Hyracotherium*," the large tapir-like pantodont *Coryphodon*, and carnivores were some of the mammals present at that time. The remains of the most spectacular of the Eocene mammals called brontotheres have been recovered from the Chadron Formation in Stark, Golden Valley, and Bowman counties (Figure 4).

Figure 4. The Eocene age elephant-size brontothere. Painting by and courtesy of Dan Varner.

Brontotheres resembled rhinoceroses and were elephant-size browsers. They are also called titanotheres or "thunder beasts" because of their immense size and were among of the largest mammals to ever live in North Dakota. They had large bony knobs on their snouts, which were probably covered with skin as in modern-day giraffes. These knobs were likely used for display or as weapons during fights to establish dominance.

Mammals of the North Dakota Scrubland and Savanna

By the early Oligocene, about 34 million years ago, worldwide climate cooling had occurred. In North Dakota the equable, humid, warm-temperate to subtropical climate that prevailed during the Paleocene and early Eocene became cooler and drier. The swampy environments, and plants and animals that inhabited them, had disappeared and an essentially treeless plain, consisting of scrubland vegetation, mostly shrubs and herbaceous plants, developed in North Dakota. By the Miocene, about 23 million years ago, true savanna habitats, dominated by grasses had become established. The Oligocene climate was seasonal and temperate. Ponds and lakes occurred on the open plain and gallery woodlands grew along the streams and rivers. The number of mammalian taxa and abundance of mammals in North Dakota dramatically increased during this time. This is reflected by the diversity of fossils found in the Brule Formation. The remains of at least 90 mammalian species in 31 families and 10 orders have been recovered from the Brule Formation in Slope and Stark counties. Many migrated here from Asia through Alaska and most were adapted for grazing. The Brule Formation mammalian fauna consisted mostly of ancestral members of families that still exist today, including dogs, camels, deer, squirrels, beavers, horses, rabbits, rhinoceroses, and mice. Remains of fish, turtles, and tortoises (some the size of the Galopogos tortoise), lizards, amphibians, birds, insects, gastropods, and bivalves in the Brule Formation indicate that these animals co-existed with the mammals.

The largest mammals living in North Dakota during the Oligocene were the giant pig-like *Archaeotherium* (Figure 5), the hippo-like *Metamynodon*, and *Subhyracodon* one of the early rhinoceroses. A sheep-like mammal, *Merycoidodon*, and the diminutive horse, *Mesohippus*, apparently roamed the open plain in large herds. *Mesohippus* resembled the modern horse except that it was much smaller, only about 60 cm tall

Figure 5. The Oligocene age giant pig-like *Archaeotherium*. Image from a mural at the North Dakota Heritage Center, Bismarck. Painting by Tom Stewart, Split Rock Studios.

Figure 6. The Oligocene age primitive horse *Mesohippus*. Painting by Heinrich Harder.

at the shoulder, and had three toes on each foot and low-crowned teeth adapted for browsing (Figure 6).

Other herbivores in the Oligocene mammalian community included the squirrel-like *Ischyromys*, the rabbit *Palaeolagus*, the beaver *Agnotocastor*, the mouse *Eumys*, the gazelle-like camel *Poebrotherium*, and tiny deer *Leptomeryx*.

Herbivores greatly outnumbered carnivores on the North Dakota Oligocene plain, as is the case today on the African savannas, although fossils of several species of carnivores have been recovered from the Brule Formation in North Dakota. Carnivores represented in the Brule Formation fossil record include the nimravid *Dinictis* (Figure 7), a saber-toothed cat-like carnivore, the weasel-like canid *Hesperocyon*, the primitive bear-dog *Daphoenus*, and the creodont *Hyaenodon*. Fossils of the insectivore *Leptictis*, distantly related to modern shrews, have also been found in the Brule Formation.

Figure 7. The Oligocene age saber-toothed cat-like *Dinictis* chasing a primitive deer-like mammal. Painting by Charles R. Knight, courtesy of the American Museum of Natural History.

Miocene age (about 23 million years ago) mammalian fossils are rare in North Dakota. Remains of a few grassland taxa have been found, including the sheep-like oreodont *Merychyus*, the small horse *Miohippus*, and the terrestrial, muskrat-size, burrowing beaver *Palaeocastor*. The largest and most spectacular of the late Miocene-early Pliocene mammals is the herbivorous, "shovel-tusked" gomphothere *Amebelodon*. These elephant-like animals had 90 cm long, flattened, spade-like tusks extending from their lower jaws. These tusks were probably used like shovels to dig up rooted water plants in rivers and ponds.

Mammals of the Great Ice Age

The fossil record of mammals in North Dakota from about 20 million years ago until about 50,000 years ago is incomplete because the sediments deposited during that time, and the fossils that would have been entombed in them, have been removed by millennia of erosion. The Great Ice Age, the Pleistocene Epoch, began about 2.6 million years ago and ended about 10,000 years ago. Continental glaciers advanced into North Dakota from Canada on several occasions during that time, and with each advance, the landscape and life in the state was altered. The record of life in North Dakota is sketchy, but fossils of some of the cold-adapted mammals that lived here at the end of the Ice Age have been found.

The mammalian megaherbivores that resided in North Dakota at the end of the Pleistocene were indeed impressive, particularly the elephant-like species. Fossils, primarily teeth, of woolly mammoths, *Mammuthus primigenius*, have been found at several localities throughout North Dakota (Figure 8).

These massive, hairy beasts, that were 3 m tall at the shoulder, are in the same family as modern elephants. The remains of a second, larger species of mammoth *(Mammuthus columbi)* have also been found here. The American mastodon, *Mammut americanum*, was one of the most common proboscideans that lived in North America at the end of the Ice Age, but few fossils of this animal have been found in North Dakota. They were about the size of the Indian elephant but were not true elephants. Unlike today's elephants, the mammoths and mastodons were covered with long, shaggy hair for insulation against the Ice Age cold.

Mammoths inhabited steppe environments whereas mastodons were woodland dwellers.

The North Dakota Ice Age mammalian community also included the huge bison *Bison latifrons* (Figure 9), the ground sloth *Megalonyx jeffersonii* (Figure 10), and the horse *Equus*.

Bison latifrons was about 25 to 50% larger than the largest bull bison that lives today. It possessed horns that spanned over 2 m, more than twice as wide as today's bison. *Bison latifrons*, unlike today's herding bison, probably lived in family groups in wooded habitats. *Megal-*

Figure 8. The Ice Age woolly mammoth *Mammuthus primigenius.* Painting by and courtesy of Dorothy S. Norton.

Figure 9. The Ice Age bison *Bison latifrons*. Painting by Fred Lahrman, courtesy of the Royal Saskatchewan Museum, Regina.

Figure 10. The Ice Age ground sloth *Megalonyx jeffersonii*. Painting by and courtesy of Dan Varner.

onyx jeffersonii was a bear-size ground sloth that grew to lengths of 2 m or more. It lived in wooded areas along the Missouri River at the end of the Ice Age. The Ice Age horse was very similar to today's horse and is also in the genus *Equus*. Like the other large mammals mentioned above, the Ice Age horse became extinct in North America at the end of the Ice Age. Whether the extinction of nearly all large mammals at the end of the Ice Age, about 13,000 years ago, was caused by climate change, disease, and habitat loss or overkill by Paleo-Indians is vigorously debated.

At the end of the Ice Age in North Dakota, the climate began to warm and the glaciers began to melt, and by about 11,000 years ago, open spruce forests were being replaced by deciduous woodlands and prairie uplands. Ponds and bogs developed in the cool and moist forested habitats. The pond and bog deposits have yielded fossils of the plants and animals that existed in the aquatic habitats. These fossils include the remains of the beaver, *Castor canadensis,* and the muskrat, *Ondatra zibethicus*; both taxa still inhabit North Dakota today.

North Dakota's Existing Mammalian Fauna

The mammalian fauna that inhabits North Dakota today became established in conjunction with amelioration of climate after the Ice Age. By about 9,000 years ago, as a result of climate warming and drier conditions, prairie ecosystems became established in North Dakota. Mammals that live in the state today began colonizing the area from throughout North America.

The Landscape and Flora of North Dakota: "An Acquired Taste"

William Jensen

Historian, author, and native son, Clay Jenkinson, has described North Dakota as ". . . an acquired taste." Upon first stepping into what is now North Dakota in October of 1804, William Clark curtly wrote ". . . the up lands naked." However, by the time they were moving upstream near present-day Williston on April 21, 1805, Lewis made the following journal entry ". . . the course of this river as far as I could see from the top of Cut Bluff was due North, it passes through a beatifull level and fertile vally about five miles in width. I think I saw about 25 miles up this river, and did not discover one tree or bush of any discription on it's borders. The vally was covered with Elk and buffaloe." Sometime in the seven months between October and April Lewis and Clark had developed a taste for the land. The landscape of North Dakota, and the northern Great Plains, is not a tropical fruit punch to be guzzled. It is a land that must be taken in slowly, over time, to be appreciated.

North Dakota is located in the geographic center of North America, ranks 17th in size compared to the other 50 states (70,594 square miles), and 48th in human population (just under 640,000 in 2007). The population is in the process of a historical shift from primarily a rural population to a 50:50 split between rural and small towns and cities. Despite these changes, farming and ranching remains a mainstay of the economy. The footprint of agriculture and, more recently, energy development (particularly oil, coal, and wind energy) have an ever increasing influence on native wildlife, including mammalian species.

In a state where approximately 89% of the surface area is in private ownership (6% public and 5% tribal ownership) land use patterns can change relatively quickly. Of particular concern for wildlife is the fragmentation and conversion of native prairie to crop production and energy development. As cereal grain commodity prices periodically surge, and more and more beef production is outsourced to other countries, there is more incentive to convert prairie grasslands, most suitable for grazing, to cropland. The future impacts of additional agricultural and energy development on native prairie, and the animal communities depending on them, are yet to be realized.

Topography

The topography of that portion of the state north and east of the Missouri River was shaped primarily by Wisconsinan glaciations about 40,000 years ago. Over the next 28,000 to 30,000 years, this mass of ice retreated and advanced several times. About 10,000 to 12,000 years ago, the glacier retreated for the last time. Glacial Lake Agassiz, formed by a giant ice dam to the north, resulted in the flat lake plain with rich soil sediments, producing the tall-grass prairie grasslands of the Red River valley (Figure 1).

Topography of the glaciated plains to the west of Glacial Lake Agassiz is often referred to as two physiographic regions: the drift prairie and the Missouri Coteau. Together they are often referred to as the Prairie Pothole Region. When glaciers carry moderate amounts of till (drift material) and it is deposited as the glacier melts, the resulting terrain is a "ground moraine." Much of eastern North Dakota was covered by ground moraine and is now referred to as drift prairie. This area is now characterized by a gently rolling landscape with shallow wetland basins and low hills. To the south and west of the drift prairie, the glaciated plains were shaped by the actions of "terminal moraines" or "end moraines." Terminal moraines formed as the edge of the glacier repeatedly melted back and moved forward. The result is a terrain now characterized by a hilly landscape speckled with numerous small seasonally to permanently flooded wetlands. These wetlands resulting from glaciers are commonly referred to as potholes. This topography is now referred to as the Missouri Coteau within the Prairie Pothole Region.

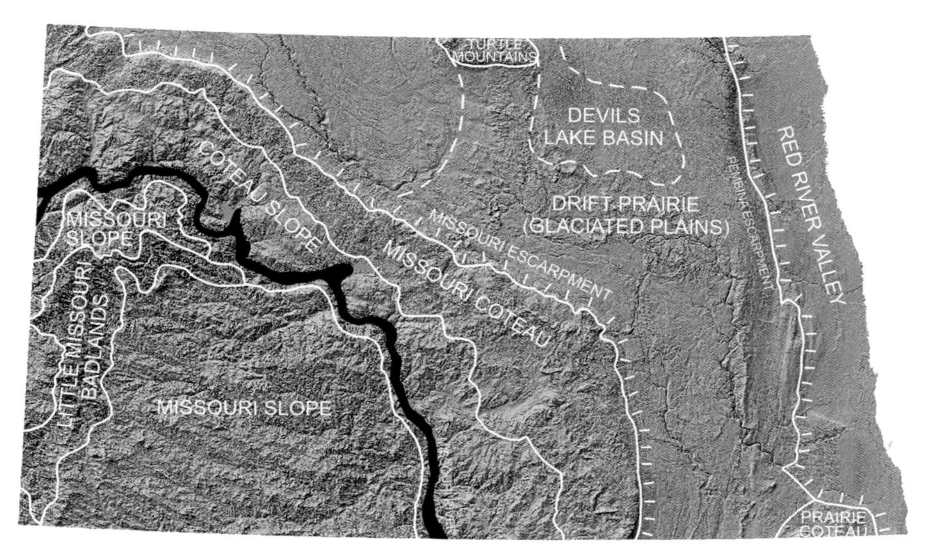

Figure 1. Map showing the main physiographic areas in North Dakota. Computer-enhanced imagery shows the influence of glaciations as recent as 13,000 years ago to the north and east of the Missouri River, and much slower erosion processes of wind and water. (Map provided by North Dakota Geological Survey)

The Turtle Mountain region is an area of younger glacial sediments, deposited on top of a harder substrate after the surrounding older sediment eroded away. The added elevation of 600 to 800 feet above the surrounding terrain now allows woodlands of aspen, paper birch, box elder, and bur oak to persist.

That portion of the state south and west of the Missouri River was not affected during the Wisconsonan glaciations, but was subject to wind and water erosion for a much longer period. Areas that resisted erosion resulted in the formation of buttes that now dot the region. The elevation and shade provided by these buttes allows small forested stands of aspen, green ash, and Rocky Mountain juniper to become established, primarily along the northern and eastern facing slopes.

The Badlands are the most striking example of water erosion in the state. Originally the Little Missouri River, like all the state's ancient river systems, flowed in a northern and easterly direction. About 600,000 years ago, glaciers blocked its course and forced it to turn to the east. The new river channel was shorter and steeper, thus causing a rapid down-cutting of the channel and creating the rugged features of steep-sided buttes, clay knobs, and arroyos. This erosion process also exposed prehistoric ocean sediments dating back more than 50 million years and allowed for exposed coal veins to be subject to prairie fires and burn, thus producing the distinctive red scoria rock from clay that had been deposited on top.

Today North Dakota is a relatively flat plain ranging in elevation from 750 feet above sea level along the Red River in the northeastern corner of the state to 3,506 feet above sea level at the top of White Butte in the southwestern corner. There are five major drainage basins in the state. The Red River of the North along the eastern edge of the state is fed by the Sheyenne River and several other tributaries. The Devils Lake basin is closed and receives water from surrounding lakes and wetlands. The Souris basin, which consists primarily of the Des Lacs and Souris rivers, loops south into the state and flows back into Canada. The James River basin flows out of the south-central part of North Dakota and joins the Missouri River in South Dakota. The Missouri River basin drains the southwestern third of the state and collects the waters of the Yellowstone, Little Missouri, Knife, Heart and Cannonball rivers.

Climate

North Dakota lies in the cool, subhumid or semiarid continental interior. Its climate is dominated by air masses originating from the Arctic, Pacific, and Gulf of Mexico. The state is characterized by hot summers, very cold winters, and sparse to moderate rainfall. Periodic drought and fire historically were the primary forces that shaped the vegetative communities in the state and the northern Great Plains in general. Based upon a review of 585 years of tree ring and weather data from central and western North Dakota, the following observations were noted: (1) The mean length of dry and wet periods were 8.8 and 8.3 years, respectively, (2) The longest dry period was 14 years (1633 to 1646), and the longest wet period was 39 years (1663 to 1702); annual precipitation rates between 1892 and 1990 were in a drought status for 35% of the years, (3) Drought conditions existed during 40% of the summer growing seasons and 53% of the early fall periods, and (4) Precipitation rates were totally independent of the previous years.

The mean annual temperature is 40°F (4°C), ranging from a mean of 7°F (-14°C) in January to 69°F (21°C) in July. The record low temperature, -60°F (-51°C), was set at Parshall (Mountrail County) on February 15,1936; the record high, 121°F (49°C), at Steele (Kidder County) on July 6, 1936; a range of more than 180°F (95°C) within a 142-day period. The average yearly precipitation (1971 to 2000) is about 16.8 inches (42.7 cm) and total annual snowfall averages 44 inches (112 cm) in central North Dakota (Burleigh County). As one travels from the northeastern quarter of the state to the southwest, mean annual temperature increases from 37°F (3°C) to 43°F (6°C), and annual precipitation rates decrease from about 20 inches (51 cm) to less than 14 inches (36 cm).

Landscape and Habitat Components

North Dakota lies within the grassland biome, characterized by a variety of grasses such as big and little bluestem, western wheatgrass, and green needlegrass. Subtle changes that occur across the landscape, due to climate and geological history of the region, produce gradual shifts in the plant communities and their composition. In addition, soils, topography, and human disturbances have also influenced these plant

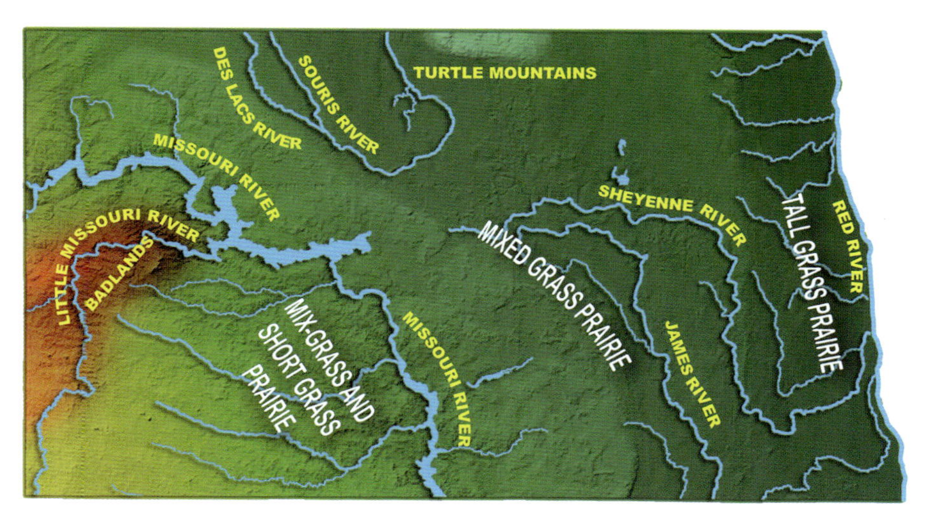

Figure 2. Map of North Dakota showing the distribution of major habitat types and river systems across the state.

communities. As a result, there is no uniformly accepted vegetation or habitat type map for the state or region. Information presented below is a blending of several vegetation mapping efforts for the region, including "natural vegetation" consisting of the native species found at the time of European settlement and what has been called the "real vegetation" consisting of all the types of vegetation currently present, including agricultural crops. North Dakota holds four general plant communities or habitat types: prairie, riparian and upland forests, wetlands, and badlands (Figure 2).

The following are descriptions of natural vegetation and topography of the state within each of the four general habitat types and representative natural areas that can be visited by those wanting to glimpse communities of the past, and brief comments about current agricultural uses and human impacts.

Prairie

Prairie is by far the largest habitat type and extends across nearly the entire state. The composition of prairie grasses and forbs varies, with taller species in the east, shorter species in the southwest, and a mix of the two types in between. Precipitation, topography, and soil types influence species composition.

Tall-grass Prairie: This habitat type encompasses the region along the eastern edge of the state (Figure 3).

The flat topography and rich soils are the result of deposits of Glacial Lake Agassiz that covered this region 10,000 years ago. Predominant natural vegetation includes big bluestem, switch grass, Indiangrass, prairie dropseed, slender wheatgrass, meadow anemone, prairie cinquefoil, wild licorice, and prairie blazing star. Common species characteristic of neighboring eastern and southern mammalian fauna include eastern cottontail, prairie deer mouse, and meadow vole. Associated mammalian species of conservation priority (SCP, a North Dakota Game and Fish Department classification) are arctic shrew, pygmy shrew, plains pocket mouse, Richardson's ground squirrel, eastern spotted skunk, and river otter. Because this area was so productive and easily converted to agricultural use, tall-grass prairie is now one of the most endangered habitat types in the world. Most of this area currently supports corn and sugar beet farming operations. Less than

Figure 3. Tall-grass prairie once covered the entire Red River basin. Now less than 2% remains in native vegetation in the state. The largest remaining stands of tall-grass prairie in North Dakota are found within the Sheyenne National Grasslands in Ransom and Richland counties. (Sandra Johnson, North Dakota Game and Fish Department)

2% of the state's surface area in this region remains in natural tall-grass prairie, with another 7% disturbed and planted back into hayland (e.g., brome grass and alfalfa). About 5% consists of lakes, wetlands and forests along the major river systems. The Sheyenne National Grasslands, in Ransom and Richland counties, is home to more than 40 sensitive species of plants and animals such as Greater prairie-chicken and the threatened western prairie fringed orchid. The Sheyenne National Grasslands holds the bulk of remaining native tall-grass prairie in the state. Tewaukon National Wildlife Refuge (NWR) in Sargent County, Englevale Wildlife Management Area (WMA) in Ransom County, and Kelly's Slough and Prairie Chicken WMA in Grand Forks County are all excellent sites for wildlife viewing.

Mixed-grass Prairie: This habitat type encompasses more than half of North Dakota, and covers the Prairie Pothole Region. Mixed-grass prairie is the transition zone between the wetter tall-grass prairie to the east and drier shortgrass prairie to the southwest. The topography of the more eastern prairie, sometimes referred to as drift prairie, is generally flat to gently rolling. The Missouri Coteau, which runs in a band across the state from northwest to southeast, is characterized by steeper topography and a high concentration of temporary and seasonal wetlands (Figure 4).

Predominant natural vegetation includes prairie junegrass, green needlegrass, needle-and-thread, blue grama, little bluestem, western wheatgrass, pasque flower, western wall-flower, gumweed, fringed sage, prairie rose, and Missouri milkvetch. Today native mixed-grass prairie makes up about 15% of the eastern Drift Prairie and about half of the Missouri Coteau. Ranching operations are more prevalent as you travel

Figure 4. Mixed-grass prairie once covered the majority of the state. Today, only about 15% of the drift prairie and more than 50% of the Missouri Coteau remains in native prairie. (Sandra Johnson, North Dakota Game and Fish Department)

west. Common mammalian species of the mixed-grass prairie include Franklin's ground squirrel, muskrat, coyote, red fox, and white-tailed deer. Associated mammalian SCP include: long-legged myotis, pygmy shrew, arctic shrew, plains pocket mouse, and Richardson's ground squirrel. During the spring and summer, this region of the state, America's duck factory, literally explodes with waterfowl and shorebirds. Audubon WMA in McLean County, Arena Lake and McKenzie Slough WMAs and Long Lake NWR in Burleigh County, Chase Lake WMA and Arrowwood NWR in Stutsman County, Lonetree WMA in Sheridan County, as well as, Des Lacs, and J. Clark Salyer NWRs in Bottineau and Renville counties, and Lostwood NWR in Burke County all provide unique hunting and wildlife viewing opportunities.

Mixed- and Short-grass Prairie: This habitat type encompasses that portion of the state south and west of the Missouri River. The rolling

Figure 5. Mixed- and short-grass prairie is primarily confined to the southwestern corner of the state. The landscape is dotted with small buttes. The buttes shown above are the Dog Tooth buttes north of Raleigh in Grant County. Today about 25% of the former mixed- and short-grass prairie remains in native vegetation. (Bill Jensen, North Dakota Game and Fish Department)

topography, dotted with isolated buttes, is the result of water and wind erosion (Figure 5).

Predominant natural vegetation includes blue grama, western wheatgrass, prairie junegrass, needle-and thread, needleleaf sedge, yarrow, gumweed, purple coneflower, silver sage, and prickly pear cactus. Common mammals include coyote, badger, and white-tailed deer. This region also marks the eastern fringe of pronghorn and mule deer range. Associated mammalian SCP include: hispid pocket mouse, sagebrush vole, black-tailed prairie dog, swift fox, and black-footed ferret. About 27% of the unglaciated mixed- and short-grass prairie remains in its natural vegetative condition (13% native prairie, 12% alkaline and barren clay pan flats, and 2% wetlands). Small isolated buttes provide unique islands of native flora that have remained unchanged except for the encroachment of exotic weeds (leafy spurge and spotted knapweed). Schnell Recreational Area (Stark County), Little Missouri National Grasslands, and Theodore Roosevelt National Park are all well worth a weekend camping trip.

Riparian and Upland Forests

North Dakota is the least forested state in the nation with less than 5% in this habitat type. As a result, forested areas add an important habitat type in providing faunal diversity. In North Dakota, forest species composition ranges from components of the northern hardwood forests to the east, oak and aspen savanna to the north, and ponderosa pine forests of the west. As with variations in prairie types, woodland species composition is also influenced by precipitation, topography, and soils.

Upland Forests: The Pembina Escarpment in Cavalier County, Turtle Mountains in Bottineau and Rolette counties, and Devils Lake in Eddy and Ramsey counties are distinctive forested areas within the eastern mixed-grass prairie. The Killdeer Mountains in Dunn County are a similar forested area within the western mixed- and short-grass prairie (Figure 6).

The distinctive feature of these areas is the steep or elevated terrain that supports an upland deciduous forest community. Predominant natural vegetation include bur oak, aspen, paper birch, green ash, American hazelnut, choke cherry, Saskatoon serviceberry, wild sarsaparilla,

Figure 6. Elevation, steep terrain, and proximity to permanent water sources allow forest communities to develop on the prairie. The Killdeer Mountains are large buttes in Dunn County that support stands of aspen, bur oak, and green ash. (Sandra Johnson, North Dakota Game and Fish Department)

marsh and bushy vetches, and pink dogbane. Characteristic common mammals of interest in upland forests include white-tailed deer, beaver, and red squirrel. Mammals less common, but found in this habitat, include snowshoe hare, least chipmunk, marten, fisher, elk, and moose. Associated mammalian SCP for these woodlands are arctic shrew, pygmy shrew, river otter, gray wolf, western small-footed myotis, long-eared myotis, and long-legged myotis. Due to steep terrain, most of this habitat remains intact; however, over half of the forest in the Turtle Mountains has been converted to agricultural uses on the United States side of the border. The Canadian side of the Turtle Mountains is primarily a provincial park and remains nearly completely forested. Pembina Hills, Cavalier and Charles V. Pulver WMAs in Cavalier County, Jay V. Wessels WMA in Pembina County, Thompson Lake WMA in Bottineau County, Wakopa WMA in Rolette County, and Killdeer WMA in Dunn

County all provide excellent wildlife hunting and viewing opportunities within upland deciduous forests. Sully's Hill National Game Preserve in Benson County offers developed hiking trails, picnic areas and interpretive programs within a native forest setting, as well as captive bison and elk herds for viewing.

Riparian Forests: Riparian habitats are those areas along streams, lakes, and river systems whose vegetation is influenced by the stream, lake, or river itself. In North Dakota, this habitat type is restricted to narrow zones along our limited number of lakes, streams, and rivers. The width of these riparian habitats can range for a few feet along ephemeral creeks to over a mile where the Missouri River spreads out. Although relatively small in total area, this habitat type is extremely productive for wildlife. Even the small islands and sandbars within river systems offer important wildlife habitat. Riparian forests also provide avenues for the dispersal of many species of forest wildlife far out into the prairie areas of the state.

Cottonwood gallery forests are a dwindling habitat type in the state. The Missouri River flood plain and adjacent breaks encompass about 700 square miles of water and 3,500 square miles of land and is the largest riparian area within the state (Figure 7).

The level topography and rich soils are the result of erosion and deposition by the Missouri River within the flood plain. Cottonwoods, willows, and green ash are the predominant natural vegetation occupying the flood plain bottoms. Juniper and deciduous trees, such as buffalo berry, green ash and burr oak, can be found in ravines and on north-facing slopes that border the flood plain. Of the 350 miles (560 kilometers [km]) of Missouri River in North Dakota, about 95 miles (150 km) remain free-flowing and unimpeded by dams. The free-flowing portions of the river include the 70 miles (110 km) between Bismarck and the Garrison Dam, and 25 miles (40 km) upstream from Williston. Portions of the Little Missouri River and other western rivers also support cottonwood forests.

Due to the elimination of periodic flooding, as a result of the Garrison and Oahe dams, the gallery cottonwood forests that Lewis and Clark viewed along the Missouri River are now being replaced by aspen, ash, elm, and oaks. Additionally, Russian olive and salt cedar (exotics) are encroaching and expanding on the Missouri River bottoms.

Figure 7. The Missouri River valley in North Dakota once supported a nearly continuous ribbon of gallery cottonwood forest. Cottonwoods require the scouring of the land by spring flooding for sapling regeneration. The confluence of the Yellowstone and Missouri rivers is still a free-flowing stretch of water that allows this unique forest community to persist. (Craig Bihrle, North Dakota Game and Fish Department)

Mammals characteristic of these areas include beaver, white-tailed deer, elk, and moose. Associated mammalian SCP for these regions include: arctic shrew, pygmy shrew, river otter, and western small-footed myotis. People interested in seeing the remaining vestiges of this unique forest community should focus on Cross Ranch State Park and Smith Grove WMA in Oliver County, Fort Mandan County Park in McLean County, Trenton WMA, Lewis and Clark WMA and Ochs Point WMA in Williams County.

Even small streams can influence the vegetation along their banks (Figure 8).

Trees found in these riparian areas, such as aspens, elms, willows, green ash, box elder and bur oak, are species not dependent upon the periodic floods that is essential for cottonwoods; however, the under-

Figure 8. Small rivers and streams allow narrow bands of riparian forest to extend out on to the prairie. The Wild Rice River in Richland County is one such river. (Craig Bihrle, North Dakota Game and Fish Department)

story component is similar. Mammals characteristic of these areas include beaver, fox squirrel, gray squirrel, raccoon, ermine, least weasel, and mink. Associated mammalian species of SCP are the arctic shrew, pygmy shrew, and river otter. Those interested in seeing nice examples of forested riparian communities should consider visiting the Sheyenne National Grasslands in Ransom and Richland counties, Fort Ransom State Park in Ransom County, Turtle River State Park in Grand Forks County, Icelandic State Park in Pembina County, and Heart Butte Dam in Grant County.

Western Conifer Stands: Early journals and photographs suggest that Rocky Mountain juniper stands were a relatively small habitat component in western North Dakota. With the suppression of fire over the past 100 years, junipers have greatly expanded their distribution on the landscape, particularly on the northern-facing slopes in the Badlands. These dense stands provide hiding and thermal cover for a variety of small mammals, mule deer, and elk. However, they may also be a limiting factor for habitat use by bighorn sheep in some areas. Bighorns prefer more open terrain in order to detect predators (Figure 9).

Isolated stands of pines, more common to Montana and Wyoming, can be found in the southern Badlands. A stand of ponderosa pine cov-

Figure 9. Rocky Mountain juniper stands are common in the Badlands, particularly along northern-facing slopes. With the suppression of fire, this habitat type has greatly expanded. (Sandra Johnson, North Dakota Game and Fish Department)

ering several thousand acres can be found along the Little Missouri River in northern Slope County. A small stand of limber pine can also be found along the west bank of the Little Missouri River north of Marmarth, in Slope County. The closest large stands of ponderosa and limber pine are in the Black Hills and North Cave Hills of South Dakota and Bighorn Mountains of Wyoming. These small forests may be of importance to some species of bats in North Dakota.

Wetlands

A wetland is a basin or low area of land that holds water for at least a portion of the year. Temporary wetlands are shallow depressions that may hold water for only a few days or weeks after the snow melts or heavy rains (Figure 10).

Common vegetation includes cordgrass, sloughgrass, willow-leaved dock, and goldenrod.

Seasonal wetlands are generally larger basins that usually hold water until the middle of July (Figure 11).

Common vegetation includes whitetop, slough sedge, and mannagrass.

Semipermanent wetlands are basins commonly 2 to 4 feet deep that generally hold water year-round (Figure 12).

Open water and submerged plants include common bladderwort, coontail, sago pondweed, and claspingleaf pondweed. Emergent vegetation around the borders of these wetlands include: bulrush, cattail, and phragmites or common reed grass. Woody vegetation, such as willows and osier-red dogwood, may also occur around semipermanent wetlands.

Permanent wetlands or lakes hold water year-round and only dry out during the most severe droughts. Vegetation in and around these

Figure 10. Temporary wetlands are shallow depressions that may hold water for only a few days or weeks after the snow melts or heavy rain. (Craig Bihrle, North Dakota Game and Fish Department)

Figure 11. Seasonal wetlands are generally larger basins that usually hold water until the middle of July. (Mark Fisher, U.S. Fish and Wildlife Service)

Figure 12. Semipermanent wetlands are basins 2 to 4 feet deep that generally hold water year-round. Permanent wetlands or lakes hold water year-round and only dry out during the most severe droughts (Pierce County). (Craig Bihrle, North Dakota Game and Fish Department)

wetlands is similar to that of the semi-permanent wetlands with the addition of trees such as cottonwood, green ash, and bur oak around the margin of some lakes. The majority of the 1.7 million wetland basins that currently exist in North Dakota fall within the glaciated mixed-grass prairie region of the state. It is estimated that between 1850 and 1985, about half the wetlands in the state had been drained. Although consolidation of wetlands has resulted in larger and deeper wetland basins, North Dakota may be down to less than 40% of her original wetland basins.

Mammals that frequent the smaller wetlands include meadow voles, short-tailed shrew, long-tailed weasel, mink, red fox, and raccoon. Muskrats require the stability of semipermanent wetlands and lakes. White-tailed deer, coyotes, and red fox frequently use cattail wetlands for cover during the winter months. These wetlands are so ubiquitous in much of eastern North Dakota that we do not appreciate their true value.

Badlands

The North Dakota Badlands encompass the area drained by the Little Missouri River along the western edge of the state (Figure 13).

The highly eroded topography is the result of relatively rapid down-cutting by the Little Missouri River when glaciers redirected the course of the river and caused rapid down-cutting. With the exposure of coal veins, underground fires repeatedly occurred. These fires produced scoria, the baked clay that gives a reddish hue to the landscape. The end result of these events is a vista of unique topographical features more characteristic of the Southwestern deserts than the northern Great Plains. Predominant natural vegetation includes cottonwood and green ash along river bottoms, and Rocky Mountain juniper, buffalo berry, western snowberry, yucca, prickly pear cactus, purple coneflower, silver sage, western wheatgrass, and little bluestem in the rougher terrain.

The southern Badlands of Bowman and Slope counties contain vegetative aspects of the big sagebrush flats more common to eastern Wyoming and Montana. Characteristic mammalian species of the southern Badlands include Ord's kangaroo rat, bushy-tailed wood rat, desert cottontail, white-tailed jackrabbit, and pronghorn; the northern Badlands support more mule deer, elk, bighorn sheep, and mountain lion. Associated mammalian SCP for the Badlands as a whole include:

Figure 13. The North Dakota Badlands encompass the area drained by the Little Missouri River. Rapid erosion produced a landscape of unique topography and scenic vistas. Today more than 90% of the Badlands remains in native vegetation. (Sandra Johnson, North Dakota Game and Fish Department)

western small-footed myotis, long-eared myotis, long-legged myotis, hispid pocket mouse, sagebrush vole, black-tailed prairie dog, black-footed ferret, and swift fox.

The steep and rugged Badlands terrain, flanking the Little Missouri River, caused many a settler to think twice before putting down a plow in this part of the state. As a result, over 93% of the land remains in natural vegetation. Fire suppression has allowed the expansion of Rocky Mountain juniper stands throughout the Badlands. The north and south units of Theodore Roosevelt National Park provide visitors with striking vistas and abundant wildlife viewing opportunities. For the more adventurous, the 1.1 million acres of the Little Missouri National Grasslands provide camping, hunting, and wildlife viewing opportunities available nowhere else in the state. The Little Missouri Badlands are truly North Dakota's crown jewel of natural areas.

Mammalian Biogeography of North Dakota

To the casual observer, North Dakota, dominated by grasslands, might appear to be biologically monotonous. However, it supports a rich mammalian fauna, a function of its geologic history, climate, and the impacts of human populations. North Dakota's diversity of 88 species of wild mammals is higher than those of neighboring Saskatchewan (75), Manitoba (80), and Minnesota (79). It is less than that of South Dakota (95) and Montana (102). The influence of latitudinal gradients and topographic relief on species diversity has been long known and aids in understanding the lower numbers of the far north and higher diversities around the Black Hills and Rocky Mountains.

With its geographic location in the center of North America, North Dakota's fauna has been influenced by neighboring ecological regions. Its proximity to the northern boreal forests of Canada has facilitated immigration of red squirrels, moose, fisher, and lynx. Riparian woodlands along the major rivers and their tributaries have served as avenues of transport for eastern species such as gray and fox squirrels, eastern chipmunks, and red-backed voles. Arriving in the Badlands from the southwest have been the desert cottontails, kangaroo rats, and pocket mice.

Human activity has had a major impact on our fauna. Gone are the black-footed ferret, plains wolf, and grizzly bear, which previously occurred in North Dakota. The bison now exists only in captive and semicaptive herds. Elk and bighorn sheep were extirpated and have only recently been reintroduced. Exotic species such as the Norway rat and house mouse arrived with steam boats and railroads. Human settlement was accompanied by expansion of the eastern cottontail, gray and fox squirrels, and red fox.

One of the earliest systems attempting to provide an ecological classification of animals and plants was the "life zone" concept developed by C. H. Merriam in the 1890s. Life zones were originally devised as temperature zones, but were later used in attempting to identify ecological communities. The concept was influenced by the rather sharp elevational zones of plants and animals in mountain ranges attributed to changes in temperature. Bailey (1926) used the concept in attempting to characterize assemblages of mammals, birds, and plants in his biological survey of North Dakota. Three zones were recognized. The warmest, the Upper Austral Zone, extended along the Missouri and Cannonball rivers as far as Fort Rice, the Yellowstone and Missouri rivers near Buford and Williston, and the Badlands along the Little Missouri River. Characteristic mammals of this zone included the bighorn sheep, bushy-tailed wood rat, Ord's kangaroo rat, black-tailed prairie dog, and black-footed ferret. The coldest zone, the Canadian Zone, was restricted to the aspen forests of the Turtle Mountains, although elements in the Pembina Hills and Killdeer Mountains were also recognized. Characteristic mammals of this zone included the red squirrel, snowshoe hare, lynx, and moose. The rest of the state was included in the Transition Zone, with humid eastern and semiarid western divisions. This zone included species such as the Richardson's ground squirrel, jumping mice, white-tailed jackrabbit, red fox, and striped skunk. The life zone concept has since largely fallen into disuse for its failure to include other ecological factors such as moisture and geologic history, but it should be recognized as an early attempt at classifying ecological associations of organisms.

Mammal Provinces of North America (Hagmeier and Stults 1964, Hagmeier 1966) were developed from existing range maps of species, and based on the tendency for range limits to be clumped and on calculation of degrees of faunal homogeneity. Characteristic assemblages of mammals were identified, resulting in 35 proposed mammal provinces for the continent. Frequently, the proposed provinces closely corresponded with existing biomes, or major ecologic regions. Under this scheme, the mammalian fauna of North Dakota is influenced by four mammal provinces. The **Alleghenian** province exerts a boreal influence from southeastern Canada and the Great Lakes states, covering much of northeastern North Dakota, including the Red River valley, the major tributaries of the Red River, the Pembina Hills, and Turtle Mountains. Thirty-nine species of North Dakota mammals, including moose, snowshoe hare, and red squirrel, were included in this province. The **Saskatchewanian** province (45 species) provides a northern Plains influence out of the Canadian Prairie Provinces and eastern Montana, and covers much of northwestern and north-central North Dakota. Species such as the white-tailed jackrabbit, pronghorn, and Richardson's ground squirrel are especially characteristic of this province. Sweeping up from the more southern Plains of South Dakota, Nebraska, and Kansas is the **Kansan** mammal province (44 North Dakota species), covering much of southwestern North Dakota. Characteristic species include Ord's kangaroo rat, desert cottontail, hispid pocket mouse, and black-tailed prairie dog. A slight southeastern influence, the **Illinoian** province (40 North Dakota species), extends from southern Minnesota barely into extreme southeastern North Dakota. Most of the state's reports of opossums have come from this area. No one province has a totally unique component of species, hence the rather large number for each and the total for all four provinces exceeding the number of species in the state.

A third approach to describing the biogeography of Plains mammals identified distributional patterns, faunal elements, and principal habitats for the mammals of the plains states of North Dakota, South Dakota, Nebraska, Kansas, and Oklahoma (Armstrong et al. 1986, J. R. Choate, Museum of the High Plains, Hays, KS, pers. comm.). Regional faunal elements based on the shapes of mapped ranges and centers of coincidence of species were recognized as follows:

- **Widespread** species (24 in North Dakota) are distributed on a continental scale, and do not reach range limits in the plains states. The big brown bat, beaver, deer mouse, and coyote are examples.
- The **Chihuahuan** faunal element (seven North Dakota species) is Southwestern, centering in northern Mexico. The western small-footed myotis, desert cottontail, and Ord's kangaroo rat are examples.
- The **Campestrian** faunal element (11 North Dakota species) centers in the high Plains of western Kansas, eastern Colorado, and the panhandles of Oklahoma and Texas. Examples include thirteen-lined ground squirrel, black-tailed prairie dog, prairie vole, and pronghorn.

- The **Eastern** faunal element (eight North Dakota species) centers in the lower Ohio Valley. Examples of eastern species include the eastern cottontail, eastern chipmunk, and gray squirrel.
- The **Austral** faunal element (one North Dakota species) centers on the Gulf Coast of Mississippi and Louisiana. The spotted skunk, rare in North Dakota, is our only example.
- The **Boreo-Cordilleran** and related **Cordilleran** and **Boreal** faunal elements (22 North Dakota species) occur across northern North America and the Rocky Mountains. North Dakota examples include the arctic shrew, snowshoe hare, least chipmunk, red squirrel, and moose.
- The **Great Basin** faunal element (five North Dakota species) centers on the Great Basin and Wyoming Basin, and includes the white-tailed jack rabbit and sagebrush vole.
- The **Neotropical** faunal element (two North Dakota species) extends north from Central America and northern South America, and has contributed the opossum and gray fox to the state's fauna.

On a broad scale, these faunal elements appear to conform fairly well to the previously discussed mammal provinces of the state. Thus, in addition to its widespread species, we can see that the mammalian fauna of North Dakota is enriched by its relative position in the center of the continent, receiving contributions from neighboring major ecological communities, especially from the north, east, and southwest.

Checklist of the Mammals of North Dakota

The following checklist includes all species known to occur in North Dakota in historic time. They are grouped by major taxonomic categories, with genera and species alphabetized. The most recognized common name(s) are included after the scientific name. Scientific nomenclature is based on Wilson and Reeder's (2005) "Mammal Species of the World: a Taxonomic and Geographic Reference." The ground squirrel genus *Spermophilus* has been recently revised (Helgen et al. 2009), and the new nomenclature is included. In cases of recent changes in scientific names, commonly used synonyms are included in parenthesis. Many species were known and utilized by Native Americans living in the region long before white settlement. Hence, known Native American names are also included.

CLASS MAMMALIA

Subclass Theria

Infraclass Metatheria

Order Didelphimorphia (Opossums)

Family Didelphidae (New World Opossums)

Didelphis virginiana (Virginia Opossum; Sinte slatanka–Lakota)

Infraclass Eutheria

Order Eulipotyphla (Soricomorpha)

Family Soricidae (Shrews; Pute hanska itunkala–Lakota; Kinikisiwapikanotcki–Ojibway))

Blarina brevicauda (Northern Short-tailed Shrew)
Sorex arcticus (Arctic Shrew)
Sorex cinereus (Masked Shrew)
Sorex haydeni (Hayden's Shrew)
Sorex hoyi (Pygmy Shrew)
Sorex merriami (Merriam's Shrew)
Sorex palustris (Northern Water Shrew)

Order Chiroptera (Bats; Hupaki glake–Lakota)

Family Vespertilionidae (Vespertilionid Bats)

Corynorhinus townsendii (Townsend's Big-eared Bat)
Eptesicus fuscus (Big Brown Bat; Ágráphiga–Mandan; Ishwatáshia–Hidatsa; Hupáhwaki´kadakena–Dakota)
Lasionycteris noctivagans (Silver-haired Bat)
Lasiurus borealis (Red Bat)
Lasiurus cinereus (Hoary Bat)
Myotis ciliolabrum (Western Small-footed Myotis)
Myotis evotis (Long-eared Myotis)
Myotis lucifugus (Little Brown Myotis)
Myotis septentrionalis (Northern Myotis)
Myotis thysanodes (Fringed Myotis)
Myotis volans (Long-legged Myotis)

Order Carnivora (Carnivores)

Family Felidae (Cats; Igmu–Lakota)

Lynx canadensis (Lynx; Inmuchota–Dakota; Igmu hota–Lakota; Wach–Arikara; Sihtachache–Hidatsa)
Lynx rufus (Bobcat; Itupa-púzi, Bidábaho pusika–Hidatsa; Mantóka–Mandan; Igmu gleska–Lakota)
Puma concolor (Mountain Lion; Immu-tanka–Dakota; Igmu watogla, Igmu tanka–Lakota; Shunta-ha^{n}ska–Mandan; Itupa-ichtia–Hidatsa; Wachtas–Arikara

CLASS MAMMALIA (continued)

Family Canidae (Dogs; Sunka–Lakota)

Canis latrans (Coyote; Mes-ta-chá-gan-es–Ojibway; Míca, Mícaksica–Dakota; Sunkmanitu, Mayasleca–Lakota; Míkasi–Omaha; Schékè–Mandan; Mótsa–Hidatsa; Stshirits pukatsh–Arikara)

Canis lupus (Gray Wolf; Shu^{n}g-tokeca–Dakota; Sunkmanitu, Caksi–Lakota; Harrata–Mandan; Tshesha–Hidatsa; Stshirita-kusa–Arikara)

Urocyon cinereoargenteus (Gray Fox)

Vulpes velox (Swift Fox; Ihoichka–Hidatsa; Ohcha - Mandan; So^{n}gina–Dakota; Ciwaku–Arikara)

Vulpes vulpes (Red Fox; Ehchokuschi–Hidatsa; Hirúttsa–Mandan; Sungila lutah–Lakota)

Family Ursidae (Bears)

Ursus americanus (American Black Bear; Wasabè–Omaha; Wachank-shica–Dakota; Mato sapa–Lakota; Konuch-katit–Arikara; Haschida–Hidatsa; Ischída–Mandan)

Ursus arctos (Grizzly Bear; Mato, Mato-chota–Dakota; Sakehanska–Lakota; Mato, Mato unknapininde–Mandan; Lachpitzi–Hidatsa; Kúnuch, Konuch-tarawis –Arikara)

Family Mustelidae

Gulo gulo (Wolverine; Eh-tupah–Hidatsa; Matóka–Mandan; Sacatungka, Skecatanka–Lakota)

Lontra canadensis (River Otter; Pta^{n}–Dakota; Petakè–Mandan; Midapóka–Hidatsa; Citapat–Arikara)

Martes americana (American Marten; Wa'anksica–Lakota)

Pekania (Martes) pennant (Fisher; Skeca–Lakota)

Martes pennanti (Fisher; Skeca–Lakota)

Mustela erminea (Short-tailed Weasel; Ermine; Mach-schipka–Mandan; Itunkasan–Lakota)

Mustela frenata (Long-tailed weasel; Ohsisa–Hidatsa; Mahchpach-piraka–Mandan)

Mustela nigripes (Black-footed Ferret; Etopta sapa–Yankton Sioux; Itopta sapa–Lakota, Nazi–Mandan; Tahu akukahak napish–Hidatsa)

Mustela nivalis (Least Weasel; Hitu^{n}ka-sa^{n}–Dakota)

Neovison (Mustela) vison (American Mink; Dokshi^{n}ca–Dakota; Ikusan–Lakota; Daktsuà, Naksua–Hidatsa; Monika suntike, Mini-gasundek–Mandan; Eruch–Arikara; Zhon-gwayzh–Ojibway)

CLASS MAMMALIA (continued)

Taxidea taxus (Badger; Choka–Dakota; Hoka–Lakota; Maté–Mandan; Amaka, Awaga Hidatsa; Sunuh-katuh–Arikara)

Family Mephitidae (Skunks)

Mephitis mephitis (Striped Skunk; Ma^{n}ka–Dakota; Maka–Lakota; Su^{n}kte–Mandan; Hohga–Hidatsa; Nichwit–Arikara)

Spilogale putorius (Eastern Spotted Skunk)

Family Procyonidae (Raccoons, Ringtails, Coatis)

Procyon lotor (Raccoon; Wica–Dakota; Wiciteglega–Lakota; Miká–Omaha; Isat–Arikara; Shunte-pusa–Mandan; Sida-buzhe–Hidatsa; Asebun–Ojibway)

Order Rodentia (Rodents)

Family Sciuridae (Squirrels; Hetkala–Lakota)

Cynomys ludovicianus (Black-tailed Prairie Dog; Pi^{n}spi^{n}sa–Dakota; Pispiza–Lakota; Achks–Arikara; Shopka–Mandan; Si^{n}hpa, Tsipá–Hidatsa)

Glaucomys sabrinus (Northern Flying Squirrel; Psinca–Lakota)

Ictidomys (Spermophilus) tridecemlineatus (Thirteen-lined Ground Squirrel; Tashnáheca–Dakota; Tshíkarani–Arikara; Naksátshi–Hidatsa; Mashedónikcha–Mandan; Miniwakao–Cheyenne)

Marmota monax (Woodchuck; Groundhog)

Neotamias (Tamias) minimus (Least Chipmunk; Sachho–Arikara; Hetkada^{n}–Mandan; Kokokshi–Hidatsa)

Poliocitellus (Spermophilus) franklinii (Franklin's Ground Squirrel)

Sciurus carolinensis (Eastern Gray Squirrel; Zicahota–Lakota)

Sciurus niger (Fox Squirrel)

Tamias striatus (Eastern Chipmunk; Hetkal–Lakota)

Tamiasciurus hudsonicus (Red Squirrel; Ahjiduhmo–Ojibway; Zica–Lakota)

Urocitellus (Spermophilus) richardsonii (Richardson's Ground Squirrel; Flickertail; Honkóta–Arikara; Pinsa–Dakota; Shopka-sop–Mandan; Tsipá sopa–Hidatsa)

Family Castoridae (Beavers)

Castor canadensis (North American Beaver; Ah-mik–Ojibway; Ah-misk–Cree; Capa–Lakota)

CLASS MAMMALIA (continued)

Family Geomyidae (Pocket Gophers; Wahinheya–Lakota)

Geomys bursarius (Plains Pocket Gopher)

Thomomys talpoides (Northern Pocket Gopher; Machtóhpka–Mandan; Mánica–Dakota; Cipans–Arikara; Kípapudè–Hidatsa)

Family Heteromyidae (Pocket Mice, Kangaroo Rats)

Chaetodipus hispidus (Hispid Pocket Mouse)

Dipodomys ordii (Ord's Kangaroo Rat)

Perognathus fasciatus (Olive-backed Pocket Mouse; Apapsá–Hidatsa; Zhizhina–Dakota)

Perognathus flavescens (Plains Pocket Mouse)

Family Dipodidae (Jumping Mice; Psipsicala–Lakota)

Zapus hudsonius (Meadow Jumping Mouse)

Zapus princeps (Western Jumping Mouse)

Family Cricetidae (New World Mice, Rats, and Voles; Itunkala, Itungtanka, Intunpsicala, Iunka–Lakota)

Lemmiscus curtatus (Sagebrush Vole)

Microtus ochrogaster (Prairie Vole)

Microtus pennsylvanicus (Meadow Vole; Hi^{n}tu^{n}ka–Dakota; Gipápuli, Bidábaho itáhu–Hidatsa; Sakch–Arikara)

Myodes (Clethrionomys) gapperi (Southern Red-backed Vole)

Neotoma cinerea (Bushy-tailed Wood Rat)

Ondatra zibethicus (Muskrat; Zih-zirukka–Hidatsa; Si^{n}kpé–Dakota; Shantshuke–Mandan; Citakh–Arikara)

Onychomys leucogaster (Northern Grasshopper Mouse; Michtika, Michtik-tak–Mandan)

Peromyscus leucopus (White-footed Mouse; Wiyashpena–Dakota)

Peromyscus maniculatus (Deer Mouse; Tepa-uti–Omaha)

Reithrodontomys megalotis (Western Harvest Mouse)

Reithrodontomys montanus (Plains Harvest Mouse)

Family Muridae (Old World Mice and Rats)

Mus musculus (House Mouse)

Rattus norvegicus (Norway Rat)

Family Erethizontidae (New World Porcupines)

Erethizon dorsata (North American Porcupine; Pahi–Mandan, Dakota; Apadin–Hidatsa; Suunu–Arikara; Pahin–Lakota)

CLASS MAMMALIA (continued)

Order Lagomorpha (Hares, Rabbits, and Pikas)

Family Leporidae (Hares and Rabbits)

Lepus americanus (Snowshoe Hare; Waziyata mastinska–Lakota)

Lepus townsendii (White-tailed Jackrabbit; Warchu–Arikara; Ma^{n}sti^{n}ska–Dakota)

Sylvilagus audubonii (Desert [Audubon's] Cottontail)

Sylvlagus floridanus (Eastern Cottontail; Wahboos–Chippewa; Ma^{n}sti^{n}-sapana–Dakota; Mo^{n}stinga–Omaha)

Sylvilagus nuttallii (Mountain [Nuttall's] Cottontail; Nis–Arikara; Itakshipisha–Hidatsa)

Order Artiodactyla (Even-toed Ungulates)

Family Cervidae (Deer; Tahca–Lakota)

Alces americanus (alces) (Moose; Muswa–Cree, Ojibway; Wesucharut–Arikara; Ta, Tabloka, Tawiyela–Lakota; Pachúptaptach–Mandan)

Cervus elaphus (canadensis) (Elk; Wapiti–Shawnee; Wah–Arikara; Ompa, O^{n}pa–Mandan; A^{n}pa^{n}–Omaha; Upa^{n}–Dakota; Hehaka, Unpan–Lakota; Madoka–Hidatsa)

Odocoileus hemionus (Mule Deer; Tsitashipisa–Hidatsa; Sinte-sapana–Dakota; Sinte sapela–Lakota; Shunte-psih–Mandan; Takatit–Arikara)

Odocoileus virginianus (White-tailed Deer; Tachtsha–Dakota; Sintehanska–Lakota; Tstita-taki–Hidatsa; Mahmanaku–Mandan; Ta-paht–Arikara)

Family Antilocapridae (Pronghorn)

Antilocapra americana (Pronghorn; Koka–Mandan; Tatókana–Dakota; Tatokala, Nigesanla–Lakota; Uchi–Hidatsa; Chka–Arikara)

Family Bovidae (Bovids)

Bison (Bos) bison (Bison; Te–Omaha; Pte–Dakota, Mandan; Pte, Tatanka–Lakota; Mité–Hidatsa; Tanaha–Arikara)

Ovis canadensis (Bighorn Sheep; Ansa-chta–Mandan; Heki^{n}skagi–Dakota; Azichtia–Hidatsa; Arikusa–Arikara)

Order Didelphimorphia—**Opossums**

Until fairly recently this order, along with six others, was included in the single order Marsupialia, the marsupials. However, recent research has resulted in the recognition of separate orders. The common term "marsupial" continues in use when referring to these mammals. In general, the marsupials are known for their unique mode of reproduction involving a very short gestation resulting in the young born in an extremely undeveloped state. The newborn is able to move into a pouch on the female's abdomen, the marsupium, and attach itself to a teat, where it remains for an extended period. This form of reproduction is in contrast to that of the more modern "placental" mammals in which the young are born in a more advanced state of development and are able to become independent more quickly.

Only one family, Didelphidae, occurs in North America, with a single species, the Virginia opossum. It has been said that when you see an opossum, you are looking at a relict of Cretaceous times, for it has many skeletal features characteristic of the earliest mammals. Didelphid ancestors occurred in North America during Eocene times, but died out in the Miocene (20 million years ago). They recolonized the continent from South America with the reopening of the Panamanian land bridge during the Pleistocene. The opossum's survival in competition with more modern northern mammals is probably due to its fairly large size and unspecialized food habits, resulting in a generalized life style allowing it to occupy a variety of habitats.

Family Didelphidae
New World Opossums

Virginia Opossum
Didelphis virginiana

Sinte slatnka—Lakota

R. W. Van Devender, American Society of Mammalogists Mammal Images Library

Description

The Virginia opossum is the only marsupial mammal in North America north of Mexico. These primitive mammals have existed almost unchanged since their Cretaceous origins 130 million years ago. Their young are born in a very undeveloped state and carried by the mother in an abdominal pouch ("marsupium") for an extended period. They exhibit numerous primitive dental and skeletal characteristics distinguishing them from the more advanced "placental" mammals. Adult Virginia opossums are about the size of a house cat, weighing 2 to 4 kg. The underfur is white and overlain with black guard hairs, giving the animal an overall grayish appearance. The ears and tail are naked. The tail is prehensile and, together with an opposable big toe, aids in tree-climbing. Average standard body measurements in the northern portion of the range are: Total Length - 759 millimeters (mm), Tail Length - 295 mm, Hind Foot - 67 mm, Ear - 50 mm. Northern specimens frequently exhibit shorter ears and tails due to frostbite damage.

Distribution

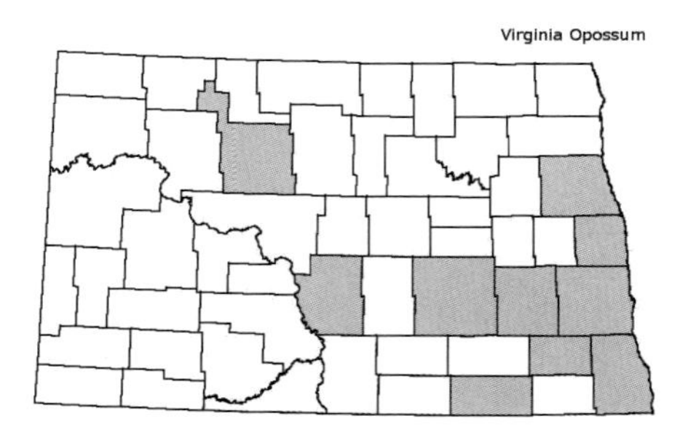

The natural distribution of the opossum includes the eastern portions of North America from southern Canada south to Nicaragua, and west into the Great Plains. It has been extensively introduced elsewhere in western United States, especially along the West Coast. Opossums appear to be recent arrivals in North Dakota. The adjacent Minnesota range includes all of the western counties and the central and southern portions of the state. There are scattered records for eastern South Dakota, but through the 1950s, there were only a few reports from southeastern North Dakota south of Hillsboro (Traill County). There has been one additional record (1991) from Bismarck (Burleigh County) cataloged into the University of North Dakota collection. There has been one additional record (1991) from Bismarck (Burleigh County), two from Grand Forks County, and scattered reports from the southeastern counties.

Habitat

Although Virginia opossums are found throughout much of the United States, including relatively arid country, they are most common in more mesic habitats, especially woody areas near rivers and streams. Furthermore, the few North Dakota records are from areas of riparian woodland adjacent to the Red and Missouri rivers.

Ecology and Behavior

The opossum is a true generalist and is able to survive in a wide variety of situations. They take a wide variety of animal and plant foods, including bird eggs and young, small mammals, frogs, clams, crayfish, insects, carrion, fruits and grains. They will den in hollow trees, under stumps and logs, abandoned burrows, or deserted buildings. This generalized life style has been thought to be one of the keys to their survival over millions of years.

Home ranges in this northern portion of the species distribution are unknown, but in more southern areas have been recorded at about 4 to 5 hectares (ha), and quite elongate when along water courses. Populations are heavily weighted toward young of the year.

Opossums are nocturnal animals. Although they can maintain their body temperature in ambient temperatures below 0°C, they frequently exhibit evidence of frostbite in northern climates and are inactive below -7°C. Opossums are usually solitary (except during the breeding season); encounters between adults are agonistic, manifested by growling and hissing, bared teeth, and tail lashing. If fighting occurs or the animal is struck by a sharp blow, it may feign death by going into a catatonic state ("playing possum").

Opossums seldom live more than two years in the wild. Predation, other than by some raptors, does not appear to be a significant cause of mortality. Hence, other factors such as starvation, accidents and weather must be regarded to be of greater significance in limiting populations.

Reproduction

There are few data on breeding seasons in the Upper Midwest. The earliest recorded breeding activity in Iowa was in early February, while January-March and May-July breeding peaks occur in New York. Considering the severity of the northern Plains climate, we may expect any reproduction in North Dakota to occur no earlier than mid-spring.

An average of six to eight extremely undeveloped young are born following a gestation of only 12 to13 days. The young crawl to the marsupium and attach to a teat, where they remain for about 60 days. Weaning takes place at about 100 days, but the young begin taking solid food prior to that time. Sexual maturity of females may be attained during the first breeding season after birth.

Status and Conservation

The distribution of the Virginia opossum has been expanding in the Upper Midwest, coinciding with human population growth and urbanization, and additional North Dakota observations may be made in the foreseeable future. However, because of the severe climate, the species will probably never become a significant component of the state's mammalian fauna.

The Virginia opossum is not regarded to be an important furbearer or game species in North Dakota. Elsewhere in the United States, opossums cause some agricultural damage through depredation on poultry, eggs, or other agricultural products. However, it is unlikely that the species will attain significant numbers to be a problem in North Dakota.

Selected References

Adams (1961), Hazard (1982), Higgins et al. (2000), Jones et al. (1983), McManus (1974), Walsh et al. (2017)

Order Eulipotyphla (Soricomorpha)—**Shrews and Moles**

Until recently, these mammals were included in the now disbanded order Insectivora. They include the shrews and moles, along with several West Indian forms. Members of this order feed primarily on insects and other small invertebrates, although they may feed on any animal material, and occasionally even seeds. This order (417 species worldwide) has retained many structural features similar to the primitive "placental" mammals of the Cretaceous.

Only one family of soricomorphs, the Soricidae (shrews), is known to inhabit North Dakota. Shrews are small, mouse-size mammals with short fur, tiny eyes, external ears buried in the fur, and long tapering noses. They are widespread, occurring over much of the world. Six species have been reported in the state.

Family Soricidae

Shrews; Pute hanska itunkala—Lakota; Kinikisiwapikanotcki—Ojibway

Northern Short-tailed Shrew
Blarina brevicauda

R. Altig, American Society of Mammalogists Mammal Images Libarary

Description

This relatively common species is North Dakota's largest shrew, easily distinguished from other members of the family. Like other shrews, it has inconspicuous eyes and ears and a long, pointed snout. However, its snout is relatively shorter and broader than that of members of the genus *Sorex*. The dorsal pelage is short, soft, and dark slate-colored; ventral pelage is somewhat lighter. Average body measurements are: Total Length - 129 mm, Tail Length - 25 mm, Hind Foot - 16 mm, Ear - 7 mm.

Distribution

The northern short-tailed shrew occurs over much of the northeastern United States and southern adjacent Canadian provinces, west

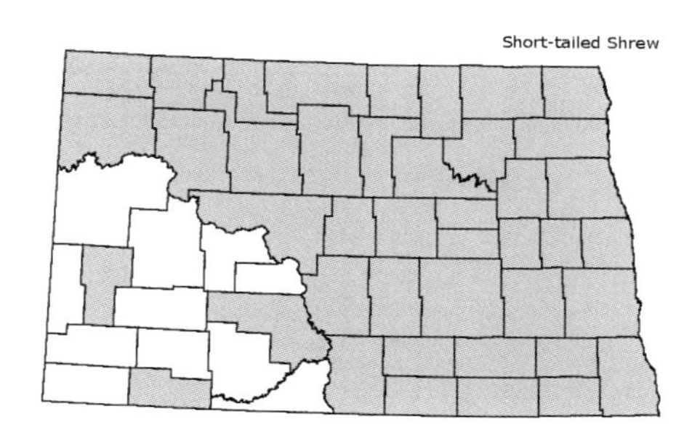

to eastern North Dakota, South Dakota, and Nebraska. In North Dakota, it primarily occurs east of the Missouri River, although specimens have been taken in Morton, Adams, and Billings counties. Its South Dakota distribution is similar, with additional reports from several western counties adjacent to the Missouri River.

Habitat

In the eastern part of its range, this shrew is primarily a forest-dwelling species. However, on the Great Plains, it occurs in a variety of habitats but seems to prefer heavy vegetation. It is common in tall-grass prairie, brushy areas, and forests, where it establishes burrows and runs in thick ground cover and leaf litter. In the prairie-forest transition of North Dakota and Minnesota, short-tailed shrews are common in nearly all habitats, including prairie, aspen parkland, and deciduous and coniferous forests.

Ecology and Behavior

Like other members of the family, the short-tailed shrew is primarily a carnivore, although some plant material may be eaten. Because of its size, it tends to prey on large invertebrates and small vertebrates, including small mammals. At least in captivity, they have been observed to kill adult meadow voles. Their ability to take larger prey is aided by venom, unique in North American mammals, which is produced by their submaxillary glands. When bitten, prey may be quickly killed or stunned, a feature which may facilitate food hoarding or storage. Short-tailed shrew bites of humans are painful, but not considered dangerous. For its weight, food consumption is significantly less than for smaller shrews. This is probably related to its larger size and lower rate of heat loss due to a proportionately smaller surface area per unit of mass compared to other shrews.

Short-tailed shrews are well-adapted for digging, and have extensive tunnel systems. Shallow tunnels, 2 to 3 cm in diameter, are constructed a few centimeters below the surface, and interconnect with deeper systems at about 50 cm. Runways and tunnels of other species are used as well. Spherical nests are constructed underground, and are lined with vegetation and fur of other small mammals.

Home ranges of short-tailed shrews average about 2.5 ha and frequently overlap with other members of the species. Hence, these shrews are not believed to be territorial. Population levels vary greatly; estimates have run from less than two to about 120 per hectare. Mortality can be very high, especially during winter, when up to 90% of a population may be decimated. Such high winter mortalities have been thought to be cold-related. A wide variety of animals prey on Blarina, including snakes, owls, hawks, ground squirrels, and small to medium-sized predatory mammals.

Like other shrews, short tailed shrews are active day and night throughout the year. They emit a variety of high-pitched vocalizations. In addition, they engage in visual communication related to agonistic behavior. Five separate postures indicating differing degrees of intimidation have been described. They also emit ultrasonic "clicks," enabling some degree of echolocation, but that ability is not as refined as that observed in bats.

Reproduction

The breeding season at the latitude of North Dakota probably runs from May through August. At least two litters of four to seven young are produced in a given breeding season. Gestation periods have been reported to be 21 to 22 days. Young are weaned at 25 days and attain sexual maturity at 45 to 50 days. Successful breeding may occur at about 2 months. Hence, it is possible that early litters may reproduce late in the summer of their birth.

Status and Conservation

The northern short-tailed shrew is common in North Dakota east of the Missouri River. It is probably absent from the western part of the state, with the exception of mesic areas near the Missouri River floodplain.

This shrew is not regarded as a harmful species, but may be important in control of insects and some rodent populations. As a carnivore, it has been known to concentrate insecticide residues in its tissues, which may have an impact on populations. Heavy grazing, or other activities reducing vegetative cover, could have a negative impact on populations.

Selected References

George et al. (1986), Hazard (1982), Higgins et al. (2000), Iverson et al. (1967)

Arctic Shrew
Sorex arcticus

Description

The arctic shrew is a medium-sized shrew, somewhat larger than Hayden's and masked shrews. Its most distinctive external characteristic is its tricolored pelage, a dark brown to black back, pale brown sides, and grayish brown underparts. This pelage is most distinctive in adults; young of the year in summer have more subdued tricoloration and, except for the larger size, could be confused with Hayden's and masked shrews. Average body measurements are: Total Length - 109 mm, Tail Length - 40 mm, Hind Foot - 14 mm, Ear - 8 mm. Body weight averages 9.1 g. The skull is longer (>17 mm) than other members of the genus in North Dakota.

Distribution

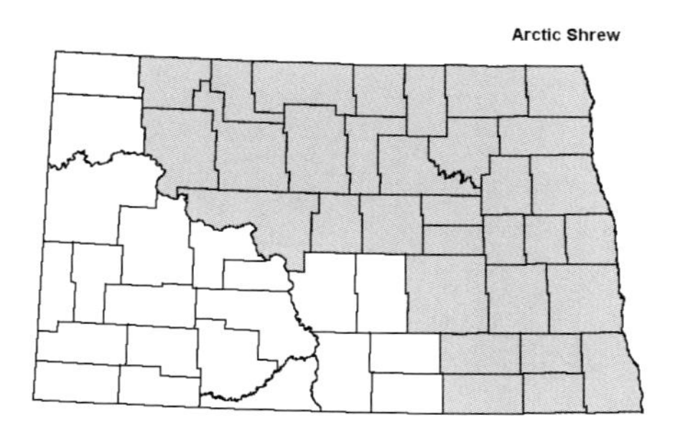

The arctic shrew occurs primarily in the boreal forested areas of North America, from the southern Yukon to the Maritime Provinces of Canada, south to Michigan, Wisconsin, Minnesota, and the eastern Dakotas. The North Dakota distribution includes the northeastern half of the state, from Divide and Williams counties to the southeastern corner. There is also a record from McKenzie County, the only one west of the Missouri River. Two subspecies are believed to occur in the state: the slightly darker, smaller *S. a. arcticus* in the northwestern corner, and the larger, lighter *S. a. laricorum* in the rest.

Habitat

In boreal regions, arctic shrews inhabit damp spruce and tamarack forests, but in the northern Great Plains, primary habitats consist of low, marshy areas with mixed cattails, forbs and willows, and along lake shores. In the prairie-forest transition of North Dakota-Minnesota, they have been found in both grassland and aspen-forested areas. Early North Dakota reports are from forested valleys, marshes and lake shores, and more recent records from Grand Forks County include edges of marshes and coulees, and lowland prairie.

Ecology and Behavior

Arctic shrews are active throughout the 24-hour period, but primarily during the hours of darkness. Typical of other shrews, they are constantly foraging for food, primarily insects, including grasshoppers, caterpillars, and larch sawflies. In some areas, larch sawflies may comprise nearly 70% of the diet. Arctic shrews have been observed to hunt grasshoppers resting on grass stems during early morning. Shrews climbed adjacent stems, and then launched themselves to the insects, grasping them with their jaws and feet.

Home ranges of arctic shrews can average about 0.5 ha. Population levels during summer and fall can run from 4 to 9 shrews per ha. Apparently arctic shrews tolerate significant overlap of home ranges, in contrast to other shrew species. Mortality is high, especially among nestlings, where it has been estimated at about 50%. Nevertheless, shrews that attain sexual maturity may live as long as 18 months.

Reproduction

The breeding season of the arctic shrew at the latitude of North Dakota runs from April through August. A litter of five to nine young is born following a 3 week gestation, and are weaned at 3 weeks of age. Two to 3 litters may be produced during a breeding season. Early litters may attain sexual maturity and breed by August.

Status and Conservation

The arctic shrew is fairly uncommon in eastern North Dakota, and absent west of the Missouri River. The North Dakota Game and Fish Department has designated this shrew as a Level III Species of Conservation Priority. As with other shrews, the arctic shrew is considered to be beneficial, consuming large numbers of insects. Indeed, it has been studied in Canada for its potential in control of larch sawflies. Low, moist areas, including wetland edges, should be protected to enhance populations of this species.

Selected References

Bailey (1926), Buckner (1957), Dyke et al. (2004), Foresman (2001), Hazard (1982), Iverson et al. (1967), Jones et al. (1983), Kirkland and Schmidt (1996), Shaughnessy (2016)

Masked Shrew
Sorex cinereus

Hayden's Shrew
Sorex haydeni

Donald Rubbelke

Description

These shrews are among the smallest of mammals, frequently weighing less than 3 grams. Until recently, masked and Hayden's shrews were considered to be separate subspecies of one species, *S. cinereus*. However, they have subsequently been split into two morphologically very similar species, distinguished only by subtle external and cranial characteristics.

Masked/Hayden's shrews have long pointed rostrums, small eyes, and teeth with reddish-brown tips. The fur is fine and very soft, brown to brownish-gray dorsally, and pale gray ventrally. The tail is relatively long and indistinctly bicolored, brown above and pale gray underneath, with a black tip (viewed from above). Average standard body measurements are for the masked shrew: Total length - 94.5 mm, Tail length - 40 mm, Hind foot - 11 mm, Ear - 6 mm; for the Hayden's shrew: Total Length - 82 mm, Tail Length - 32 mm, Hind Foot - 11 mm, Ear - 8 mm. *S. cinereus* is darker brown, and has a longer tail with more tuft.

Distribution

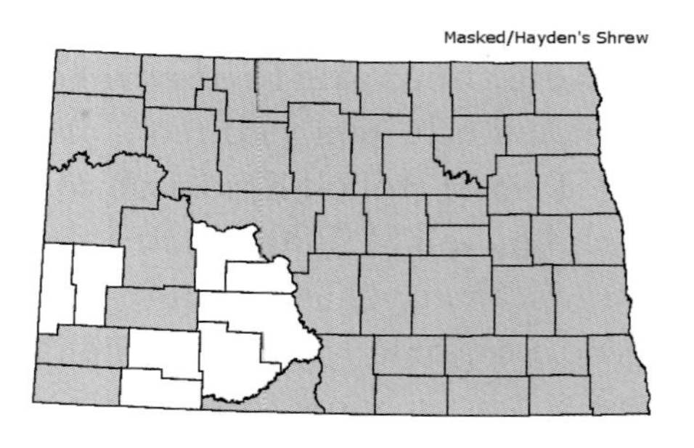

The masked shrew has the most extensive distribution of shrews in North America, occurring in the northern half of the continent. It has also been recorded in eastern Siberia. Hayden's shrew is found throughout the prairies of the Dakotas, eastern Montana, western Minnesota, Nebraska and Iowa, and as far south as northern Kansas and Missouri. The northern limit is southern Alberta, Saskatchewan, and Manitoba. Because of the recent splitting into two species and subtle taxonomic differences, local collections have not been analyzed to determine the exact distribution of masked and Hayden's shrews in North Dakota. In western North Dakota, scattered records have been reported near Selfridge and Cannonball (Sioux County), and Bowman, Dunn, McKenzie, Slope, and Stark counties. We suspect that Hayden's shrew is the more common of the two in North Dakota, occurring statewide in mesic prairie habitats. The masked shrew may occur in wooded and riparian areas, especially in the eastern half of the state.

Habitat

The masked shrew occurs in a wide variety of habitats, from dense forests to prairies, to arctic tundra. Moisture appears to be a limiting factor, with more mesic sites being preferred. Also preferred are areas with significant ground cover, including herbaceous vegetation, leaves, logs, stumps, and brush piles. During periods of high moisture, they may colonize open fields. Although some habitat overlap seems to occur, the Hayden's shrew prefers drier, grassier habitats than those used by masked shrews. Thus Hayden's shrew may be regarded as more of a habitat specialist.

Ecology and Behavior

These two shrews, like other members of the genus, are active day and night, with the greatest amount of activity occurring during the hours of darkness. Foraging bouts occur throughout the day and night, with intervening periods of a few hours of inactivity. This pattern reflects their extremely high metabolic rates. A shrew may consume up to 75% of its body weight in a single 24-hour period, and can starve to death in a matter of a few hours. Foods are primarily insects (adults, pupae, larvae, and eggs), earthworms, and other small invertebrates. Small vertebrates, such as baby mice, may also be consumed, as well as carrion and seeds. Shrews forage by burrowing through loose soil, grass, leaves, and other above-ground material, and through the underground tunnels of other species.

Shrews do not hibernate and are active throughout the year, foraging under the snow. In northern climates, winter survival may be contingent on adequate snow cover protecting shrews and other small mammals from temperature extremes.

Population densities may run as high as 25 per hectare, and may fluctuate greatly from year to year. Home ranges average about 0.5 ha. Mortality, especially among young, can be very high. Common causes of mortality include starvation, cold, and predation. Mammalian predators may kill shrews, but not eat them because of their strong musky odor. Longevity may be as high as 14 to 16 months, but most probably do not survive their first year.

Shrews are highly vocal, and will engage in vigorous squeaking, visual display, and ritualized fighting when an intruder is detected. They are also known to emit ultrasound which functions in echolocation.

Reproduction

Breeding in the northern Plains occurs from April to October. A female may produce a litter of two to 10 young following a gestation period of about 18 days. Two to three litters per year may be produced. Young are weaned at about 20 days and mature rapidly. Early litters may breed by late summer, but those born later attain sexual maturity the following spring.

Status and Conservation

Both species are regarded as common in relatively moist habitats, but may undergo large annual fluctuations in numbers. Neither species has been reported from southwestern North Dakota in recent years.

Neither species has any known negative economic impact. Both consume large numbers of insects. Moist areas, such as wetlands, low areas, and riparian areas are important to shrews, particularly to maintain source populations during dry periods. Maintenance of undisturbed areas with a good ground cover including litter, brush piles, and downed woody materials is beneficial. Heavy grazing, wetland drainage, and application of pesticides and insecticides likely have an adverse effect on shrew populations.

Selected References

Bailey (1926), Foresman (2001), Foresman and Jensen (1992), Genoways and Jones (1972), Seabloom et al. (1978), Svihovec (1967), van Zyll de Jong (1980), Whitaker (2004)

Pygmy Shrew

Sorex hoyi

Description

Until recently, this shrew had been regarded as under a separate genus, *Microsorex*. It is now placed in the genus *Sorex*, along with the other long-tailed shrews. Weighing only about 2 grams, the pygmy shrew is the smallest mammal in North America, and one of the smallest in the world. Its size and grayish-brown color make it difficult to distinguish externally from the closely related masked and Hayden's shrews. Average body measurements are: Total Length - 84 mm, Tail Length - 30 mm, Hind Foot - 10 mm, Ear - 4 mm.

Distribution

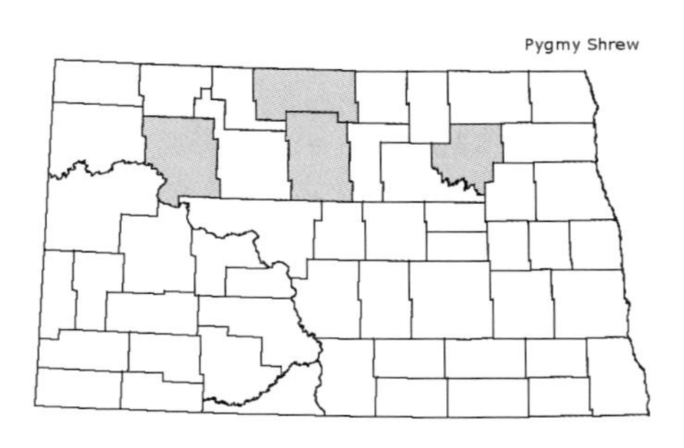

The pygmy shrew is primarily a boreal animal, occurring over much of Alaska and Canada. The distribution dips into eastern Washington, northern Idaho, western and extreme northeastern Montana, North and South Dakota east of the Missouri River, and the Great Lakes states. In Minnesota, it occurs in approximately the northeastern two-thirds of the state, but is absent from the western tier of counties adjacent to the Dakotas. The specific North Dakota distribution is problematic, with the only records being from Devils Lake (Ramsey County), J. Clark Salyer National Refuge (McHenry County), the Turtle Mountains (Bottineau County), and Mountrail County. Because of the subtleties of identification, however, it may be important to re-examine cranial specimens from existing North Dakota collections.

Habitat

This shrew is usually found in varied forest habitats, swamps, grassy clearings, and floodplains. In spring, they use swamps and marshes, and move to upland areas in late summer. They appear to prefer moist soils, tunnels of small mammals, leaf litter, roots, and stumps.

Ecology and Behavior

Pygmy shrews are insectivorous, feeding on ants, flies, spiders, small beetles, grubs, and caterpillars. Earthworms are also taken. Carrion feeding has been observed in captives.

Population densities appear to be low compared to other shrews, about 0.5/ha. It is not known if these shrews are territorial. Known predators include some hawks and garter snakes. Maximum longevity is probably 15 to 16 months.

Reproduction

Very little is known of reproduction in this shrew. Elsewhere in the species range, pregnant females with five to seven embryos have been found during July and August. Based on scanty evidence, it is believed that only one litter per year is produced, which may explain reports of relatively low population densities.

Status and Conservation

The pygmy shrew is rare in North Dakota. The North Dakota Game and Fish Department has designated it as a Level II Species of Conservation Priority. Its status may be clarified with additional surveys in forested river valleys in the eastern part of the state.

This species is undoubtedly of little economic significance. In North Dakota, its importance may lay in its contribution to the state's faunal diversity.

Selected References

Bailey (1926), Datta et al. (2018), Dyke et al. (2004), Foresman (2001), Hazard (1982), Higgins et al. (2000), Long (1974)

Merriam's Shrew

Sorex merriami

Description

This shrew can be distinguished from other North Dakota long-tailed shrews by its pale color. The dorsal color is light gray, while the sides and ventral surface are whitish to buffy. Average standard body measurements are: Total Length - 96 mm, Tail Length - 36 mm, Hind Foot - 12 mm, Ear - 5 mm. The skull is similar to that of *S. cinereus*/ *S. haydeni*, but has a greater maxillary breadth and subtle differences in the upper incisors.

Distribution

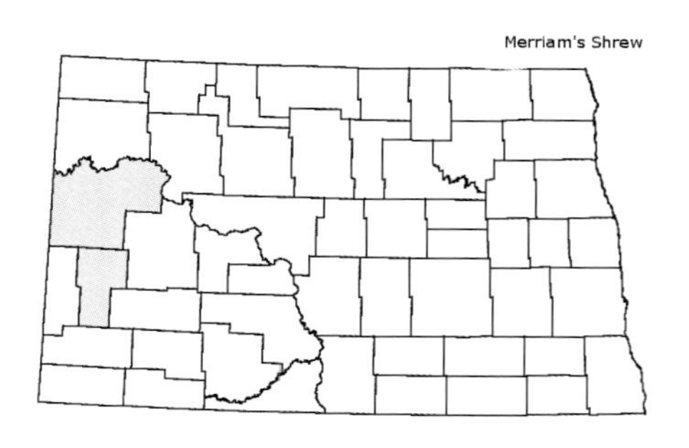

Merriam's shrew is distributed from the western Dakotas and Nebraska to southeastern British Columbia, eastern Washington, Oregon, and California, and south to Arizona and New Mexico. For many years, the species was known in the Dakotas only by a single specimen from Butte County, South Dakota, and a partial specimen from Billings County, North Dakota. Recently, two additional specimens were collected from Billings and McKenzie counties, thus verifying the continued presence of the species in North Dakota. A few additional records occur from central and eastern Montana.

Habitat

Merriam's shrews are adapted to arid shortgrass prairie and sagebrush in the northern Great Plains. On the Great Plains, they have been found in shrub-steppe, open dry grasslands, disturbed roadsides, and grassy openings in ponderosa pine. This species may be one of the most xeric-adapted shrews in North America. The species is sometimes associated with runways of the sagebrush vole, *Lemmiscus curtatus*.

Ecology and Behavior

Very little is known about the natural history of Merriam's shrew. There appears to be a close association with the sagebrush vole. Food habits include caterpillars, spiders, beetles and grubs, and crickets. Owls are the only known predators.

Reproduction

Breeding may occur from mid March to early July. Litter size, based on a few necropsy records, probably ranges from five to seven.

Status and Conservation

Merriam's shrew must be regarded as extremely rare in North Dakota. This species, because of its rare status, is very vulnerable to overgrazing or other practices damaging to shortgrass prairie and sagebrush.

Selected References

Armstrong and Jones (1971), Bailey (1926), Foresman (2001), Mullican (1994), Shaughnessy and Woodman (2015)

Northern Water Shrew
Sorex palustris

Donald Rubbelke

Description

The northern water shrew is the largest of North Dakota's long-tailed shrews. The back and sides are black, the underparts silvery gray, and there is a conspicuous fringe of stiff hairs on the hind feet that aid in swimming. Average standard measurements are: Total Length - 149 mm, Tail Length - 67 mm, Hind Foot - 19 mm, Ear - 9 mm. *Sorex palustris* is the largest of our long-tailed shrews, with males averaging 15.4 g and females 12.3 g.

Donald Rubbelke

Distribution

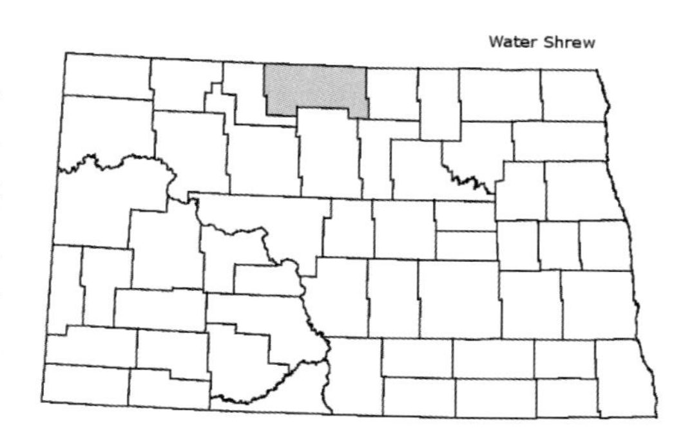

The northern water shew occurs throughout the cooler boreal and montane regions of North America. Regionally, in Manitoba there are southerly records from boreal outliers in the Porcupine, Duck, and Riding Mountains, and the Carberry Sandhills. There are records from northern Minnesota, including the eastern portions of counties adjacent to the Red River. There is just one confirmed record from Marshall County in northeastern South Dakota.

Until recently, northern water shrews had not been reported in North Dakota. Bailey (1962) discussed records from extreme northeastern South Dakota and near Winnipeg, Manitoba, which indicated their possible occurrence along the Red River and its tributaries. Then, in 2018, two specimens were recorded in the forested Turtle Mountains (Bottineau County), the first records for the state.

Habitat

As the name implies, these shrews live along unpolluted streams, lake margins, and wetlands having rocks, logs, crevices, and overhanging banks. Beaver and muskrat activity and associated lodges, dams, and vegetation is also associated with water shrews. The two recent North Dakota specimens were collected in separate semi-permanent wetlands in the Turtle Mountains. Consequently, further surveys may reveal additional records in the lakes, ponds, and streams of the Turtle Mountains, Pembina Hills, and tributaries of the Red River.

Nests are constructed in the cavities of overhanging stream or lake banks. Beaver dams and lodges and hollow logs may also be used. Nesting materials include grass, moss, and sticks.

Ecology and Behavior

These shrews are capable swimmers and divers, entering the water to capture prey and escape predators. However, to remain under water they must be constantly swimming, due to a film of air covering their fur, increasing buoyancy and causing them to float like a cork when they stop paddling. They can remain in the water only a few minutes before having to return to land to groom and dry their fur.

Foods of water shrews are primarily aquatic insects, but slugs, snails, earthworms, and small fish are also important in their diet. Shrews will actively hunt and feed for about thirty minutes, followed by an hour of rest. Peaks of activity occur between sunset and 11 p.m. and shortly before sunrise.

Water shrews appear to be solitary and do not exhibit a dominance hierarchy or occur in social groups. Encounters between individuals has been antagonistic, and both sexes are equally prone to fight.

Like other small mammals, longevity is short and probably does not exceed 18 months. Predators of water shrews have been recorded as including predatory fish, snakes, weasels, and hawks and owls.

Reproduction

Water shrews mature earlier than other members of the genus. In males, testicular enlargement begins in late December, and females exhibit ovarian activity in January. First pregnancies have been reported in February. Reproductive activity continues into August. First-year shrews tend to delay reproduction until their second winter.

Females produce two or three litters per year. Length of gestation is unknown, but for most shrews gestation averages about 21 days. Litter size, based on embryo counts, averages 6 (3–10).

Status and Conservation

The northern water shrew has no known economic significance. Its apparent preference for clear lakes and streams may provide an indication of habitat quality.

Currently, the northern water shrew must be considered as rare in North Dakota, with just one confirmed report over the past 100+ years. Future collecting efforts in northeastern streams and lakes may clarify its status in the state. Its documented presence in North Dakota makes a significant contribution to the state's faunal diversity.

Selected References

Beneski and Stinson (1987), Datta et al. (2018), Foresman (2012), Hazard (1982), Higgins et al. (2000), Rubbelke (1984), Sasmal et al. (2018), Wrigley et al. (1979)

Order Chiroptera—
Bats

Hupaki glake—Lakota

The bats are the only mammals which have attained true flight. The forelimbs, especially the digits, are greatly elongated and form a bony support for the wing membrane. In many groups, an additional membrane (uropatagium) extends between the hind limbs, forming an additional control surface in flight.

While many bats have good vision, most emit ultrahigh frequency sounds for echolocation of obstacles and prey. Most species are insectivorous, but some tropical species specialize on pollen, fruit, small vertebrates, and blood. Bats are long-lived (up to 30 years), especially considering their size. Secure roosting sites, along with their ability to hibernate during severe seasons, appear to be major factors contributing to their longevity. However, human-caused factors, including pesticide use and habitat destruction, have resulted in population declines. Wind energy facilities have been recently shown to cause high mortalities, especially to tree bats *(Lasionycteris, Lasiurus)*. Barotrauma, lung damage resulting from sudden changes in air pressure, occurs when bats fly near rotating wind turbine blades.

This is the second largest order of mammals, with more than 850 species. It is primarily a tropical order, and species diversity decreases rapidly in the northern hemisphere. Those that do occur in northern regions survive by migrating to suitable hibernating sites during winter periods when food supplies are unavailable. For many years, the assumption was that North Dakota bats were forced to migrate out of the state, due to its severe winter climate and lack of suitable hibernation sites (caves). Those in the east were thought to migrate to southern Minnesota caves and woodlands, and western bats to caves in South Dakota's Black Hills. Recently, however, six species of western North

Dakota bats were documented to hibernate in mud and rock caves and crevices of the North Dakota badlands. One family, the Vespertilionidae, occurs seasonally in North Dakota. Eleven species have been reported in the state. Several species are colonial, the most common being the little brown bat and big brown bat.

Family Vespertilionidae
Vespertilionid Bats

Townsend's Big-eared Bat
Corynorhinus (Plecotus) townsendii

Dan Brockman

Description

The Townsend's big-eared bat is easily distinguished from other regional bats by its extremely large ears. They are joined across the forehead at the base, and when laid back, extend to the middle of the body. There are also prominent facial glands between the nostril and the eye. The pelage color can range from brown to slate. Average standard body measurements are: Total Length - 102 mm, Tail Length - 46 mm, Hind Foot - 11 mm, Ear - 36 mm.

Distribution

This bat occurs from southern British Columbia south through much of the western United States, into central Mexico. There are also isolated southeastern populations of separate subspecies in Missouri,

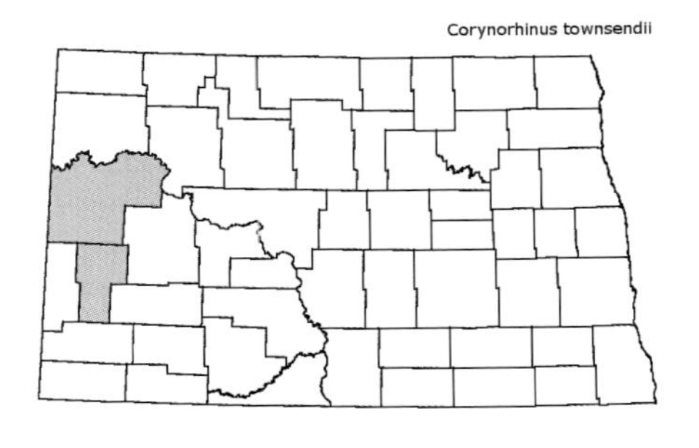

Arkansas, Oklahoma, Kentucky, and the Virginias. Until very recently, there were no known records from North Dakota, but its occurrence was suspected because of records in adjacent counties in extreme northwestern South Dakota and southeastern Montana. In June 2009, seven females, two of which were visibly pregnant, were live-captured in McKenzie County (Cori Lausen, Birchdale Ecological Ltd., Kaslo, BC, Canada, personal communication). More recent records have been documented in Billings County.

Habitat

Throughout much of its range, Townsend's big-eared bat is regarded as a habitat generalist, but it is most commonly associated with mesic deciduous and coniferous forests. In semiarid portions of eastern Montana, it occurs in Rocky Mountain juniper-limber pine vegetation, while in South Dakota, most reports are from the forested Black Hills.

Spring and summer roosting habitat includes the warmer portions of caves and mines, and buildings. The North Dakota observations were made in cottonwood/willow habitat near water and close to log buildings. Winter hibernation occurs in relatively cool but thermally stable sites in caves and mines. We had assumed that this bat migrated to the caves and mines of the Black Hills for hibernation. However, a recent survey documented these bats overwintering in small caves and deep crevices in the North Dakota badlands.

Ecology and Behavior

This is a relatively sedentary bat, not known to make long-distance migrations. Recorded movements from maternity roosts to hibernacula have ranged from 3 to 70 km. The recent North Dakota observations could imply much greater potential migration distances, the nearest large caves and mine shafts being in South Dakota's Black Hills, a distance of over 370 km. Alternatively, North Dakota's Badlands contain numerous small caves in eroded mudstone, which might provide suitable hibernacula for this and other bat species.

During spring and summer, males are solitary while females assemble in maternity colonies of usually under 100 bats. Throughout hibernation, these bats roost singly or in small clusters of up to 58 individuals.

Townsend's big-eared bat forages later than many other bats, reaching its peak of activity nearly two hours after sunset. The species seems to prefer forest edges and riparian zones within 3 km of the roost. Foraging is principally on small moths, but other insects, e.g., beetles, lacewings, flies, wasps, etc., are taken as well.

Mortality of young-of-the-year bats runs about 5% prior to hibernation, and 50% over the first year. Annual survival of adults is approximately 80%, with a maximum known longevity of 16 years.

Reproduction

The breeding season extends from October through late February. Young females are believed to be capable of reproduction during their first year, but young males are not. As with other regional bats, delayed fertilization occurs, with parturition beginning in late May following a gestation of 56 to 100 days. Young are fledged at 2.5 to 3 weeks, and weaned by 6 weeks of age.

Status and Conservation

The U. S. Forest Service lists the Townsend's big-eared bat as a Sensitive Species, and the states of South Dakota and Wyoming list it as a species of concern. Two eastern subspecies are listed by the U. S. Fish and Wildlife Service as Endangered. Until 2009, the species was not known to occur in North Dakota, thus it has not received conservation attention. The recent southwestern North Dakota observation warrants continued monitoring to determine if this bat is a regular resident of the state.

Selected References

Barnhart and Gilam (2016), Barnhart and Gilam (2017), Cockrum and Cross (1964), Foresman (2001), Higgins (2000), Jones et al. (1982), Kunz and Martin (1982), Lausen and Barclay (2006), Nelson et al. (2015), Schmidt (2003), Sherwin et al. (2000), Sherwin et al. (2003), U. S. Fish and Wildlife Service (2004)

Big Brown Bat

Eptesicus fuscus

Ágráphiga – Mandan; Ishwatáshia – Hidatsa; Hupáhwaki´kadakena – Dakota

G. L. Twiest, American Society of Mammalogists Mammal Images Library

Description

Except for its much larger size, the big brown bat superficially resembles the genus *Myotis* in North Dakota. It is the second largest bat in the region, averaging over 20 grams; only the hoary bat is larger. The dorsal fur is uniformly brown, becoming lighter on the underside. The wing and tail membranes are nearly naked. The ears are short and rounded, with a broad tragus. Average standard body measurements are: Total Length - 121 mm, Tail Length - 46 mm, Hind Foot - 11 mm, Ear - 18 mm.

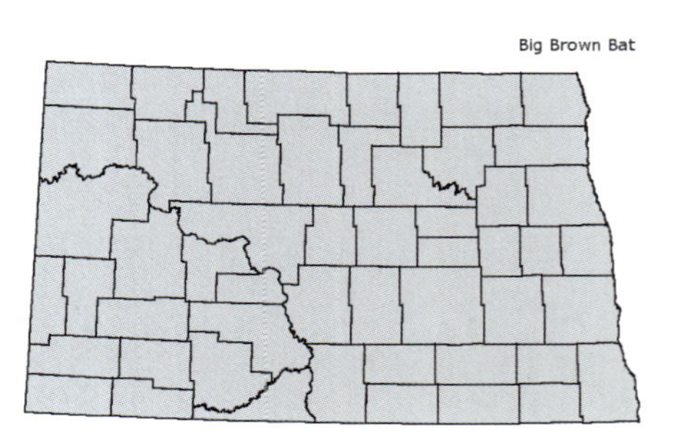

Distribution

This widely distributed species occurs over much of North America, from southern Canada, through all of the United States and much of Mexico, into Central America and northern South America. In North Dakota, it has been recorded throughout the state.

Habitat

The big brown bat is a habitat generalist, foraging over both land and water, in both rural and urban environments. Its use of a particular habitat seems more associated with insect abundance than a specific habitat setting. In some areas, it appears to be more abundant in open deciduous forest having a closed canopy.

Following hibernation, females form maternity colonies, usually of 25 to 75 individuals, while males are generally solitary. Some males may roost in small groups or with females. Maternity colonies occur mainly in buildings, but also in hollow trees and rock crevices. When they occur in buildings, relatively tall, older structures having temperatures 8 to 10° above ambient are preferred. When roosting in rock crevices, preferred sites have small openings and vertical orientation, and are inaccessible from above. Large trees and snags also provide sites for maternity roosts. Winter hibernation habitat is also diverse, including caves, storm sewers and buildings. As with several other bat species, the big brown bat has been recently documented overwintering in the North Dakota badlands.

Ecology and Behavior

The big brown bat forages over a wide variety of habitats, ranging out 1 to 2 km from the roost. Foraging begins shortly after sunset and continues throughout the night. Small beetles dominate the diet, and include a number of significant agricultural pests. Flies, moths, bugs, wasps, and other insects are also taken, but these are considered to be minor foods.

Dispersal from summer roosts may begin as early as August, but bats do not arrive at their hibernacula before November. North Dakota bats are not known to hibernate in the state, and are thought to disperse into Minnesota and South Dakota to find adequate winter roosts. Recent observations of hibernation in rocky areas of southeastern Alberta, however, raise the possibility of hibernacula occurring in mudstone caves of North Dakota's Badlands.

These bats are capable of hibernating under cooler, drier conditions than other temperate species, such as the little brown bat. They tend to hibernate singly, or in small clusters, wedging themselves into small crevices or under rocks in the hibernaculum.

Significant factors causing mortality in big brown bats include insufficient fat storage prior to hibernation, weather, accidents, and predation. Important predators include grackles, kestrels, owls, weasels, cats and rats. They may also be vulnerable to accumulation of insecticides in their tissues sufficient to cause death. Big brown bats can carry rabies, St. Louis encephalitis, and *Histoplasma*. As with other members of the family, big brown bats are relatively long-lived, the record being 19 years.

Reproduction

Copulation occurs any time from September to March, followed by delayed fertilization upon arousal from hibernation. Following a 60-day gestation, a single young is born between May and July. As with a number of other bat species, the young fledge at 18 to 35 days, usually prior to weaning. Adult females cease lactating between 32 and 40 days post-partum. Young males tend to become sexually mature by fall; however, not all young females breed during their first year.

Status and Conservation

The big brown bat is common throughout North Dakota. This species is not limited by habitat type, provided adequate roost sites and insect prey are available. It has been cited as somewhat unique among bats for its hard-bodied insect diet, which includes many important agricultural pests. Protection and enhancement of roost sites, including man-made structures, along with limitation of pesticide use are important practices in maintaining the species. Because of its potential as a carrier of several human diseases, direct contact is discouraged.

Selected References

Bailey (1926), Barnhart and Gilam (2017), Cryan et al. (2001), Genoways and Jones (1972), Goehring (1972), Hibbard (1972), Jones and Genoways (1966), Kurta and Baker (1990), Lausen and Barclay (2006), Svihovec (1967)

Silver-haired Bat

Lasionyteris noctivagans

R. W. Van Devender, American Society of Mammalogists Mammal Images Library

Description

The silver-haired bat is relatively easily distinguished from other North Dakota bat species. It is a medium-sized bat, covered with dark brown to black fur with many hairs having silvery white tips, giving it a frosted appearance. Another distinguishing feature is fur covering the basal half of the interfemoral membrane. Average standard body measurements are: Total Length - 101 mm, Tail Length - 41 mm, Hind Foot - 9 mm, Ear - 15 mm.

Distribution

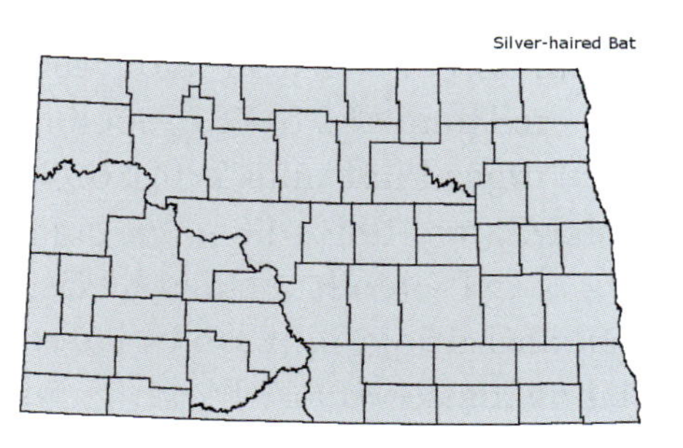

This bat is widely distributed over much of North America from southeastern Alaska and southern Canada throughout most of the United States, and as far south as northern Mexico. There are scattered records from most of North Dakota.

Habitat

The silver-haired bat, along with the red bat and hoary bat, has been referred to as a "tree bat." It prefers deciduous forested areas with nearby water bodies for foraging. This bat does not usually use caves for hibernation. Hence, North Dakota bats probably migrate several hundred miles to a climate mild enough for hibernation. However, the silver-haired bat has been recently documented overwintering in the North Dakota badlands, indicating availability of some suitable hibernacula.

Ecology and Behavior

This is a solitary species, typically roosting in the tree canopy or under loose bark. It has also been known to roost in buildings, wood piles, rock crevices, and occasionally in caves. In South Dakota's Black Hills, silver-haired bats prefer hollow dead snags of ponderosa pine for maternity colonies, where they roost in groups of six to 55. Winter hibernation roosts include hollow trees, loose tree bark, rock crevices, and caves. There is only one record of cave roosting for hibernation in Minnesota, and reports indicate that most silver-haired bats migrate out of that state for the winter.

Little is known about the migratory patterns of the silver-haired bat, but they probably arrive in North Dakota during April and May, coincident with emergence of flying insects. One specimen in the University of North Dakota collection was taken on May 24. Departure patterns are also unknown, but there is a fall specimen in the UND collection dated September 30. Museum records indicate that the species winters in the eastern and southeastern United States.

Foraging occurs over water bodies and along forest edges. Feeding appears to be opportunistic, with a wide variety of insect prey taken. This bat becomes active later than other species and follows a bimodal pattern of activity, with peaks about three hours and seven hours after sunset.

Although this is a very common bat, its numbers tend to fluctuate greatly, both annually and geographically. It is relatively short-lived for a bat species, with an average longevity of about two years and a maximum longevity of 12 years. This relatively short longevity could be attributed to its solitary nature, the vulnerability of its summer and hibernation roosts to disturbance, predation, and the vagaries of climate. Recent research indicates that this species, along with other "tree bats," is especially vulnerable to mortality associated with wind energy facilities.

Reproduction

The reproductive pattern follows other temperate members of the family. Mating is thought to occur in the fall, with storage of sperm in the female during the hibernation period. Ovulation occurs during April and May, followed by fertilization (delayed fertilization). Gestation is 50 to 60 days, followed by parturition in June and July. Females usually give birth to two young, each weighing about 2 grams. At North Dakota's latitude, young bats have been observed to begin flying by late July.

Status and Conservation

Although its numbers can fluctuate greatly from year to year, the silver-haired bat is one of the more common species in North Dakota. Reports of the species seem to be most frequent during spring and fall migration.

The silver-haired bat is a known carrier of rabies, so caution should be used when encountering or handling bats.

Deforestation, resulting in loss of roosting habitat, and insecticide use are probably the most significant factors limiting this bat in North Dakota. Practices protecting older forests, including dead and dying trees, would enhance roosting habitat for the species. Research indicates the likely need for mitigation efforts to minimize mortality from barotrauma in the vicinity of wind energy facilities.

Selected References

Arnett et al. (2008), Baerwald et al. (2008), Bailey (1926), Barnhart and Gilam (2016), Cryan (2003), Hazard (1982), Jones et al. (1983), Jones et al. (1973), Kunz (1982), Mattson et al. (1996)

Red Bat
Lasiurus borealis

R. Altig, American Society of Mammalogists Mammal Images Library

Description

This medium-size bat (8 to 14 g) is one of the most easily identified of North Dakota's bats. It has a distinctive reddish coloration, washed with white hairs, and a buff-colored patch on the shoulder. The interfemoral membrane is entirely covered with fur, along with the basal portions of the wings. The ears are short and rounded, and have a triangular tragus. Average standard body measurements are: Total Length - 113 mm, Tail Length - 48 mm, Hind Foot - 9 mm, Ear - 12 mm.

Distribution

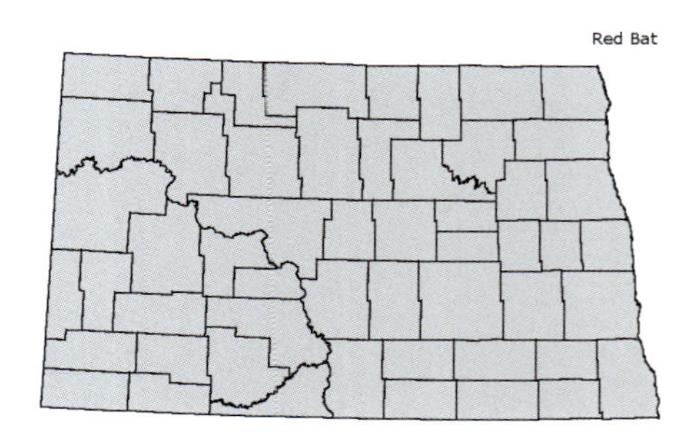

The red bat is widely distributed from southern Canada south throughout much of the United States (except the mountainous West), Mexico, Central America, and South America as far south as Argentina and Chile. In North Dakota, there are records from throughout the state, and there are numerous records from neighboring Minnesota. It is more sparsely distributed in South Dakota, and is believed to be the least common bat species in the Black Hills. There are only two reports of specimens collected in Montana, where it is considered rare.

Habitat

This bat roosts mainly in forest edges near water. These bats will roost at varying heights from low shrubs to the upper canopy. American elm seems to be a preferred roosting tree, but a wide variety of other deciduous trees and shrubs are used as well. Red bats seem to prefer heavy shade and cover from above and to all sides, but not below. Such sites are thought to not only provide concealment from predators, but also a more stable thermal environment.

Little is known about winter habitat requirements of the red bat. They have been known to swarm around cave entrances with other species, but do not usually undergo hibernation in these sites. Winter habitat seems to consist of forested areas in southern states having daytime temperatures warm enough to allow occasional foraging.

Ecology and Behavior

The red bat is a solitary species, roosting singly or in family groups. Although these bats are not colonial, there seems to be some vocal communication around favored roosting sites. Rather than utilizing loose bark and cavities, as in other tree-roosting bats, this bat hangs from twigs and leaf petioles and blends in well with the foliage.

As with other solitary bats, little is known of the red bat's migratory patterns. The earliest known arrival in the northern Great Plains is mid-April. This probably coincides with emergence of primary insect prey. Departure dates are unknown, but are likely in September or October. Museum records indicate that the species winters in the southeastern United States.

Nightly foraging begins shortly after sunset, with a secondary peak occurring just before sunrise. They do not range over long distances, but rather forage over a favored area within about 1,000 meters from the roost. These bats feed on a wide variety of larger insects, but seem to prefer moths.

The red bat, along with other tree bats, is subject to different patterns of mortality than the colonial species. A variety of avian predators and some mammals are known to take red bats. The blue jay is probably the most important predator on these bats, especially their young. This species has one of the higher incidences of rabies among North American bats. As with the other "tree bats," red bats appear to be susceptible to mortality from barotrauma associated with wind turbines.

Reproduction

As with other temperate zone members of the family, this bat breeds in late summer and fall and stores sperm over winter (delayed fertilization). Following an 80- to 90-day gestation, parturition occurs during mid- to late June. The litter size ranges from one to five, averaging about three. This is the highest known litter size for North American bats. Young are weaned at 4 to 6 weeks, and achieve flight status between 3 and 6 weeks.

Status and Conservation

The red bat is considered a common summer resident throughout North Dakota, especially in wooded areas.

As with other tree bats, this species requires woodland for roosting, and generally prefers water areas having large numbers of flying insects. As an insect-eating species, it is believed to be vulnerable to insecticide use. As a potential rabies carrier, human contact with the species should be avoided.

Selected References

Arnett et al. (2008), Baerwald et al. (2008), Bailey (1926), Barnhart and Gilam (2016), Cryan (2003), Foresman (2001), Hazard (1982), Higgins et al. (2000), Jones et al. (1983), Shump and Shump (1982a)

Hoary Bat
Lasiurus cinereus

D. Riepe, American Society of Mammalogists Mammal Images Library

Description

The hoary bat is easily identified by its large size, coloration, and pelage. Weighing over 20 grams, its fur is a mixture of dark brown and grayish with hairs tipped with white, giving it a frosted appearance. There is a white patch on the shoulder and a yellowish patch on the throat. Like the smaller red bat, and unlike the smaller silver-haired bat, the interfemoral membrane is completely furred. The ears are short and rounded, and have a short, blunt tragus. Average standard body measurements are: Total Length - 139 mm, Tail Length - 55 mm, Hind Foot - 13 mm, Ear - 19 mm.

Distribution

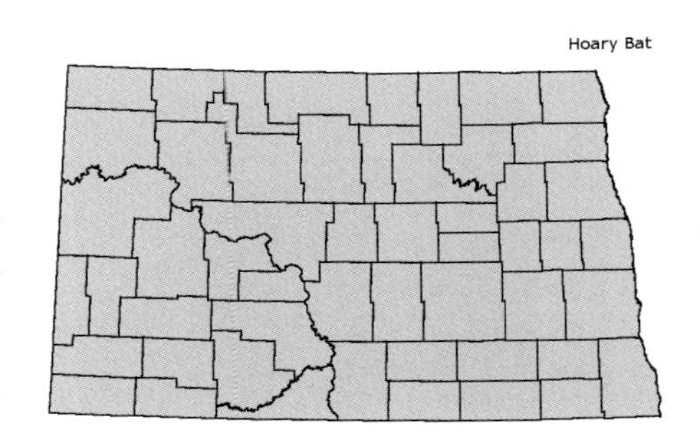

This is our most widely distributed bat, ranging from the tree line in Canada south to Argentina and Chile. It occurs in all of the lower 48 states, and some of the Carribean and north Atlantic islands. In North Dakota, there have been widely scattered records over the state.

Habitat

Tree foliage provides the primary roosting sites for the hoary bat, but they have also been observed in tree cavities, squirrel nests, and on the sides of buildings. They tend to prefer to roost at woodland edges, at least 3 meters above the ground, and well covered from above by foliage.

Ecology and Behavior

Like the red bat, the hoary bat is solitary, mainly roosting along the edges of open woodlands. Their roosting sites among tree foliage are well-concealed from above, presumably protection from avian predation.

The species appears to exhibit sexual differentiation in migration patterns and summer range. Museum records indicate that hoary bats mainly winter in California, Mexico, and the southeastern states. However, during the warm months males are predominant in the western states, females in the east. Both sexes occur in the Plains states, including North Dakota. Some bats may winter in relatively northern states, going into hibernation, while those that migrate further south remain relatively active during the colder months.

Nightly foraging of these bats peaks around 3 to 4 hours after sunset. In wintering areas with a relatively warm climate, they may be active during the day, feeding on insects. Though solitary while roosting, they are known to forage in groups for insects, and also may associate with other bat species while foraging. They seem to prefer moths, but a variety of other larger insects are also taken.

Little is known about mortality in hoary bats, but predation by hawks and owls has been suspected. High winds have been known to dislodge females with attached young, subjecting them to potential predation on the ground. Impalement on barbed wire has also been noted. Like the other "tree bats," hoary bats appear to be especially vulnerable to wind turbine fatalities. Hoary bats have had a relatively high incidence of rabies, up to 17 to 25%.

Reproduction

Copulation may occur during fall migration, followed by delayed fertilization characteristic of other bat species in the family. Most reported parturition dates range from mid-June to early July, following a 90-day gestation. However, one South Dakota report estimated a parturition date of May 15. Typical litter size is two but can run from one to four. Lactating females have been observed in the northern Plains states from mid-July through the first week of August. The young fledge at about 3 weeks of age, but are not weaned until 7 weeks. They have a relatively slow rate of growth, which may be compensated for by migration to warmer climates which facilitate winter foraging.

Status and Conservation

Although not often observed, this bat is regarded as relatively common in North Dakota. Wooded areas having an abundance of flying insects comprise essential habitat for this tree bat. As with other North Dakota bats, it is vulnerable to indiscriminate insecticide use. As a carrier of rabies, human contact should be avoided.

Selected References

Arnett et al. (2008), Baerwald et al. (2008), Bailey (1926), Cryan (2003), Foresman (2001), Hibbard (1963), Hibbard (1972), Jones et al. (1983), Mullican (1999), Shump and Shump (1982b)

Western Small-footed Myotis

Myotis ciliolabrum

Merlin D. Tuttle, Bat Conservation International

Description

This bat has undergone a series of taxonomic changes in recent years. It was originally known as *M. subulatus*, then *M leibii*, and finally separated into two distinct species, the eastern small-footed myotis *(M. leibii)* and the western small-footed myotis *(M. ciliolabrum)*. Weighing less than 8 grams, it is North Dakota's smallest bat. The species can be distinguished from other related bats by its coppery coat coloration, blackish mask, ears and membranes, and small feet (<9 mm) and forearm (<34 mm). Standard body measurements are: Total Length - 87 mm, Tail Length - 38 mm, Hind Foot - 8 mm, Ear - 14 mm.

Distribution

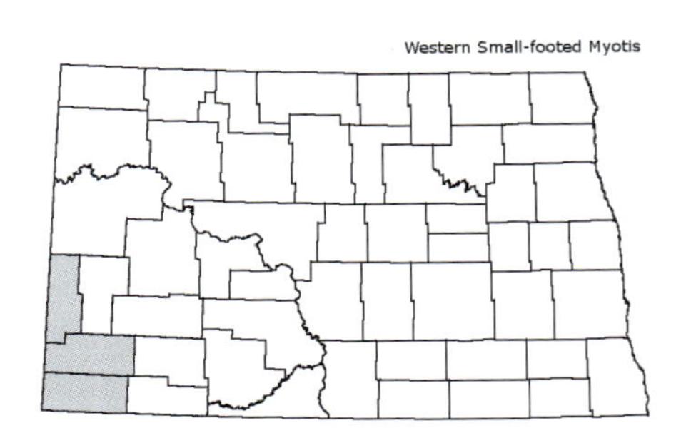

The western small-footed myotis occurs throughout much of western North America from central British Columbia, Alberta, and Saskatchewan, south well into Mexico, and east to the southwestern corner of North Dakota, the southwestern half of South Dakota, and western Nebraska and Kansas. In North Dakota, all records have been from Badlands areas of Golden Valley, Slope, and Bowman counties.

Habitat

Preferred habitats appear to be semiarid and arid badlands and desert regions, but bats have also been taken near water bodies and deciduous and coniferous forests. North Dakota records have been from the rugged Badlands adjacent to the Little Missouri River, and also from trees and buildings. Summer roosts are typically in rocky crevices, holes in the sides of buttes, under tree bark, or in buildings. Winter hibernation is in caves and mine shafts. mine shafts. As with several other species, this bat has been recently documented overwintering in the North Dakota badlands.

Ecology and Behavior

These bats probably arrive in North Dakota during April and May, establishing roosts and small maternal colonies. Unlike the little brown bat, this bat is relatively solitary, roosting singly or in very small groups. Nightly foraging begins around dusk, and peaks from 10 p.m. to 2 a.m.. The western small-footed myotis feeds on a variety of night-flying insects, especially moths.

During the fall, these and other North Dakota bats may migrate to caves and mine shafts in the Black Hills of South Dakota for hibernation. Recent records of hibernation in southeastern Alberta habitats similar to the North Dakota Badlands raise the possibility of overwintering in the state. This species does not aggregate in large groups. They usually hibernate singly or in clusters of two; less than 12 bats are normally found in a cave.

Because of its solitary nature, little is known of the population ecology of this bat. Causes of mortality are probably similar to those of the little brown bat, i.e. limited predation, accidents, and human disturbance including pesticide use. As with other bats, it appears to be quite long-lived, the record for a known-age individual being 12 years.

Reproduction

Like the little brown bat, breeding takes place in the fall, just prior to hibernation, with ovulation and fertilization occurring upon arousal in the spring (delayed fertilization). A single young is born during June-July. North Dakota collection records document pregnant females between June 17 and 29.

Status and Conservation

Because of the solitary nature of the species and few collection records, this bat might appear to be relatively uncommon in southwestern North Dakota. However, surveys indicate that it may be the commonest member of the genus in the Badlands along the Little Missouri River.

These bats do not aggregate in large enough numbers to be considered to be pests. Neither have they been documented as carriers of rabies or other diseases. Because of its occurrence in a very limited portion of North Dakota, the species merits careful scrutiny and has been designated a species of conservation priority in the state.

Selected References

Barnhart and Gilam (2016), Barnhart and Gilam (2017), Genoways and Jones (1972), Holloway and Barclay (2001), Jones et al. (1983), Jones and Genoways (1966), Jones and Stanley (1962), Lausen and Barclay (2006), Seabloom et al. (1978), Svihovec (1967)

Long-eared Myotis

Myotis evotis

Merlin D. Tuttle, Bat Conservation International

Description

As the common name implies, the ears of this bat distinguish it from other myotis bats in North Dakota. When laid forward, they extend >5 mm beyond the tip of the nose. A minute fringe of hairs along the edge of the tail membrane (uropatagium) also aids in identification of the species. Its brownish coat color, and black ears and wing membranes are similar to most other members of the genus. Average standard body measurements are: Total Length - 86 mm, Tail Length - 44 mm, Hind Foot - 9 mm, Ear - 18 mm.

Distribution

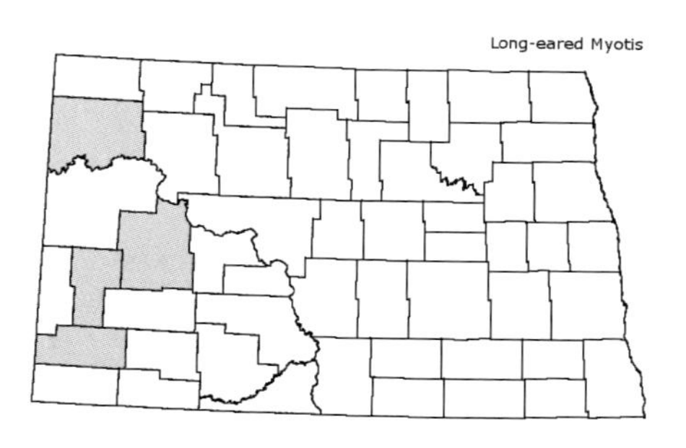

The long-eared myotis occurs from central British Columbia, Alberta, and southwestern Saskatchewan through much of the western United States, south into Baja California, and as far east as extreme western North Dakota and South Dakota. In North Dakota, there are records from Grinnell in southeastern Williams County, the Killdeer Mountains of Dunn County, Billings County along the Little Missouri River, and Slope County.

Habitat

This bat appears to be primarily a forest dweller. It has been reported in ponderosa pine woodlands of the Badlands areas of the Dakotas and southeastern Montana, although it has also been reported in agricultural areas, sagebrush and shrublands. Summer roosts include buildings, hollow trees, rock crevices, or other protected sites. Foraging frequently occurs near water areas where there are abundant insects. Caves and mine shafts are preferred hibernacula, but this bat has been recently documented overwintering in the North Dakota badlands.

Ecology and Behavior

The spring arrival of this relatively solitary bat in North Dakota is unknown, but probably coincides with the emergence of flying insects. Females establish small maternity colonies in buildings or other protected places, while males and non-reproductive females roost singly or in small groups.

Long-eared myotis forage later than do other myotis bats, frequently after dark, and they continue well past midnight. A variety of nocturnal insects are taken, primarily moths.

Practically nothing is known of hibernation sites or behavior of this species, but these western North Dakota bats may migrate to hibernacula in caves and mine shafts in the South Dakota Black Hills. Recent observations of hibernation in southeastern Alberta raise the possibility of the species overwintering in the North Dakota Badlands.

Reproduction

There are practically no data on reproduction in the long-eared myotis, but it probably follows a similar pattern to that exhibited by the little brown bat. In southeastern Montana, pregnant females were observed during early July.

Status and Conservation

Based on reports and collections, the long-eared myotis is believed to be uncommon to rare in western North Dakota. It has been designated a Level III Species of Conservation Priority in the state.

Management efforts for this species at its eastern range limit should entail protection of forest land in southwestern North Dakota, especially ponderosa pine and juniper woodland. As with other bats, use of insecticides around these areas and nearby water bodies may prove harmful to the species.

Selected References

Bailey (1926), Barnhart and Gilam (2017), Dyke et al. (2004), Foresman (2001), Genoways and Jones (1972), Higgins et al. (2000), Jones and Genoways (1966), Jones et al. (1973), Lausen and Barclay (2006), Manning and Jones (1989)

Little Brown Bat
Little Brown Myotis
Myotis lucifugus

R. K. Laval, American Society of Mammalogists Mammal Images Library

Description

The little brown bat is one of the most common and well-known bats in North America. It is a small bat, coppery brown above, lighter below, and has a dark spot on the shoulder. The wings and interfemoral membrane (uropatagium) on the tail are naked. Average standard body measurements are: Total Length - 91 mm, Tail Length - 38 mm, Hind Foot - 10 mm, Ear - 15 mm. The tragus (a fleshy flap projecting up from the base of the ear) is short and blunt. This bat is quite similar to two other myotis occurring in North Dakota: the northern myotis and the long-legged myotis. Refer to those species' accounts for distinguishing characteristics.

Two subspecies of *M. lucifugus* occur in North Dakota: *M. l. lucifugus* in the eastern half of the state and *M. l. carissima* in the west. The eastern subspecies is darker and slightly smaller than that occurring in the west.

Distribution

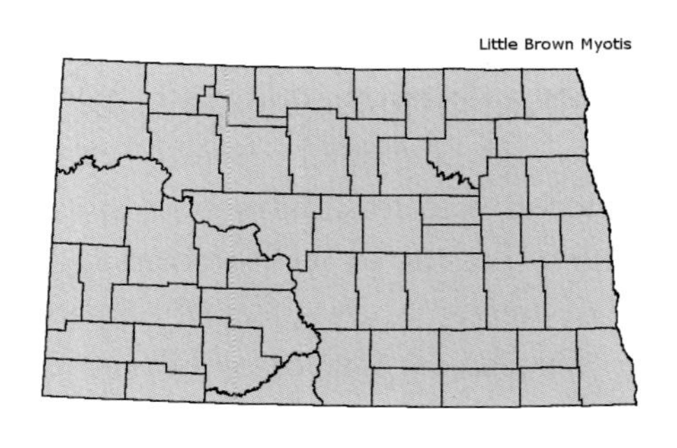

This species is widely distributed throughout much of North America, occurring from the tree line in Alaska and Canada south well into Mexico. Little brown bats occur throughout North Dakota, wherever there are suitable roosts (usually buildings) and water. Caves and mine shafts are preferred hibernacula, but this bat has been recently documented overwintering in the North Dakota badlands.

Habitat

During the warm months, female little brown bats establish traditional roosts which may be used for many years. These roosts are usually in buildings (attics, barns, sheds, etc.), but they may also be found in other protected sites such as under bridges or in hollow trees. These concentrations of bats may number more than 1,000. Males and nonbreeding females may assemble in small groups, or singly, in other protected sites such as under eaves of buildings or rock outcrops. Roosts are typically near water, such as lakes, streams or stock ponds, where there are large concentrations of flying insects.

The cold months of the year are spent in a hibernation site (hibernaculum), usually a cave or mine shaft. Requisite conditions in the hibernaculum are temperatures above freezing and high humidity (≥90%). The nearest known sites for North Dakota *Myotis* are caves in central/southern Minnesota for eastern populations, and the Black Hills of South Dakota for bats living in the west.

Ecology and Behavior

Little brown bats probably arrive in North Dakota during April and May, coincident with the appearance of flying insects. They emerge from the roost, and commence their nightly foraging about dusk. Foraging continues for at least five hours, with bats flying at heights of 3 to 10 m in the vicinity of water bodies, woodland edges, street lights, or wherever flying insects are likely to congregate. During foraging, these bats may consume prodigious numbers of flying insects, up to the equivalent of 1,200 mosquitoes per hour. Following the nightly foraging bout, they return to the roost, where they enter into a light torpor.

During September-October, little brown bats depart North Dakota, migrating to their winter hibernacula. Banding returns indicate that they are capable of migrating several hundred miles. Upon arrival at the hibernaculum, they aggregate in large numbers, mate, and enter into hibernation. Hibernation is not continuous; bats arouse periodically during the winter months and engage in some movement in the hibernaculum.

The mating system of the little brown bat is random and promiscuous. There is no evidence of an organized social structure or territorial behavior. They are quite vocal and emit loud calls, especially when arousing from torpor. Bats of this family (Vespertilionidae) are known for their ability to echolocate using ultrasonic pulses. This echolocation is used in location of prey and in avoiding obstacles.

Little brown bats are very long-lived for their size. Average longevity is 2 to 3 years; 10 years is common, and the known record for an individual in the wild is over 31 years. Such longevities have been attributed to roosting in relatively secure places, seasonal exploitation of abundant food resources, and hibernation during the severe months. A wide variety of predators will take bats, but none seem significant in controlling populations. Accidents and human disturbance, including use of pesticides, appear to be important limiting factors. Drastic population declines in some areas have been attributed to pesticide use and human disturbance of hibernacula. Little brown bats can carry rabies, but the incidence is low (<1%), and there is no clear relationship to rabies in other species of wildlife.

Reproduction

Most little brown bats mate just prior to entering hibernation. In a process known as "delayed fertilization," sperm are stored as a semisolid mass in the female during hibernation. Upon arousal in the spring, the female ovulates and fertilization occurs, followed by a 50- to 60-day gestation. A single young is born following arrival at the nursery roost.

The young is able to cling to its mother, even during flight. Growth is rapid, with weaning and capability of flight occurring at about 3 weeks.

Status and Conservation

The little brown bat is one of the most common bats in North Dakota, and may be found wherever there are suitable roosts and nearby water bodies.

Occasionally, little brown bats establish roosts in occupied buildings, where they are regarded as pests because of their vocalizations, accumulation of excrement, and potential for disease transmission. Recommendations for mitigation or control in such situations are available through Bat Conservation International (http://www.batcon.org/).

Because of its potential for control of flying insect pests, this species is generally regarded as beneficial, and efforts have increased to encourage populations. In some areas, erection of bat houses has been successful in establishment of new roosts. Plans are available from Bat Conservation International and the North Dakota Game and Fish Department. The use of insecticides in the vicinity of known bat roosts should be discouraged.

Selected References

Bailey (1926), Fenton and Barclay (1980), Genoways and Jones (1972), Hazard (1982), Jones et al. (1983), Jones and Genoways (1966), Seabloom et al. (1978), Svihovec (1967)

Northern Myotis
Myotis septentrionalis

Merlin D. Tuttle, Bat Conservation International

Description

Once known as a subspecies of Keen's little brown bat *(M. keenii)*, this bat is now recognized as a separate species. A significant character of this small bat is its relatively long ears, tragus, and tail. The ears, when laid forward, extend just beyond the tip of the nose. They are not as long as those of the long-eared myotis, however, which extend up to 5 mm. Another character is its long, pointed tragus (10 to 12 mm), in contrast to a shorter (7 to 8 mm), blunt tragus in the little brown bat. The tail is also longer than that of the little brown bat. Average standard body measurements are: Total Length - 94 mm, Tail Length - 40 mm, Hind Foot - 9 mm, Ear - 17 mm. Many of the distinguishing features of the myotis bats are relatively subtle, and care should be used in identification.

Distribution

This bat occurs over much of Canada east of British Columbia, and throughout most of the eastern half of the United States. There is

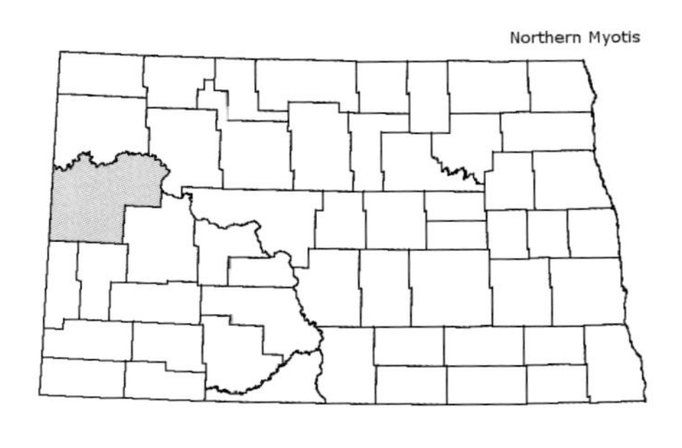

a single record from northeastern Montana, and scattered records from South Dakota, primarily from the Badlands and Black Hills. Minnesota records are from the eastern half of the state. Its North Dakota distribution is problematic. There is a marginal record from Fort Buford (McKenzie County), but it has not been known to be collected in the state in recent years. This bat is uncommon near its western range limits, and there have been few collections from treeless prairies.

Habitat

Summer habitat for the northern myotis appears to be wooded areas, where it frequently roosts singly or in small clusters behind the loose bark of over mature and decaying trees. It has also been known to roost behind shingles, in caves, and other protected sites. Small maternity colonies of less than 60 bats occur in similar sites, and also in buildings. As with other myotis bats, hibernation is in caves and mine shafts.

Ecology and Behavior

This relatively solitary bat forages intermittently through the night, but has two peaks, just after dusk and again before sunrise. It not only feeds on flying insects, but also gleans them off of surfaces such as tree bark and leaves. This opportunistic feeding allows a more varied diet, and utilization of prey that may be able to detect ultrasound in echolocation. A wide variety of insect prey is taken depending on location, season, and individual preferences.

As with other myotis bats, the nearest hibernacula for North Dakota summer residents are probably in central Minnesota and the Black Hills of South Dakota. These bats have been known to hibernate in Minnesota as far north as St. Cloud.

Little is known of mortality or its causes in the northern myotis, but an extreme longevity of 18.5 years has been recorded.

Reproduction

Reproductive patterns appear to be similar to those of other myotis bats (see account of *M. lucifugus*). Copulations have been observed around the time of arrival at the hibernaculum in late summer or early fall. Delayed fertilization is believed to occur, with parturition occurring from mid-May to July. Flying subadults have been captured as early as July.

Status and Conservation

In Minnesota, this bat has been listed as a species of special concern. It is also on the "blue list" in Alberta as an at risk species and has been recently listed as threatened in the United States. The paucity of records near its western range limits, together with the lack of any recent records from North Dakota suggests that it is a very rare bat in the state. As with other forest bats protection of woodlands and limitation on use of insecticides are recommended.

Selected References

Barnhart and Gilam (2016), Caceres and Barclay (2000), Foresman (2001), Hall (1981), Hazard (1982), Higgins et al. (2000), Jones et al. (1983)

Fringed Myotis
Myotis thysanodes

Description

This bat is similar to the long-eared myotis, except for being larger but with a smaller ear size. The range of standard body measurements are: Total Length - 16-20 mm: Length of Forearm - 40-47 mm. A diagnostic feature is a distinct fringe of stiff hair along the posterior edge of the uropatagium (interfemoral membrane). Color can vary from yellowish-brown to darker olivaceous.

Distribution

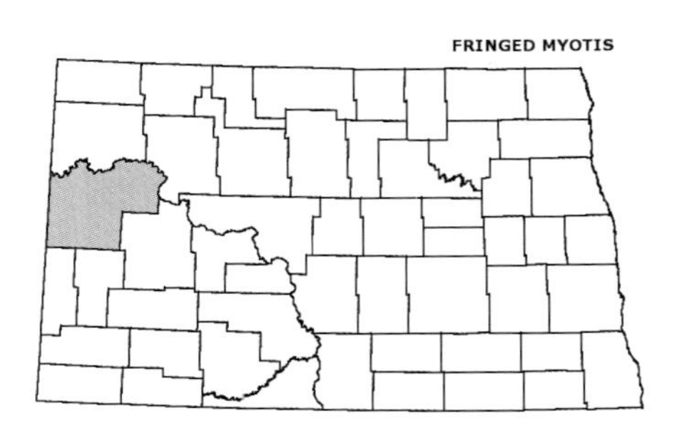

The fringed myotis is a western bat, occurring from British Columbia south into Mexico (Veracruz and Chiapas). A disjunct population occurs in the Black Hills of South Dakota and Wyoming, and the Nebraska panhandle. It was unknown in North Dakota until 2012, when an adult male was captured in the North Unit of Theodore Roosevelt National Park (McKenzie County).

Habitat

Fringed myotis bats are found in a variety of habitats. They appear to be mostly found in dry habitats (deserts, grasslands) interspersed with mature forests (ponderosa pine, pinyon-juniper, oak). Maternity roosting sites and hibernacula include caves, mines, and buildings. Sides of ceiling joists are among preferred roosting sites, where bats gather in tight clusters. Roosting under bark and in hollow tree snags is also common. The one North Dakota record was taken along a creek near the campground of the North Unit of Theodore Roosevelt National park.

Ecology and Behavior

Based on capture rates, fringed myotis appear to active from shortly before sunset until 4–5 hours after sunset, with peak activity during the first 1–2 hours. These bats have relatively short, rounded wings, resulting in slow but highly maneuverable flight enabling them to negotiate around the tree canopy. Food habits analysis indicates that beetles and moths are preferred. Because of their relatively small wing size, fringed myotis may have limited size foraging areas. One study estimated foraging area of about 38 ha.

Limited wing size may also limit migration distance from summer roosts to hibernacula. Unless there are suitable hibernacula in North Dakota, any fringed myotis in the state would likely have to migrate to South Dakota's Black Hills or Cave Hills.

Like other bats, fringed myotis may live up to 11 years. Roosting in secure location seems to assure long life spans. Mortality of neonates has been estimated at less than 1%, a striking contrast to other small mammals. Predation appears to be inconsequential, although some predators may cause abandonment of roost sites. There have been a few reports of rabies, but the disease is of minimal threat to humans and none to the species. White-nose syndrome, which has devastated bat populations in the eastern United States, has not yet reached the Northern Plains region. However, its appearance in neighboring Minnesota may be cause for future concern.

Reproduction

Like other myotis bats, copulation probably occurs in the fall after females have left the maternity roost. Delayed fertilization occurs with sperm stored until emergence from hibernation. Ovulation, fertilization, and implantation takes place in April and May, followed by a 50-60-day gestation. Young are born in late June to early July. Females produce only one young per year, compensation for their overall longevity. The young are capable of limited flight at 2+ weeks, and adult flight at about 3 weeks.

Status and Conservation

Several state and federal agencies have recognized the fringed myotis as a sensitive species that is rare and possibly threatened. However, little is known about the status of populations in the United States. Even less is known of its status in North Dakota, with only one confirmed observation to date. However, recent growth in the state's energy development (especially oil and wind) has stimulated increased interest in bats, resulting in serious surveys. Two new species (Townsend's big-eared bat, fringed myotis) have been identified in North Dakota, along with several range extensions.

As with other bat species, primary threats to the fringed myotis are roost disturbance, habitat alteration, and toxic chemicals. Additional recent concerns include wind farm development (causing barotrauma) and the possibility of white-nose syndrome.

Selected References

Foresman (2012), Higgins et al. (2000), Keinath (2004), Nelson et al. (2015), O'Farrell and Studier (1980), Panella (2013), Tigner and Stukel (2003)

Long-legged Myotis

Myotis volans

Merlin D. Tuttle, Bat Conservation International

Description

The long-legged myotis is of medium size when compared to its close relatives. Overall, it is very similar in appearance to the little brown bat. As the common name implies, it has a longer tibia than other myotis bats, but reliable identification requires careful examination of other characters as well. Other identifying features include its short, rounded ears, which do not reach the tip of the nose when laid forward, a keeled calcar (see glossary), and fairly dense fur on the underside of the wing extending about to a line connecting the knee and elbow. Average standard body measurements are: Total Length - 97 mm, Tail Length - 44 mm, Hind Foot - 9 mm, Ear - 14 mm.

Distribution

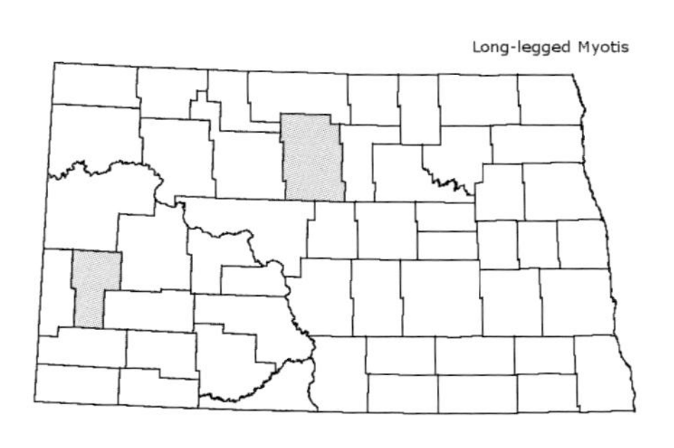

This is a western North American bat, occurring from southeastern Alaska to central Mexico. Its eastern distributional limits include the western Dakotas and Nebraska. There are only two North Dakota records. The first North Dakota record was a bat collected in 1965 along the Little Missouri River in Billings County, near Medora. The other records are from the vicinity of Granville in McHenry County. These two reports constitute the most northeasterly records of the species.

Habitat

Coniferous forests appear to comprise the primary habitat of the long-legged myotis, but they have also been reported along riparian woodland, grasslands, and badlands. The two North Dakota records are from cottonwood river bottoms and from a house in agricultural land. Roosts include loose tree bark, rock crevices, and buildings. Typically, roosts comprise only a few bats, but maternity roosts can be quite large, up to 180. One of the North Dakota records is from a maternity colony in an attic comprising about 60 individuals. Hibernation is in caves and mine shafts; hence, as with other myotis bats, this species undergoes fall migration from North Dakota, presumably to the Black Hills of South Dakota.

Ecology and Behavior

This bat is active throughout the night, but appears to have an activity peak several hours after sunset. A large variety of soft-bodied flying insects are taken during foraging. As with other myotis bats, the long-legged myotis is quite long-lived, the record being 21 years.

Reproduction

The reproductive strategy of this bat is similar to its close relatives, i.e., fall breeding, delayed fertilization, and parturition during spring and summer. Pregnant females have been captured from mid-April to mid-August. Females are thought to reach sexual maturity during their first year, but males may be more variable.

Status and Conservation

With only two known observations in North Dakota, the long-legged myotis must be considered to be rare to uncommon. The North Dakota Game and Fish Department has designated the long-legged myotis as a Level III Species of Conservation Priority. Additional surveys in the state together with careful re-examination of existing collections may result in reassessment of its status.

Management strategies for this species, as well as its relatives, entails protection of conifer forests and riparian woodland in the western part of the state. Insecticide use in such areas should be limited.

Selected References

Barnhart and Gilam (2016), Foresman (2001), Genoways (1967), Genoways and Jones (1972), Higgins (2000), Jones et al. (1983), Jones and Genoways (1966), Warner and Czaplewski (1984)

Order Carnivora—Carnivores

The term "carnivore" implies flesh-eating, and is partially true for this relatively large order. However, members of this order exhibit a wide variation in food habits, ranging from the almost exclusively vegetarian giant pandas, insectivorous meerkats, fish-eating seals, and the omnivorous raccoons, skunks, and black bears, to the cats and wolves which feed primarily on flesh. In North Dakota, carnivores range in size from the tiny least weasel (40 to 60 g) to the black bear (60 to 100 kg).

All of our local carnivores have tooth and jaw characteristics which are adaptations for the capture and consumption of prey. Prominent canine teeth enable capture and holding of prey. The incisors aid in stripping muscle from bone, while the carnassials (the last upper premolar and first lower molar) are shearing teeth providing for cutting connective tissue and cracking bone. The jaws are greatly strengthened and hinged, and do not provide for lateral movement. Other carnivorous adaptations include binocular vision, an enlarged brain, elongated limbs, and (in the cats) retractile claws for grasping prey.

The carnivores are widely distributed on all continents except Antarctica. There are 240 species worldwide and 57 species in North America north of Mexico. Historically, 23 species have occurred in North Dakota, but only about 16 species are currently resident in the state. North Dakota families include the Felidae (cats), Canidae (dogs), Ursidae (bears), Mustelidae (weasels and allies), Mephitidae (skunks), and Procyonidae (raccoons).

Family Felidae
Cats; Igmu—Lakota

Canada Lynx
Lynx canadensis
Inmuchota–Dakota; Igmu hota–Lakota;
Wach–Arikara; Sihtachache–Hidatsa

L. L. Master, American Society of Mammalogists Mammal Images Library

Description

Weighing from 7 to 16 kg, the Canada lynx is about twice the size of a large house cat. It has relatively long legs, a stubby tail, and large feet. The pelage is usually pale gray, mixed with brown, and lacks the spotting of the related bobcat *(L. rufus)*. The face has ruffs on either side, there are tufts on the ears, and the tail has a black tip which completely encircles it. Ranges of standard body measurements are: Total Length - 875 to 1,000 mm, Tail Length - 100 to 120 mm, Hind Foot - 215 to 250 mm, Ear - 70 to 80 mm. Lynx can be distinguished from bobcats by their more pronounced tufted ears, longer fur, lack of spots, larger feet, and a black tip completely encircling the tail.

Distribution

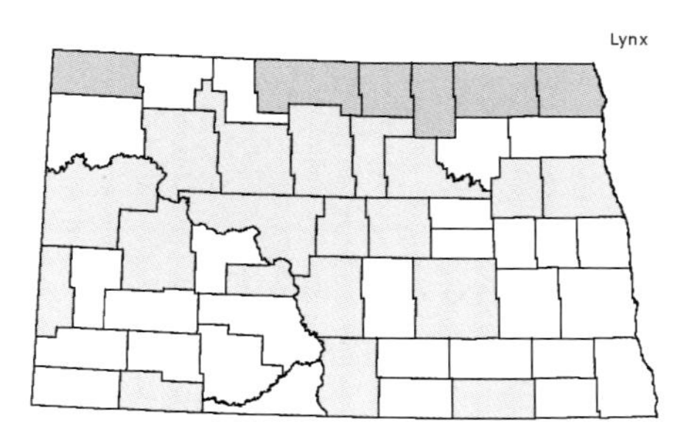

The Canada lynx is an inhabitant of the boreal forested areas of North America, predominately throughout much of Alaska and Canada. Historically, it ranged through the Rocky Mountains as far south as Nevada, Utah, and Colorado. There are scattered records from the Great Plains, including the Dakotas, Nebraska, and Iowa, and throughout the Midwest and northeastern states. It was largely extirpated from much of the conterminous United States, following white settlement.

In North Dakota, Canada lynx originally occurred over much of the state, ranging as far south as the Little Missouri Badlands in the west and Stutsman County in the east. Early fur traders regularly reported taking fairly large numbers of lynx, but some may have come out of Canada. Most reports came out of forested areas near the Canadian border, such as the Turtle Mountains and Pembina Hills. More recently, there have been scattered reports from throughout most of the state, but especially the heavily forested areas of the Pembina Hills, Turtle Mountains, and the Souris River. Most reports were from 1962 to 1963 and 1972 to 1973, when lynx may have dispersed out of Canada, possibly resulting from changes in snowshoe hare populations, forest fires, spraying, or disease.

Habitat

Typical habitat for Canada lynx is deep boreal forest, where they depend on snowshoe hares and other species for their prey. They are not well-adapted to other habitats, although they may inhabit farming country if it is interspersed with extensive forest. They are occasionally found in settled areas, but this is in response to dispersal from their primary range following crashes of snowshoe hare populations. As indicated above, most North Dakota reports have been from the aspen-dominated forests of the Turtle Mountains and Pembina Hills, and riparian forests adjacent to major rivers.

Ecology and Behavior

Lynx are solitary animals, and tend not to have social interactions other than during the breeding season and when mothers are raising their kittens. Scent marking is used to effect mutual avoidance. They are nocturnal, although most active during the twilight hours corresponding to the activity patterns of their prey, usually rabbits and snowshoe hares. They den by day in any convenient shelter, such as deadfalls or overhangs.

Canada lynx are highly specialized predators, depending heavily on snowshoe hares *(Lepus americanus)* for 30 to 100% of their diet. Following a cyclic peak of hares, lynx will take up to an average of 1.2 hares per day. Other prey taken opportunistically includes red and flying squirrels, mice and voles, ground squirrels, beaver, muskrats, grouse, and other birds. Occasionally deer and moose young are taken, but these comprise a minor part of the diet. When snowshoe hares are abundant, they may comprise nearly 100% of the diet, but during population declines lynx switch to alternate prey. When prey are abundant, they may be cached, but lynx do not always return to the cache site.

Home ranges are typically from 16 to 20 km^2, but can vary greatly. A transitory population in Minnesota resulting from dispersal out of Canada had unusually large home ranges of 51 to 122 km^2 for females and 145 to 243 km^2 for males. Lynx may or may not segregate their home ranges from each other. In Minnesota, male home ranges did not overlap, and overlapped with females very little. Daily movements can be highly variable, a function of prey density and snow conditions. During periods of hare abundance, daily movements may average nearly 3 km, but during scarcity this may increase to more than 5 km. Dispersal involves juveniles departing from their mothers' home ranges, or adults and juveniles dispersing in response to declines in hare populations. Lynx may disperse for great distances, often more than 500 km, and one case of 1,100 km has been recorded. Thus, occasional reports of lynx far from their normal range should not be surprising.

Canada lynx populations have long been known to fluctuate synchronously with the eight- to 11-year snowshoe hare cycle, lagging slightly behind the peaks and troughs of hare abundance. During these cycles, lynx populations can vary by three to 17 times, from 1 per 5 to 10 km^2 during highs to one per 50 to 70 km^2 during lows. These crashes result from poor recruitment, increased adult mortality, and increased dispersal. Reduced recruitment results from low pregnancy rates, decreased litter sizes, and poor survival of kittens. Other than trapping and hunting, starvation, predation, disease, and road-kills are major specific causes of mortality. Starvation of kittens is predominant during population declines. Coyotes, wolves, and domestic dogs have been known to prey on lynx, but infanticide by other lynx may comprise the most significant predation loss. Annual mortality among kittens can be as high as 95% during periods of population decline, while overall non-trapping mortality in a population can range from 34 to 68%. Adult longevity in the wild probably does not exceed five years.

Reproduction

Canada lynx, like other North American cats, are polygamous. The breeding season is short, typically between mid-March and early April, but may vary geographically and in response to prey availability. Females may breed during their first year if prey is abundant, but males are not capable of breeding until their second winter. The gestation period is about 63 days, with parturition typically occurring in mid to late May. Litters range from one to five, averaging 3.7, but the average litter size decreases during periods of prey scarcity. During these periods, some females may only reproduce on alternate years.

Newborn kittens are blind and helpless, weighing about 200 g. Their rate of development is contingent on food abundance. With abundant food, their eyes are open at 2 weeks, they emerge from the den at 33 to 42 days, and begin taking solid food soon after. They are weaned at 12 weeks. During food scarcity, their development is retarded and few survive. This is a major immediate cause of lynx population crashes. Maximum weight is not reached until nearly 2 years. Kittens remain with their mother until 9 to 10 months of age.

Status and Conservation

The International Union for the Conservation of Nature (IUCN) has listed the Canada lynx as threatened in Canada and the United States. While legally trapped for fur in Canada, there is concern that data are insufficient to guarantee sustained harvests and healthy populations. Lynx are more scarce in the contiguous United States and vulnerable to human-induced factors such as recreational use of national forests and snow compaction on trails, which favors competing species. Lynx are listed as Threatened in the contiguous United States.

In neighboring Minnesota, there have been 92 verified reports of lynx between 2000 and 2004. There have been no reports of lynx in North Dakota since the 1970s. The species is listed as a furbearer in the state, with a closed season.

Continued monitoring in North Dakota, combined with verification of reports of sightings, is necessary to determine if lynx may be dispersing into the state from Canada and Minnesota. Any dispersal into the state will undoubtedly coincide with fluctuations of snowshoe hare populations.

Selected References

Anderson and Lovallo (2003), Bailey (1926), Gunderson (1978), Hall (1981), Hazard (1982), Jones et al. (1983), Mech (1980), Quinn and Parker (1987), Tumlison (1987)

Bobcat

Lynx rufus

Itupa-púzi, Bidábaho pusika–Hidatsa;
Mantóka–Mandan; Igmu gleska–Lakota

J. O. Wolff, American Society of Mammalogists Mammal Images Library

Description

The bobcat, about twice the size of a house cat, is similar in body size and proportions to the lynx. Weights typically range from 7 to 16 kg, with males averaging about 11 kg and females about 7 kg. Compared to the lynx, the ear tufts and facial ruffs are smaller, the feet are smaller, the black tip on the tail is incomplete, and the fur has well-defined spots. Overall coloration is more reddish than gray. Ranges of standard body measurements are: (Males) Total Length - 475 to 1,252 mm, Tail Length - 108 to 201 mm, Hind Foot - 135 to 223 mm, Ear - 37 to 85 mm; (Females) Total Length - 610 to 1,092 mm, Tail Length - 90 to 171 mm, Hind Foot - 125 to 190 mm, Ear - 32 to 80 mm). Males typically average about 11 kg in weight, and females about 7 kg.

Two subspecies of bobcats are known to occur in North Dakota. A smaller, darker *L. r. rufus* occurs east of the Missouri coteau, and a larger, paler *L. r. palescens* is found in the western portions of the state.

Distribution

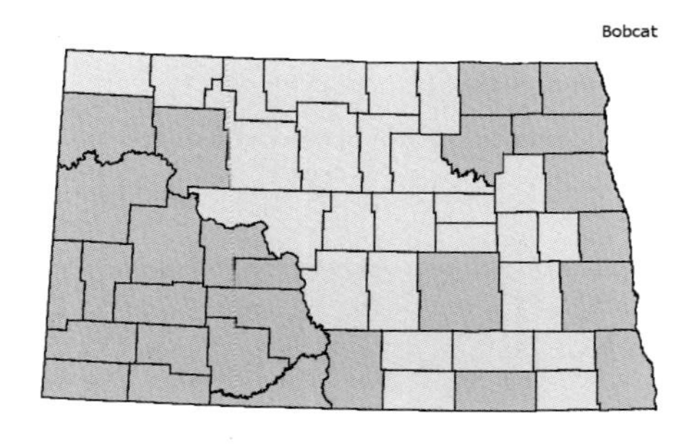

The bobcat occurs throughout southern Canada, the contiguous United States, and into Mexico as far south as the Isthmus of Tehuantepec. It has been largely extirpated from much of the midwestern United States, a result of habitat change due to agriculture, along with human exploitation. During the past 100 years, however, its range has expanded, probably in response to forest clearing, the retreat of competing lynx populations, and harvest regulation.

Bobcats are uncommon in eastern North Dakota, with only a few scattered recent records from counties with wooded areas near lakes and waterways. They are fairly common, but sparse, in the western part of the state, predominately in counties adjacent to the Missouri, Heart, Cannonball, and Little Missouri rivers. They can also be expected in the wooded Killdeer Mountains of Dunn County.

Habitat

As their widespread distribution implies, bobcats live in a wide variety of habitats, including Southern bottomland forests, deserts, northern boreal forests, and the tropics. In general, they prefer rough, broken country interspersed with dense cover. Areas used most heavily have dense understory vegetation and high prey populations. Habitat use may shift seasonally in response to climatic variation, prey availability, and metabolic demand associated with reproduction. Preferred habitat in Minnesota is second-growth forest with dense undercover, interspersed with clearings and swamps. These situations provide protection from severe weather, adequate den sites, cover, and prey resources. The few eastern North Dakota records are associated with wooded areas adjacent to the Red River and its tributaries, while in the west they are primarily found in timbered river bottoms and the rough country of the Badlands. Similarly, in southeastern Montana, bobcats select cover types associated with the broken terrain of badlands ecosystems.

Ecology and Behavior

Like lynx, bobcats are solitary, and avoid direct social interactions, except females with kittens and males and females during the breeding season. Mutual avoidance is accomplished by frequent scent marking of their home ranges with feces, urine, and anal gland secretions. Bobcats may be active at all hours, but peaks occur during early morning and early evening, probably related to activity patterns of their major prey. During their active periods, bobcats crisscross likely hunting habitat, and then sit and watch from lookouts adjacent to game trails to ambush or stalk prey.

Bobcats are strict carnivores, depending on rabbits and hares for a major part of their diet. Squirrels, voles, and mice are also taken, as well as larger rodents and occasional deer. In western North Dakota, foods in bobcat stomachs by frequency of occurrence were cottontail (44.4%), deer (14.8%), and small mammals (29.6%).

Home ranges can vary greatly as functions of habitat quality, prey density, season, population density, and sex. Ranges increase in size with decreased prey densities, and female home ranges are typically smaller than those of males. Females further reduce their home ranges during the breeding season. Recorded home ranges in the Midwest have varied from 7 to 61 km^2. Although bobcats engage in frequent scent marking, there may be a high degree of home range overlap, especially among males and between males and females. Daily movements within the home range can range from 2 to 10 km in males, and from 1 to 7 km in females. Dispersing young have traveled as far as 182 km.

Being somewhat territorial, bobcat population densities have a rough negative correlation with home range size. In boreal forests, densities may be somewhat cyclic, following the fluctuations of snowshoe hare populations. In the Rocky Mountain states and Midwest, estimates have ranged from four to 10 per 100 km^2. Bobcats have few natural enemies and, in the absence of man, starvation may be the principal cause of mortality. Kittens may be killed by mountain lions, coyotes, foxes, owls, and adult male bobcats. Human-caused mortality (legal and illegal hunting and trapping, and road kills) may account for more than 80% of the mortality in a population. Average annual survival can range from 19% to >80%, depending on natural conditions and human factors. Maximum longevity has been recorded as 15.5 years in the wild and 32 years in captivity; however, few survive in the wild past about 4 years.

Reproduction

Like the other cats, bobcats are polygamous breeders. Females are seasonally polyestrous; those females failing to become pregnant early in the season may breed again in late spring or early summer. Breeding seasons have been reported to begin as early as December and end as late as April, and it has been suggested that southern bobcats may breed at any time of the year. Specific breeding seasons apparently vary with latitude, longitude, altitude, climate, and availability of prey. However, most breeding is in March and April. Females may breed during their first year, but the majority do not reproduce until they are 2 years old. Males mature more slowly and do not breed until they are yearlings. Gestation periods average about 63 days, varying from 50 to 70 days. Most litters are born between mid-May and mid-June. Litter sizes average about two or three (range one to seven).

At birth, kittens weigh 280 to 340 g, and are blind and helpless. Their eyes open at 9 to 18 days. They emerge from the den at 33 to 42 days, and begin taking solid food, but they are not fully weaned until 2 months. Maximum weight is not reached until about 500 days for males and 700 days for females. Young remain with the mother until the next breeding season, and continue to remain in the natal home range for several more months before dispersing.

Status and Conservation

Until the 1960's, there was little public concern over the status of bobcats, and in many states, they were bountied as vermin. Only a few states protected the species with closed seasons. With the adoption of CITES (Convention on International Trade in Endangered Species of Wild Fauna and Flora) in 1975, demand for bobcat pelts increased sharply. Bobcat management shifted to protection against overexploitation. The bobcat was listed in Appendix II of CITES, whereby member countries were required to demonstrate that international trade would not threaten the survival of the species before harvest was allowed. In spite of increased harvests, bobcat populations in the United States appear to be stable and possibly increasing in states where they had been nearly extirpated. As of 1996, 37 states allowed harvest of bobcats.

Bobcats are classified as furbearers in North Dakota but, because of their sparse populations, have been economically unimportant to the fur trade. Currently, hunting and trapping of bobcats is allowed only south and west of the Missouri Rriver, and in 2001–2002, only 12 pelts were purchased by fur buyers in the state. While sparse populations appear to be relatively stable in southwestern North Dakota, the bobcat continues to be very rare in the east.

Management strategies generally begin with assessment of population status, a very difficult task when dealing with sparse and widely dispersed predator populations, such as the bobcat. In North Dakota, the rural mail carrier surveys, harvest monitoring, hunter/trapper questionnaires, and monitoring of pelts sold have been employed for many years in an attempt to follow population trends. Results, however, may be inaccurate and imprecise. Other more intensive approaches have been tried in other states, such as use of telemetry, mark-recapture, snow tracking, scent posts, and hair snares. Feasibility of such approaches may be questionable in states like North Dakota, where the bobcat is of little economic or recreational significance, and does not appear to be threatened.

Habitat management may be feasible in areas managed for timber production, especially if coupled with management for other species such as ruffed grouse or deer. Ideal habitat conditions will favor the small mammal prey base, and could involve clear-cutting small blocks of timber interspersed with forested areas, and maintaining a mosaic of forest types with early to mid-successional stages. Thus, habitat management considerations could be applicable to the Turtle Mountains and Pembina Hills areas of North Dakota.

Depredations on domestic animals occasionally occur, but are uncommon. Reports of depredations include sheep, lambs, young pigs, and poultry. Any approaches to control should involve removal of individual problem animals.

Selected References

Adams (1961), Anderson and Lovallo (2003), Bailey (1926), Fuller et al. (1985), Giddings et al. (1990), Hazard (1982), Jones et al. (1983), Lariviere and Walton (1997), Rolley (1987), Rollings (1945), Seabloom et al. (1978), Trevor et al. (1989)

Mountain Lion (Cougar)

Puma (Felis) concolor

Shunta-ha^{n}ska – Mandan; Itupa-ichtia – Hidatsa; Wachtas – Arikara; Igmu watogla, Igmu tanka – Lakota

R. H. Barrett, American Society of Mammalogists Mammal Images Library

Description

Weighing up to 120 kg, the mountain lion is the largest of North Dakota's cats. Adults are buffy to pale brown above, and paler below. The long, cylindrical tail is tipped with black. The muzzle stripes and area behind the ears are also black. Adult males are about 50% heavier than females. Ranges of standard body measurements are: Total Length - 1,800 to 2,590 mm, Tail Length - 700 to 900 mm, Hind Foot - 254 to 295 mm, Ear - 105 mm.

Distribution

Historically, the mountain lion was the most widely distributed terrestrial mammal in the Western Hemisphere, occurring from northern British Columbia south to southern Argentina and Chile. It ranged throughout much of North America from the Pacific to the Atlantic coast. Subsequently, extirpations resulting from hunting and changes in

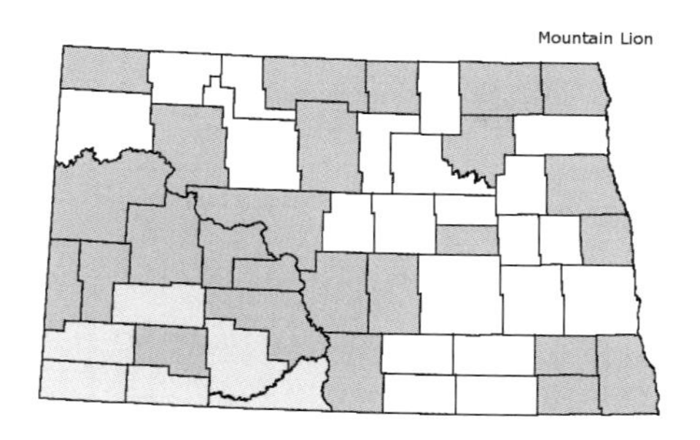

land use have reduced the core distribution of mountain lions by over 60% in North America. Currently, core populations occur largely in remote areas of the mountainous West. In recent years, however, lions seem to have expanded into inhabited areas, and human populations have moved into mountain lion habitat, resulting in increased numbers of attacks on humans.

Mountain lions have occurred throughout most of the Great Plains, but apparently were never numerous. They were extirpated from Nebraska in the 1890s, but there have been several recent reports from that state. In neighboring South Dakota, a core population in the Black Hills appears to be expanding eastward over much of the western half of the state. In Manitoba, the lack of any recent museum specimens resulted in the province being omitted from the northern distributional limits of the species until a specimen was taken near Winnipeg in 1973.

Historically found west of the Missouri River, there were no early records of mountain lions in eastern North Dakota. Even early fur traders such as Alexander Henry who worked in the Red River country made no mention of their presence in the early 1800s. At one time, however, they were relatively common in the Badlands and Killdeer Mountains south and west of the Missouri River. Mountain lions disappeared from the state by the early 20th century. However, sightings resumed during the 1950s and have subsequently increased significantly. Most observations have been in or near the badlands (Golden Valley, Billings, McKenzie, and Dunn counties), but there have been additional reports from Divide, Mountrail, McHenry, Rolette, Traill, and Grand Forks counties.

Habitat

The historic distribution of the mountain lion throughout the Western Hemisphere reflects its ability to exist in a wide variety of habitat types. In general, its occurrence seems limited by three factors: availability of prey, stalking cover, and human interference. Nearly half of the North Dakota reports from recent years have been from the vicinity of the heavily forested Turtle Mountains and Killdeer Mountains, and the rugged Badlands. In Manitoba, reports in the province were grouped as 40% wilderness (heavily wooded regions with few roads or habitations), 30% mixed land (towns and agricultural land interspersed with large forest tracts), and 30% farmland (croplands with small woodlots and forest cover along rivers).

Ecology and Behavior

Mountain lions are solitary, associating with others only for a few days during the breeding season and when females have dependent young. They are primarily nocturnal, but may also be active during the day depending on the availability of prey. Females with kittens may emerge to feed earlier in the evening than males, females, or females with juveniles, possibly to avoid contact, remaining closer to their prey caches, or because of their greater energetic requirements.

These cats are obligate carnivores, feeding almost exclusively on vertebrate prey. Deer make up the bulk of the diet in most areas, up to 75% in winter and 60% in summer. However, they are opportunistic and will capture other species, such as rabbits and ground squirrels. In ranching areas, cattle make up less than 1% of the diet. Estimates indicate that a mountain lion will kill the equivalent of one deer per 4 to 14 days, 14 to 20 deer per year. Proportionately, more old bucks and young fawns are taken than are found in the prey population. Killing behavior involves stalking to about 15 m of the prey, leaping on its back, and biting below the base of the skull. The prey is then dragged to a secluded area and consumed. Portions not immediately eaten are cached under leaves or brush for later consumption.

Attacks on humans have been rare, but increasing in recent years. During the period 1890 to 2004, there were 20 fatal attacks and 88 nonfatal attacks in the United States and Canada. Of these 108 attacks, 50 were between 1991 and 2004.

Mountain lions are wide-ranging animals. Home ranges of males vary from about 190 to >800 km^2, and females from about 70 to >200 km^2. Male home ranges do not overlap, but female home ranges may overlap with those of other males and females. Home ranges are marked by "scrapes," constructed by pulling soil and other debris into a small pile. Urine and feces are often incorporated into the scrape. In addition to marking of the home range, scrapes made by females may also communicate their reproductive status. Mountain lions may also communicate by leaving scratch marks on trees and logs, and leaving scent on objects by rubbing their cheeks, which contain scent glands.

Population densities typically range from 5 to 49 lions per 1,000 km^2. Adult females usually outnumber adult males. The transient component of the population consists of dispersing yearlings of about 18 months. These transients may disperse up to 500 km from their natal home ranges. Their survivorship is low, and this cohort is the most likely to have conflicts with humans. Survival of kittens under a year of age is about 70% in unhunted populations, and after the first year annual survival is 88 to 95%. Mountain lions have lived as long as 20 years in captivity, but probably do not survive longer than 12 or 13 years in the wild. Hunting is likely the main cause of mortality. They have few natural enemies, but kittens are vulnerable to predation and cannibalism by adult males. Accidental deaths, such as road kills, are becoming more common. Some fatalities occur during encounters with large ungulate prey, and some succumb to starvation following injuries to limbs or teeth. Some workers have claimed that, because of their solitary habits, lions are fairly free of parasites and diseases, but they are susceptible to many agents affecting domestic cats. However, epizootic diseases are probably not a threat to wild populations.

Reproduction

Mountain lions are polygamous breeders, with a single male tending all the females whose home ranges overlap with his own. Estrus females attract males by vocalizing and scenting nearby objects. The male accompanies the female for a short time, and then resumes his solitary behavior. Males do not participate in rearing of young. Females may breed annually, but more typically produce litters every 18 to 24 months.

Females may come into estrus during any season but, in the Northern Hemisphere, most births occur between April and September. Estrous cycles last about eight days, and females associate with males for two to five days. The gestation period is about 92 days. The average litter size is about three, ranging from one to six. Newborn young have their eyes and ears closed, are well-furred, and weigh an average of 400 g. Their eyes and ears open between 1 and 2 weeks. They may begin accompanying their mother to kill sites at 2 months, and are weaned shortly thereafter, when they weigh between 3 and 4 kg. Young remain with the mother up to two years, and then disperse. Females attain sexual maturity at 2 to 3 years, but probably do not mate until they have established their own home ranges. Females have been known to remain reproductively active until 12 years old, and males as long as 20 years.

Status and Conservation

The mountain lion's population is estimated at less than 50,000 breeding individuals, and has been listed by the IUCN as "Near Threatened." It is listed as a big-game species in South Dakota and Montana, and is a "Special Concern" species in Minnesota.

In North Dakota, increased numbers of reports from recent years, including observations of juveniles in the Turtle Mountains and Badlands, indicate that the state has small numbers of breeding mountain lions. There were 41 confirmed reports of mountain lions in the state during the period 2001 to 2005. Until 2005, the mountain lion was listed as a furbearer in the state, with a closed season. In 2005, an experimental season was opened, currently allowing a harvest of 10 animals in the Badlands, and an unlimited harvest in the east.

A habitat suitability map has been developed, identifying the North Dakota Badlands and Missouri River "Breaklands" as having sufficient suitable habitat to support a small population of 45 to 74 resident adults. High-quality habitat has been defined as >50% stalking/concealment cover per 2.6 km^2, >50% slope, and perennial streams or shoreline within a 2.6 km^2 area. Other than the Badlands and Missouri River Breaklands, only portions of the Turtle Mountains (Bottineau, Rolette counties) and the Pembina Gorge (Cavalier, Pembina counties) contain suitable mountain lion habitat. These are considered too small to support a resident population, but may provide temporary sites for transient animals.

Mountain lions have never been important in the fur trade, but are valued as game animals. Depredation issues are also significant to the livestock industry, and there have been increasing numbers of lion-human interactions, some involving fatalities. Some constituencies press for conservation and restoration for esthetic reasons alone.

Management strategies for mountain lions include: monitoring, affording some measure of protection, permitting hunting, and controlling problem animals. Conservation programs often consider metapopulations, habitat fragmentation, and development of corridors connecting habitat islands. Depredation issues are approached by removal of problem animals by professional trappers, harvesting populations to more acceptable levels and education to alter husbandry practices.

In North Dakota, mountain lions are monitored by recording and verification of sightings by professionals and the general public. They have been afforded protection by closed season status as furbearers but, as noted above, a limited season was opened in order to acquire biological data on the species in the state. However, if the recent trends of increased reports continue, issues of depredation, human encounters, and sport hunting may require additional approaches to management of this large carnivore.

Selected References

Bailey (1926), Currier (1983), Hazard (1982), Higgins et al. (2000), Jones et al. (1983), Lindzey (1987), Nero and Wrigley (1977), Pierce and Bleich (2003)

Family Canidae
Dogs

Coyote
Canis latrans

Mes-ta-chá-gan-es–Ojibway; Míca, Mícaksica–Dakota; Sunkmanitu, Mayasleca–Lakota; Míkasi–Omaha; Schékè–Mandan; Mótsa–Hidatsa; Stshirits pukatsh–Arikara

Harold Umber

Description

The coyote, sometimes called a brush wolf or prairie wolf, is a mid-sized canid, smaller than a gray wolf and larger than a red fox. It has an elongated, pointed muzzle with a narrower nose pad (<25mm) than gray wolves. The ears are relatively large and pointed, and are held erect and forward. The legs are moderately long, and they have a long and heavily furred bushy tail. Eyes have round pupils with yellow irises. Pelage color can be widely variable, but typically their dorsal fur ranges from buffy-gray to pale-yellow; underparts are paler. The legs and sides may be tawny and the tip of the tail is black. Coyotes stand about 460 to 510 mm at the shoulder and weigh 9 to 16 kg. Males are slightly larger than females. Average measurements are Total Length - 1,230 mm, Tail Length - 328 mm, Hind Foot - 199 mm, Ear - 103 mm.

Distribution

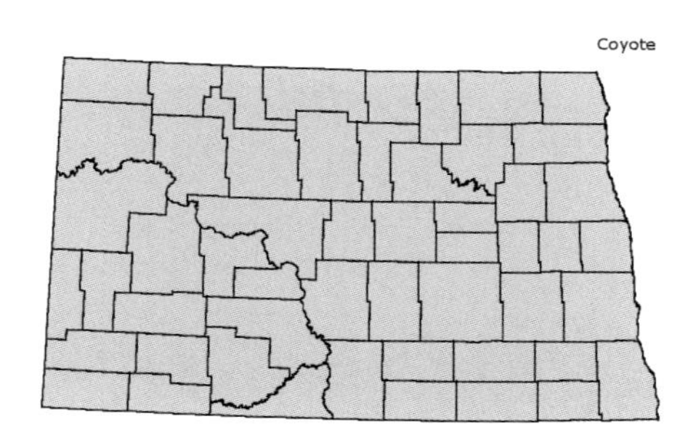

Coyotes are nearctic canids; originally they inhabited grasslands and relatively open wooded areas of western and central North America, ranging east to the western edge of the eastern deciduous forest. Their distribution expanded, partly in response to the demise of gray wolf populations, and in response to forest opening as livestock production was established. Currently, coyotes occur from Costa Rica to the North Slope of Alaska and throughout much of the continental Unites States and Canada except northern Quebec and all of Newfoundland and Labrador. Colonization in the Southeast (e.g., Florida, Georgia) occurred largely as a result of translocations by humans. In North Dakota, much of the state coyote distribution and abundance has varied historically, but they now occur throughout the state.

Habitat

Coyotes are highly adaptable and occupy diverse habitats in North Dakota, from badlands to areas of mixed grassland/cropland, but mainly open grasslands. Because coyotes are so adaptable, in most areas habitat does not limit their distribution.

Ecology and behavior

The scientific name for coyote, *Canis latrans*, means "barking dog," and coyotes are one of the few wild animals whose vocalizations are commonly heard. At night, coyotes both howl (a high, quavering cry) and emit a series of short, high-pitched yips. Howling is used to communicate with other coyotes in the area.

Coyotes are primarily nocturnal but may be active during the day. Generally, activity and movements such as foraging and territorial maintenance are most common at night. Daytime activity may increase during the breeding season and pups are more active during the day than are adults. Coyotes will use underground dens for birthing young, but resting sites during other seasons tend to be sheltered areas above ground.

The basic unit in coyote social organization is a family occupying a territory. Adult females, mated pairs, or groups of coyotes may defend territories, but defense is most common during the pup-rearing period. Families often occupy territories in the same area for many consecutive years. In North Dakota, typical coyote families (mated pair, pups, and sometimes one associated adult) will occupy mostly contiguous and non-overlapping territories. Coyotes are less likely to form packs than are wolves, and in North Dakota, they often forage in pairs or threesomes.

Coyotes, although essentially carnivorous, are versatile in their eating habits and are considered opportunistic predators. Their diet changes throughout the year in response to available food sources. Mammalian prey species are important year-round, and are supplemented in spring, summer, and fall with plants, insects, and birds. Some prey items are cached for later consumption. Small mammals are the primary prey source during the summer. In winter, larger prey including deer, livestock carrion, and rabbits constitute the majority of the diet. Coyotes will eat adult birds and eggs, but they are not considered to have as great an impact on nesting birds as do red foxes. Prey abundance affects both density and reproduction of coyotes. The prey of coyotes seems to determine group size, with larger prey items (e.g., ungulates) requiring a larger hunting group. In North Dakota, smaller food items (rabbits, fawns, rodents) are readily available and used by coyotes, thus precluding the need for larger hunting groups.

Home range size is inversely related to coyote abundance; family territory sizes in a low-density population in North Dakota averaged 61 km^2 in spring-summer and 30 km^2 during fall-winter. Home ranges of territorial individuals ranged from 5 to 80 km^2 and included most of the family territory.

It is difficult to estimate coyote population densities, but we do know that they are highly variable. Densities of 2 to 4/10 km^2 are common over much of the range. However, densities are lower in North Dakota, with estimates of 3 to 7 adults/100 km^2. Coyotes are most vulnerable during their first year, as indicated by higher mortality rates for juveniles (40 to 70%) than adults (30 to 50%). Human-caused deaths account for most mortality, but diseases and parasites, including distemper, canine hepatitis, and sarcoptic mange, are important at times.

Reproduction

Breeding takes place from January to March. Mated pairs may produce pups each year, and both the male and female raise the pups, occasionally with help from an unmated adult (yearling offspring). Gestation averages 63 days and mean litter size is about six pups, but litter size is influenced by population density and food availability. The percentage of yearlings that breed varies among region and years, but in North Dakota, typically both sexes breed their first year. Dispersal takes place in autumn and winter, when the pups are 6 to 9 months old. Dispersal distances can be as great as 60 to 80 km. Not all pups disperse in their first year, which may explain the presence of adult helpers at some dens.

Status and Conservation

Coyote populations in North Dakota have varied widely in the past 150 years and will likely experience variable populations in the future. Changes in fur value or disease outbreaks could cause major population changes in the future. A recent outbreak of sarcoptic mange has been associated with a downward trend of coyote numbers, but evidence suggests that this is a temporary phenomenon. Intensive controls including bounties, denning (i.e., killing pups at dens), shooting from aircraft, and application of toxicants have been significant factors controlling coyote populations since white settlement. Evidence strongly suggests that human-caused mortality in the early 1900s severely suppressed coyote populations in many parts of North Dakota (especially the east) and by the 1950s, coyote populations had reached their lowest level since settlement. Populations recovered by the 1980s.

Coyotes are a valuable fur resource and desired by hunters and trappers, but they also are significant as predators of livestock and poultry. Coyote management has largely been designed to reduce populations; however, due to their fecundity, behavior, and adaptability, those attempts are often difficult. Most do not prey on livestock, but once they learn that young livestock are easy prey, depredation may become a problem. Most depredation problems can be solved by targeting and removing the coyotes that are responsible. Measures to prevent livestock depredations include proper disposal of livestock carcasses, use of guard animals, and confining livestock during birthing.

In North Dakota, coyotes are managed as a furbearer with regulated hunting and trapping seasons. It is illegal to use poison as a control method. The trapping season is restricted to the fall and winter months, but the hunting season is open year-round. This liberal hunting season allows landowners to remove problem coyotes without having to obtain a special permit.

Selected References

Allen et al. (1987), Beckoff (1977), Bekoff and Gese (2003), Brillhart and Kaufmann (1994), Hazard (1982), Knowlton (1972), Lewis et al. (1994), Sargeant et al. (1987, 1993), Voigt and Berg (1987)

Gray Wolf

Canis lupus

Shu^{n}g-tokeca–Dakota; Sunkmanitu tanka, Caksi–Lakota; Harrata–Mandan; Tshesha–Hidatsa; Stshirita-kusa–Arikara

L. L. Master, American Society of Mammalogists Mammal Images Library

Description

The gray wolf, also called the timber, tundra, or plains wolf, is the largest wild canid in North America. The face can be distinguished from that of dogs and coyotes by the wide tufts of hair projecting down and outward from below the ears. Typically, wolves have white fur around the mouth, whereas domestic dogs usually have black fur in this area. Wolves have a broader snout than coyotes, and their ears are held erect and forward and are relatively shorter. The fur is usually grizzled gray in color but can vary widely from pure white through mottled gray and brown to black. Dorsal hairs are generally longer and darker than ventral hairs. The tail is heavily furred and long. They have moderately long legs. Adult males weigh from 20 to 80 kg and adult females weigh from 18 to 55 kg. Body size varies greatly depending on subspecies. Ranges of standard measurements are Total Length - 1,003 to 1,046 mm, Tail Length - 356 to 500 mm, Hind Foot - 220 to 310 mm.

Distribution

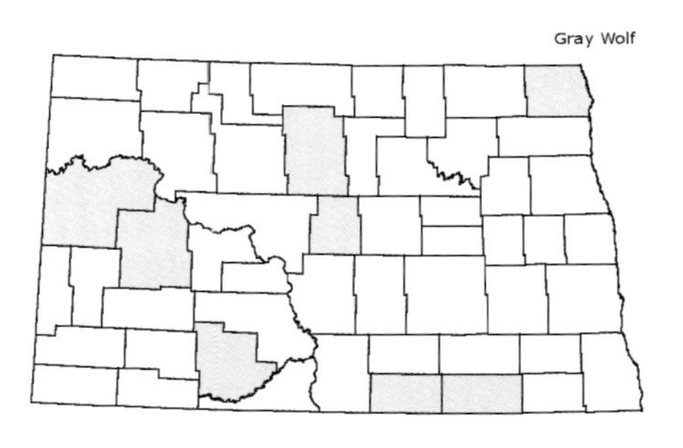

Historically, gray wolves were found throughout the Northern Hemisphere, north of 120° N latitude. In the United States, they were found everywhere except parts of California and Arizona, and in the southeastern states, which were occupied by the related red wolf. Reports suggest that wolves were abundant throughout the northern Great Plains, both on the prairie and in forested areas. Deliberate control by European settlers restricted their present range in North America. By the 20th century, the gray wolf had nearly disappeared from the eastern United States, except for portions of the Appalachians, and the northern Great Lakes region. Gray wolves once occurred throughout North Dakota, but they were extirpated by the late 1800s. Scattered sightings of wolves have been regularly reported throughout the state, likely wanderers out of Minnesota and Canada, but there are no recent records of breeding wolves in the state.

Habitat

Gray wolves are very adaptable and not habitat specific; they are considered habitat generalists. They once occurred in nearly all major habitats, including prairies, tundra, coniferous forests, and mixed coniferous-deciduous forests. Only deserts and mountain tops are avoided. Habitat use by wolves is influenced by availability and abundance of prey. They are predators of large mammals, and abundance of prey species and lack of human persecution are good predictors of where wolves can survive.

Ecology and Behavior

Wolves are social canids, functioning in packs that are typically family units of five to eight animals. Pack members are strongly tied to each other and order is maintained through a dominance hierarchy with the adult (alpha) male dominant to the alpha female and pups, and the alpha female dominant to the pups. In large packs, separate hierarchies of males and females may develop, but the alpha male usually is the pack leader. Mated pairs typically stay together for life. The alpha female is often the only mated female in the pack. Young are usually born in underground dens, but hollow logs, crevices in rocks, caves, or abandoned beaver lodges have been used as den sites. Wolves may dig their own den or enlarge and use burrows of other mammals. Typically, dens face south and are in well-drained areas close to water.

Wolves communicate with each other through postures, facial expressions, vocalizations, and scents. Communications help to keep order among pack members and reaffirm dominance relationships. Vocalizations include growls, barks, and howls, functioning as communication among pack members or between packs. Postures and facial expressions are used for greetings among pack members and as expressions of dominance and subordination. Howling functions to assemble pack members and advertise territorial boundaries. Scent marking is usually performed by high-ranking males and is believed to define and advertise territorial boundaries, and to bring new pairs together to establish new packs.

The pack maintains a large home range, but its size varies considerably from area to area, depending on the type and density of prey and the size of the pack Home range sizes of 130 to 3,374 km^2 have been reported. Daily movements within a home range for foraging and territorial maintenance vary from a few kilometers to 72 km.

Packs are cooperative in their hunting behavior, which allows them to take large prey such as bison, moose, elk, and deer. Other prey resources include rabbits, hares, beaver, small rodents, and occasionally berries, fruits, and carrion. Predation rates are variable, but packs have been estimated to take one big-game animal every three days during winter to one per five days during the summer.

Human activity is the major mortality factor for wolves in most areas. Human-caused mortality includes legal and illegal harvest and vehicle collisions. Other mortality factors include disease, malnutrition, and intraspecific aggression (usually related to low prey density).

Reproduction

Breeding takes place from January to April depending on latitude. Courtship occurs between pack members, but not all females produce pups. The dominant pair usually breeds successfully, but occasionally subordinates do as well. Young are born after a 63-day gestation period and litter sizes average six pups (range 1 to 11). Young are blind and deaf at birth. They are weaned at approximately 8 to 10 weeks. A pack generally has one to three females of breeding age (22 months), and other adult pack members assist in rearing young.

Wolves may disperse from their pack at any age, but young animals (10 to 20 months) do so more frequently. Most dispersing wolves establish new territories or join packs within 50 to 100 km of their natal pack. Dispersal movements can be extensive; one dispersing wolf traveled a straight-line distance of 206 km in two months. Dispersing wolves frequently attempt to establish new packs.

Status and Conservation

The gray wolf was listed as endangered in 1973, but has since recovered in some portions of the United States. Healthy and expanding wolf populations in the upper Great Lakes region and the successful reintroduction of wolves in Montana and Idaho have allowed the U.S. Fish and Wildlife Service to propose reclassification of the gray wolf from endangered to threatened in some parts of the country and remove the species from the Endangered Species list in other states.

Wolves once occurred throughout North Dakota, but today, only an occasional wolf wanders into the state from Minnesota, Montana, and Canada. Currently no breeding wolves are known to occur in the state. In 1994, a suspected wolf den with tracks of an adult wolf and pups was discovered in the Turtle Mountains, possibly representing the first wolves born in the state since the 1930s. Those wolves disappeared for unknown reasons. With the expanding wolf population in Minnesota, wolves may eventually pioneer areas in the Turtle Mountains or Pembina Hills. However, biologists estimate that at least 500 km^2 is needed to ensure the viability of a wolf population because of its large territorial requirements. It is unlikely that a recovery of wolves will occur in North Dakota in the near future.

Selected References

Carbyn (1987), Hall (1981), Johnson and Sargeant (1977), Mech (1970, 1974), Mech and Frenzel (1971), Paquet and Carbyn 2003, Peterson et al. (1984)

Gray Fox
Urocyon cinereoargenteus

David Strand

Description
The gray fox is somewhat smaller in stature than the red fox, having shorter legs and extremities, but it is slightly more stout and similar in weight. It has a narrow muzzle (typically shorter than red fox), and relatively large and pointed ears that are held erect and forward. The dorsal fur has a grizzled gray appearance from individual guard hairs having bands of white, gray, and black; underfur is gray. The tip of the tail is black. The throat, front of ears, lower cheeks, and chest are white; underparts are buff. Gray fox fur is coarser than the luxurious texture of red foxes, yet the pelt is beautifully marked and can be valuable depending on the fur market. Males are slightly larger than females. Ranges of standard measurements are Total Length - 80 to 1,130 mm, Tail Length - 275 to 430 mm, Hind Foot - 100 to 150 mm, Ear - 65 to 81 mm, Weight - 3.9 to 5.2 kg.

Distribution

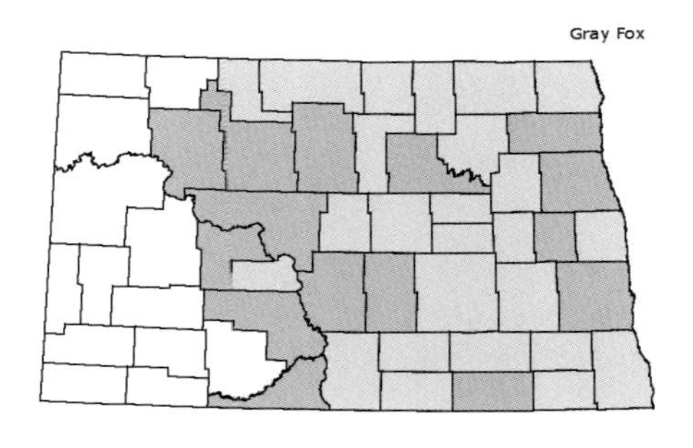

The gray fox is found from extreme southern Ontario and Quebec, southward throughout the eastern United States, through Central America to northern Venezuela and Colombia. The northwestern range extends into eastern North Dakota and southern Manitoba, eastern and portions of western South Dakota, and eastern Nebraska. It is found in parts of Wyoming, Colorado, Utah, Oregon, and New Mexico; it is absent from Washington, Idaho, and Montana.

In North Dakota, gray foxes are uncommon, occurring primarily in the eastern half of the state, largely in riparian and wooded or brushy areas. It appears to be a newcomer to North Dakota, with no reports of the species from the early surveys. Extreme western records of gray foxes in the state are from Mountrail, Ward, McLean, Mercer, Morton, and Sioux counties.

Ecology and behavior

The gray fox is primarily considered a deciduous woodland species, typically using areas with a greater proportion of wooded habits than areas used by red and swift foxes. It also thrives in mixed agricultural/woodland landscapes, although brush and tree cover are important characteristics of areas they occupy. Gray foxes, unique among North American canids, are well-known for their ability to climb trees to forage, rest, or escape from predators. They are well-adapted to climbing, with long claws and short legs, and have been observed climbing to heights of 18 meters in vertical branchless trees by grasping the truck with their front feet and pushing with their hind legs. Once in trees, they are able to jump from branch to branch. To descend trees they will back down vertical trunks or run head first down slanted trunks.

Less is known about the social ecology of gray foxes than red foxes. They are typically monogamous, living in family groups consisting of the mated pair with their young. There are no reports of a third "helper" adult in raising pups, but there are reports of two females rearing pups in the same den. Both sexes raise their pups, although females likely play a larger role. Gray fox families are territorial; home ranges of adjacent family groups may overlap, but core areas are exclusive. Adult foxes are solitary in their movements, foraging, and territorial maintenance. Gray foxes are mainly nocturnal, but may be active during daylight hours.

Reported home range sizes are quite variable, ranging from <1 to >6 km^2, and differences are believed to be related to quantity and quality of habitats. Home range size also varies according to sex, geographic location, and season. Males tend to have larger home ranges than females, foxes in the western United States tend to use smaller areas than those in the east, and both males and females tend to use larger areas in the fall and winter than spring and summer.

Gray foxes are considered more omnivorous than other North American canids. Mammals, including small rodents, sciurids, and rabbits, comprise the greatest portion of their diets, but in the summer, invertebrates such as grasshoppers and beetles are important. During the fall, fruits and berries may dominate their diets. They supplement these with a wide variety of other foods, such as plants, birds, eggs, and carrion.

Dens may be used throughout the year, but most use occurs for pup rearing. Dens may be hollow logs or trees, rocky outcrops, underground burrows (dug by other animals), wood piles, or heavily brushed areas. Activity is concentrated near the rearing den while pups are young. Human activity near a den will often cause adults to move pups to a new den site.

Information about gray fox dispersal is limited and variable. Timing of dispersal by young gray foxes is highly variable, ranging from 3 to 10 months of age. Male juveniles are more likely to disperse significant distances; distances up to 135 km have been reported.

Gray foxes are harvested throughout most of their distribution, and in many areas, this is the primary cause of mortality. Gray foxes are also depredated by golden eagles, bobcats, mountain lions and coyotes. Diseases, such as distemper and rabies, can also be a significant source of mortality.

Reproduction

The gray fox breeds from January to mid-May, but breeding likely peaks in February or the first week of March. The gestation period averages 53 days. One litter per year is born between late March and mid-May. A litter can have between one and 10 pups, but three to five is most common.

Pups are born with their eyes closed and their thin fur is black. Their eyes open at 9 to 12 days of age. Pups may emerge from the den at about a month and venture away from the den area at about 3 months of age. The family group breaks up in late summer or early fall. The young continue to grow until they are about 18 months old, but they are able to breed before they are a year old.

Status and Conservation

Because they are uncommon, little is known about gray fox populations in North Dakota. Wooded riparian habitats along parts of the Red, Sheyenne, James, and Missouri rivers are the likely habitats for sustained gray fox populations.

The gray fox is classified as a furbearer, and is managed with a regulated hunting and trapping season each year. Its rarity limits its value in the fur trade, but its striking appearance gives gray foxes value as a watchable resource. Land management practices that maintain brushy and forested cover types with high prey populations and suitable denning sites are beneficial.

Selected References

Adams (1961), Chamberlain and Leopold (2000), Cypher (2003), Fritzell (1987), Fritzell and Haroldson (1982), Hazard (1982), Hall (1981), Schantz (1950)

Swift Fox

Vulpes velox

Ihoichka–Hidatsa; Ohcha–Mandan; So^{n}gina–Dakota; Ciwaku–Arikara

George Rohde

Description

The swift fox is the smallest canid native to North Dakota. It is about the size of a house cat, having an average weight of 2.4 kg. Coloration is grizzled dark grayish-buff across the back, upper tail and sides extending to orangish buff across the lower sides, legs, and the ventral surface of the tail. The ventral fur is white with some buff on the chest. Swift foxes have distinctive black patches on each side of the muzzle, and rounded and erect ears. The bushy tail is black-tipped, in contrast to the white-tipped tail of red foxes. Average measurements (male, female, respectively) are Total Length - 800, 788 mm, Tail Length - 284, 280 mm, Hind Foot - 126,120 mm, Ear - 65, 62 mm.

Distribution

The swift fox is native to short-grass and mixed-grass prairies of the Great Plains in North America. On the northern limit of their his-

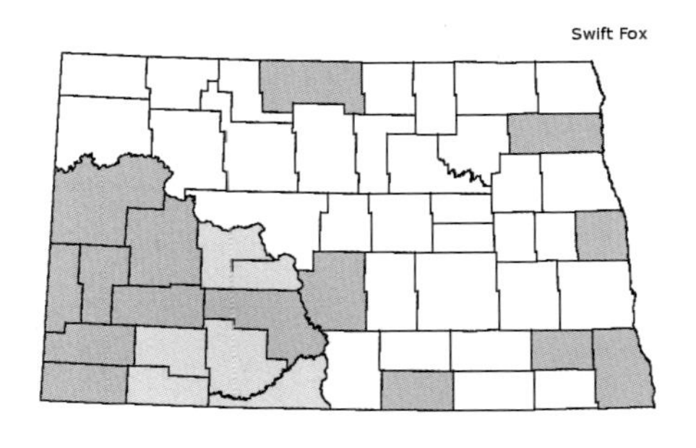

toric range, they were present in the southern prairies of Alberta, Saskatchewan, and Manitoba. The southern distribution extended to New Mexico and Texas. Records exist for all or parts of Montana, Wyoming, North Dakota, South Dakota, Nebraska, Kansas, Colorado, and Oklahoma. Following white settlement, the species declined until by 1900, it was rare throughout its historic range. Presently, healthy populations occur in Colorado, Kansas, Wyoming, Oklahoma, and New Mexico. Reintroductions have occurred in the prairies of Alberta, Saskatchewan, Montana, and South Dakota.

Prior to the mid- to late-1800s, swift foxes were likely common in most of North Dakota except for tall-grass prairie habitats along the eastern border of the state. Swift foxes have been scarce in the state since Bailey's (1926) Biological Survey of North Dakota, and in recent years were considered extirpated. However, recent confirmed sightings have become more numerous, especially in the west, possibly indicating some recovery of the species in the state.

Habitat

Swift foxes prefer shortgrass and mixed-grass prairies in gentle rolling to level terrain. Sparse vegetation, allowing for both good mobility and visibility seems to be an important characteristic of areas used by swift foxes. Additional favorable conditions likely include low potential for contact with human activities and low densities of predators such as coyotes.

Ecology and behavior

Swift foxes are primarily nocturnal, although limited daytime activity may occur near den sites. Social structure revolves around the monogamous mated pair. Occasionally, trios or groups of two males and two to three females may be found; such groups include one breeding female and non-breeding "helpers." Both members of the pair provide for the young, and young foxes remain with the adults for four to six months, longer than other North American canids.

Unlike most other canids, swift foxes use dens throughout the year. They will either excavate their own dens or use enlarged burrows of other animals. Dens serve to provide escape cover from predators, protection from extreme weather conditions in both summer and winter, and shelter for raising young. Notably, the distribution and density of dens, which are considered important components of swift fox habitat, may influence predation rates and population growth.

Like other canids, swift foxes, are opportunistic foragers, feeding on a wide variety of mammals, birds, arthropods, plants, and carrion. Mammals, especially lagomorphs and rodents usually dominate swift fox diets. Insects and birds become important food items in late summer and early fall; several studies have reported use of carrion throughout the year. The generalist foraging behavior of swift foxes makes food an unlikely limiting factor.

Estimates of average home range sizes for swift fox are variable, ranging from approximately 7 to 35 km^2. Early studies suggested that swift foxes were not territorial, although more recent data have provided evidence of territoriality. Members of family groups forage separately. Swift foxes tend to travel greater distances in the breeding season, which may increase potential encounters with coyotes and thus vulnerability to coyote predation. Young foxes may disperse from their natal areas from September to January. Males are more likely to disperse, and disperse significantly farther (9.4 km versus 2.1 km) than females. Occasionally, a yearling (most often female) will remain in the parents' home range and assist in rearing the next year's pups.

Swift fox population densities tend to be sparse, estimated at about 2/10 km^2. Coyotes have been identified as the major cause of mortality, although they typically do not consume the foxes they kill. Other predators of swift foxes include are golden eagles and American badgers. Mortality factors associated with human activities include vehicle collision, secondary poisoning, shooting, and trapping.

Reproduction

Swift foxes are monestrous, with breeding beginning in late December or early January in the southern portion of their range, to March in the northern portion of their range. Gestation is estimated to be 51 days. Average litter sizes of two to six have been reported based on counts of pups at natal dens. Pups will emerge from their dens at about 5 weeks. Both male and female parents will provide food for the young.

Status and Conservation

Until recently, swift foxes presently were considered extirpated from North Dakota. From 1970 to 2011 there were only four confirmed records from Golden Valley, McIntosh, Ransom, and Slope counties, and one unconfirmed record from Mercer County, probably representing dispersing young from South Dakota and Montana. However, from 2011 to 2017, there have been 12 additional confirmed reports, possibly indicative of recover. Swift foxes are listed as furbearers in North Dakota, but the season is closed.

Although there are habitats that are likely suitable for swift foxes in North Dakota, their long absence from the state may be due to the lack of colonists to re-establish viable populations. Increases in coyote and especially red fox populations may impair the ability of swift foxes to disperse into and reoccupy suitable habitats. Several efforts to reintroduce swift foxes into the northern parts of their historic range have succeeded, including reintroductions to southern Alberta and Saskatchewan, the Blackfoot reservation in Montana, and two locations in South Dakota (Badlands National Park and Bad River Ranch near Pierre). Swift foxes might find their way back to North Dakota on their own as distribution expands from South Dakota and Montana reintroduction efforts.

Selected References

Allardyce and Sovada (2003), Bailey (1926), Cypher (2003), Egoscue (1979), Hillman and Sharps (1978), Hines (1991), Jones et al. (1983), Kilgore (1969), Scott-Brown et al. (1987), Sovada and Carbyn (2003), Sovada et al. (1998)

Red Fox

Vulpes vulpes

Sungila lutah–Lakota; Ehchokuschi–Hidatsa; Hirúttsa–Mandan

George Rohde

Description

The red fox is a mid-sized carnivore resembling a small dog. It has an elongated, pointed muzzle, relatively large and pointed ears that are held erect and forward, moderately long legs, and a long and heavily furred bushy tail. The color of the dorsal fur ranges from pale yellowish red to deep reddish brown, with color becoming darker on the back. The tail is also reddish, but mixed with black and distinctively tipped with white. There are several atypical color variants in wild populations; the most common are "silver" and "cross." Silver foxes are black, but guard hairs are frosted with white, giving them a silver appearance. Cross foxes have a dark brown band down their back and across their shoulders. All color variants maintain the distinguishing white tip on their tails. Average measurements (male, female, respectively) are: Total

Length - 1,026, 973 mm, Tail Length - 382, 362 mm, Hind Foot - 167, 158 mm, Ear - 91, 76 mm, Weight - 5.5, 4.7 kg.

Distribution

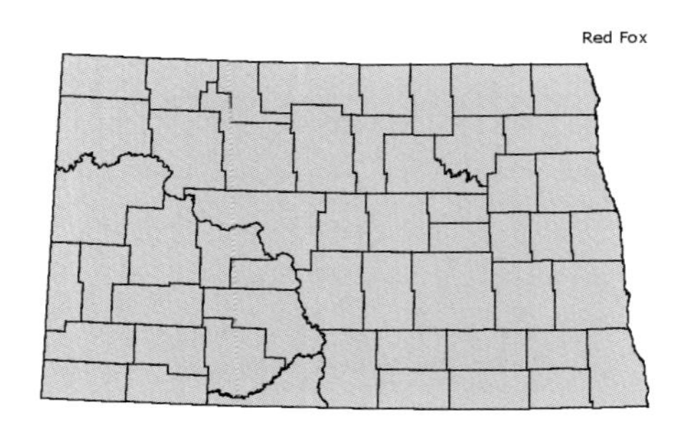

Except for the gray wolf, the red fox is the most widely distributed carnivore in the world; its distribution includes nearly all of the Northern Hemisphere. It occurs throughout the United States, except in the desert Southwest and the driest parts of the Great Plains.

In North Dakota, red foxes are found throughout the state, but their distribution and abundance has varied historically. Prior to white settlement, a period when foxes were competing with gray wolves, coyotes and other predators, red fox populations were low. Early accounts in North Dakota indicated that they were locally common, but less numerous than gray wolves. Red foxes were sparse to absent through much of the state from the 1890s to the mid-1930s. In the mid-1930s, they were observed through much of eastern and central North Dakota, and by 1944, high populations existed in most of the state, except the southwest. Bounty records suggest that red foxes became common in the southwest around the mid-1950s.

Habitat

Red foxes are found in a wide variety of habitats from forest to arctic tundra, open prairie, and cropland. Red foxes are well suited to agricultural development and thrive in intensive agricultural areas of North Dakota.

Red foxes forage primarily in upland habitats, but wet meadow edges of wetlands are particularly attractive, likely because of the diverse and abundant food resources found in these habitats.

Ecology and behavior

Red foxes live in family groups consisting of a mated pair and occasionally a third adult, usually a yearling female offspring that has delayed dispersal. Fox families are highly territorial in the spring and summer, occupying well-defined, discrete territories ranging in size from 3 to 21 km^2. Territory size is inversely related to population size, with territories expanding and contracting as population size fluctuates. Each adult fox of a pair will thoroughly cover the family territory over the course of two weeks. However, areas of intense activity within the territory might change throughout the year.

Foxes are largely carnivorous, although they opportunistically take advantage of a variety of food resources. They consume more small mammals (e.g., voles, mice, rabbits, hares, and ground squirrels) than any other food, although they supplement these with a wide variety of other foods, such as plants, birds, eggs, insects, and carrion. Red foxes take the eggs and young of ground-nesting birds, and prey selectively on nesting hens. During winter months, the diet consists mostly of rodents, including mice and voles, and they can consume significant quantities of sunflower seeds left in fields after harvest.

Red foxes are mainly nocturnal but may be active during daylight hours. Both sexes actively raise their pups, but they are solitary in their movements, foraging, and territorial maintenance. Dens are used for pup rearing, but not during other periods of the year. Typically, natal dens are refurbished holes that were dug by other species (e.g., badgers), but dens may also be dug by the foxes. Natal dens may be used for many years and can be complex in structure. Activity is concentrated near the rearing den while pups are young. Commonly, more than one rearing den will be used during pup rearing. As pups become more mobile, litters may be split and two dens may be used concurrently. Human activity near a den will often cause adults to move pups to a new den site.

Dispersal includes both young foxes leaving their natal areas and adult foxes emigrating, sometimes from territories where they have successfully raised a litter. Dispersal distances of more than 300 km have been reported in North Dakota.

Red fox populations can vary greatly. In North Dakota, aerial censuses have estimated densities as high as 17 families per township. Changes in fox populations are related to dynamics of the canid community, direct human influences on canid populations, and habitat changes. There is substantial evidence of an inverse relationship between population densities of gray wolves and coyotes, and coyotes and red foxes. Thus, the extirpation (by humans) of gray wolves in North

Dakota favored coyote population expansion, and the subsequent control of coyotes favored red fox expansion. At the same time, agricultural development provided a food base and habitat conditions more suitable to red foxes than presettlement conditions, which were best suited to wide-ranging canids such as the wolf and coyote.

Red foxes have been known to live up to 8.6 years in the wild, but few survive to 6 years. Most mortality is the result of hunting, trapping, or vehicle collisions. Recently, however, secondary infections and weakened condition resulting from sarcoptic mange have been major factors contributing to fox mortality in North Dakota. The species is also an important reservoir for rabies in some portions of North America, but not North Dakota. Coyotes will occasionally chase and sometime kill foxes, and there is evidence of coyotes killing fox pups at dens.

Reproduction

Red foxes breed mostly in January and February. In North Dakota, most yearling and adult foxes breed each year, but older foxes have greater reproductive performance. Females are monestrous and in North Dakota they typically bear four to six pups in March or April after a 53-day gestation period. Pups emerge from the dens in early May (about 5 weeks old) and initially remain near the den; distances ventured from the den gradually increase until dispersal begins in August. Once pups are weaned, both the male and female provide food for the young. Dispersal of pups peaks during October and November and continues through January.

Status and Conservation

Fox populations in North Dakota have varied widely in the past 150 years and likely will continue to be dynamic in the future. Increases in coyote numbers, rising fur value resulting in increasing harvest, or disease outbreaks could cause major population changes. An outbreak of sarcoptic mange in the 1990s has been associated with a downward trend of fox numbers, but evidence from mange outbreaks in other states suggests this is a temporary phenomenon. Fox numbers in North Dakota are slowly beginning to recover. Favorable habitat conditions in North Dakota and low fur prices foster growth in the population of red foxes in the state. Red foxes are abundant enough in North Dakota to allow hunting and trapping seasons. Currently there is a year-round season, which allows people to remove foxes that are killing poultry or causing other problems.

Red foxes are a valuable fur resource and desired by hunters and trappers. Conversely, foxes are recognized as significant predators of breeding birds, particularly upland nesting ducks. Unfortunately, to effectively manage foxes to benefit ducks, foxes must be controlled at the time fur is least valuable and when pups are being reared, which negatively affects the potential harvest for fur. Thus, within wildlife agencies, management strategies might conflict because of differing objectives. Ultimately, clear goals must be identified and effective management must be based on sound information and take into account divergent goals.

Selected References

Adams (1961), Allen (1984), Bailey (1926), Cypher (2003), Johnson and Sargeant (1977), Johnson et al. (1989), Lariviere and Pasitschniak-Arts (1996), Phillips et al. (2003), Sargeant (1972, 1982), Sargeant et al. (1984, 1986, 1987, 1993), Voigt (1987)

Family Ursidae
Bears

American Black Bear
Ursus americanus

Wasabè–Omaha; Wachank-shica–Dakota; Mato sapa–Lakota; Konuch-katit–Arikara; Haschida–Hidatsa; Ischída–Mandan

R. Riewe, American Society of Mammalogists Mammal Images Library

Description

The black bear is large and heavily built, with short powerful legs, erect ears, and a very short tail. Unlike many other regional mammals, it is plantigrade, i.e., walking on the soles of its feet. Coloration is usually black with a brown snout, but occasionally brown, cinnamon, bluish, or even beige-colored bears have been reported. White marks are sometimes seen on the chest. Weights of adult males usually range from 60 to 140 kg, and may be nearly double those of adult females (40 to 70 kg). Occasionally, an adult male may reach 300 kg. Representative ranges of body measurements are: (Males) Total Length - 1,375 to 1,780 mm, Tail Length - 90 to 125 mm, Hind Foot - 215 to 280 mm; (Females) Total Length - 1,270 to 1,475 mm, Tail Length - 80 to 115 mm, Hind Foot - 190 to 240 mm.

Distribution

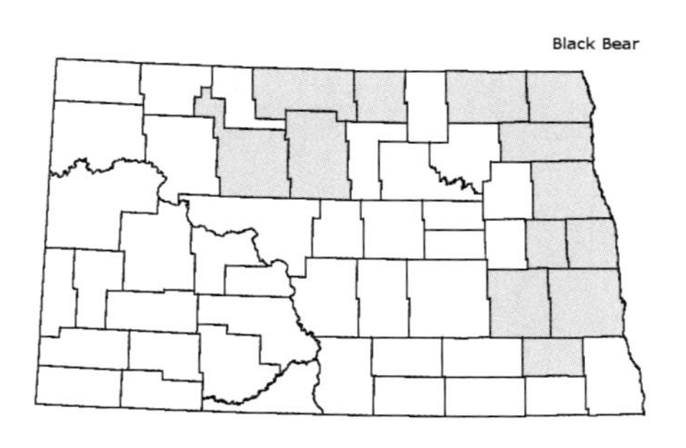

The historic distribution of the black bear included nearly all of the forested areas of North America, including Mexico. Black bears were never as widely distributed in the northern Great Plains as were grizzly bears. Currently, they are found in all Canadian provinces, except Prince Edward Island, and throughout forested areas of the United States except for the arid Southwest. They also occur in 12 Mexican states encompassing the Sierra Madre Occidental and Sierra Madre Oriental.

In neighboring South Dakota, black bears occur only in the Black Hills, and they do not appear to occur in eastern Montana. In North Dakota, they originally ranged over much of the forested areas of the state, but were most abundant in the Turtle Mountains, Pembina Hills, Red River valley, and the forested tributaries of the Red River. Their greatest abundance was in the Red River valley, where during the period from 1800 to 1808, fur trader Alexander Henry reported 906 bears taken from the region. North Dakota Game and Fish Department files indicate numerous recent reports from the Red River corridor, Pembina Hills, and Turtle Mountain areas.

Habitat

Black bears are quite adaptable, but are most abundant in relatively inaccessible country with dense understory cover and abundant food sources. Dense cover is preferred for concealment, thermal protection, and bedding. In neighboring Minnesota, they usually inhabit well-wooded areas, swamps, and cutover areas having an abundant supply of berries. Habitat selection varies seasonally, depending on availability of food. Bears will readily inhabit areas occupied by humans, especially if sources of human food and garbage are available.

Ecology and Behavior

Black bears are omnivorous, consuming a wide variety of foods, especially plants. Foods are dominated by new growth of grasses and forbs in the spring, shrub and tree-borne fruits in summer, and mast such as acorns in the fall. Animal matter is taken opportunistically, consisting mainly of beetles and colonial insects, but does not seem to constitute more than about 10% of the diet.

Activity peaks generally occur during the morning and evening twilight hours, but this can be influenced by mating activity and food availability. Greatest overall levels of activity are during the warm months of June through September, a period of increased feeding. With the onset of autumn, black bears become increasingly inactive, and between October and early January, they seek out dens in which to spend winter dormancy. Dens may be under downed timber or rock piles, or in banks, caves, or culverts. During dormancy, metabolism drops, but body temperature is maintained and bears may be easily aroused with disturbance.

Black bears have a solitary life style, except for females with cubs, during the mating season, and in aggregations around food sources. Black bears engage in marking behavior involving rubbing, biting, and clawing certain trees ("bear trees"). This behavior implies territoriality, as all marking is done by residents rather than transients. Only females are territorial, but a number of reports indicate that bears exhibit little tendency to avoid each other and that home ranges frequently overlap. In Minnesota, female territories average about 10 km^2; adult males during the mating season have overlapping ranges averaging 75 km^2. During late summer and early fall, bears may engage in extensive foraging movements outside of the home range prior to returning for denning. Yearlings undergo dispersal during early summer, females remain around their mother's home range, but males engage in extensive movements up to 200 km or more.

Population densities have averaged one bear per 1 to 11 km^2 in various parts of North America. In northeastern Minnesota, populations have been estimated at one bear per 4.1 to 6.3 km^2. Black bears are long lived, surviving up to 23 years in the wild. Most mortality is human-caused, including hunting, poaching, roadkills, and depredation control. Infanticide by adult males may also occur. Mortality increases during years of food shortage, involving mostly young bears and those forced to move long distances in foraging. Very old bears may suffer from serious dental problems, resulting in starvation.

Reproduction

The peak breeding occurs during June and July, but may extend into September. Breeding is promiscuous, with multiple males visiting females over a four- to seven-week period. Breeding associations may last two to five days, and more than one male may be present. Black bears are induced ovulators, requiring the stimulus of copulation before ovulation occurs.

Black bears have delayed implantation, the early stage embryo (blastocyst) floating in the uterus until it implants during November or December. Therefore, while the total gestation period may be seven to eight months, most development occurs during the final six to eight weeks. Parturition occurs during January or February, during denning. Newborn cubs are altricial, blind, toothless, poorly furred, and weighing 225 to 280 g. Litter size is normally two, but litters with as many as five cubs have been reported. Cubs remain with the mother for about 16 months, and then disperse. Because of this extended time with the mother, females only breed every two to three years. Sexual maturity is attained between 2 and 7 years, usually at 3 to 4 years.

Status and Conservation

Black bears continue to occupy much of their historic range, but population densities vary greatly. In a number of western states and Canada, they are relegated to nearly pest status, with thousands killed annually, while in much of the East and Southeast, they survive only in remote enclaves. They have been extirpated from some midwestern states, and are listed as threatened in South Dakota, Florida, and Louisiana. In North Dakota, the black bear has become a rare species, in contrast to its abundance during the days of the early fur trade. The North Dakota Game and Fish Department recorded only 11 reports between 1974 and 2004, most in the northeastern third of the state. These bears are likely wanderers from adjacent Minnesota and Manitoba. The black bear is listed as a furbearer in North Dakota, with a continuing closed season.

In North Dakota, the black bear became a rare species, in contrast to its abundance during the days of the early fur trade. The North Dakota Game and Fish Department recorded only 11 reports between 1974 and 2004, most in the northeastern third of the state. These bears were likely wanderers from adjacent Minnesota and Manitoba. However, 28 confirmed reports between 2011 and 2017 may indicate a comeback of the species as a resident of the state. The black bear is listed as a furbearer in North Dakota, with a continuing closed season.

Although the black bear is abundant and hunted in neighboring Minnesota, and throughout much of the West, its rare status in North Dakota over the past century makes it unlikely to become a harvested species in the foreseeable future. In states such as Minnesota, where bears are common, their active foraging habits, combined with human encroachment on habitat, has resulted in numerous human-bear conflicts. Steps applying to maintaining the species in North Dakota include habitat management for hard and soft mast production, den sites, and escape cover; protection by laws and regulations; implementation of nuisance control policies; and an information and education program, particularly regarding nuisance situations.

Selected References

Bailey (1926), Coues (1897), Foresman (2001), Hazard (1982), Higgins et al. (2000), Jones et al. (1983), Jonkel (1978), Lariviere (2001), Pelton (2003), Rogers (1987)

Family Mustelidae
Weasels and Allies

Wolverine
Gulo gulo

Sacatungka, Skecatanka – Lakota; Eh-tupah – Hidatsa; Mató-ka – Mandan; Nag-gwy-gway – Ojibway

Jeffrey C. Lewis, U.S. Dept. of Transportation

Description

The wolverine is North America's largest mustelid. It is about as heavy as a medium-sized dog, males weighing 11-18 kg. It has a compact body, broad head, short legs, and a bushy tail. Its coloration varies from brown to black, with yellowish to brownish bands from the shoulder to the sides and converging at the rump. There may also be white to yellowish throat markings. Average standard measurements (male, female, respectively) are: Total Length - 1,000, 901 mm; Tail Length - 231,223 mm; Hind Foot - 184, 160 mm; Ear - 55, 45 mm. The wolverine skull is massive, and capable of crushing the bones of any large ungulate.

Distribution

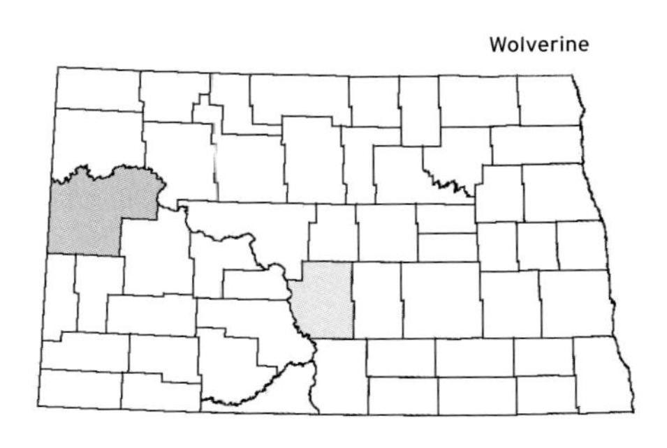

The wolverine is circumpolar in distribution, occurring in boreal zones of Eurasia and North America. Historically, wolverines ranged in North America in Alaska and northern Canada, the northern tier of states, south through the Rocky Mountains to New Mexico and Arizona.

The wolverine was apparently extirpated from much of its original range, but in recent years, various undocumented sightings in the upper Midwest as far south as Iowa and Nebraska suggest that there may be a slow recovery.

Originally, wolverines ranged over northeastern North Dakota, especially the forested areas. Early fur records indicate skins from the Park River, Hair Hills (Pembina Escarpment), Pembina River, Grand Forks, and Fort Union. There were no known records over the past century. In recent years, however, there have been scattered unverified reports from the badlands and north-central portions of the state. In addition, two relatively recent incidents are worthy of note. In May, 2016, a rancher killed a male wolverine near Alexander (McKenzie County). When a necropsy was performed, it was revealed that a transmitter had been implanted in Wyoming in 2008. The animal was last detected in Colorado in 2012, after which the battery failed. Wolverines are capable of traveling vast distances! The only other recent report for North Dakota was a similar circumstance in 1971, when a Burleigh County rancher allegedly killed a wolverine in his pasture and had it mounted. Sightings in the state would most likely be in the vicinity of the Pembina Hills or Turtle Mountains and involve wanderers out of Canada or northern Minnesota.

Habitat

The wolverine is primarily an animal of the far north in North America, inhabiting taiga and tundra regions of Canada, Alaska, and the mountainous west. However, historically, it apparently also inhabited aspen parkland and prairie habitats.

Wolverines tend to prefer isolated, rugged wilderness habitats. Deep snow conditions appear to be prerequisite for spring denning and successful rearing of young.

Ecology and Behavior

Wolverines are hunters and scavengers, admirably adapted to both life styles. Stomach and scat analyses have revealed remains of a wide variety of mammals and birds, including moose, elk, deer, voles, ground squirrels, porcupines, magpies, hawks, and even fish. In short, they will consume anything animal, even some insects and plants. They have been observed following wolves and lynx, presumably to scavenge the remains of their kills. Although remains of domestic cow and horse were detected in Montana, these were attributed to the baits of commercial trappers.

As with most members of the mustelid family, adults are solitary except during the breeding season when males and females come in contact for a few days. Daily activity is mostly nocturnal, although they are occasionally active in daylight hours. Typical activity periods are for 3–4 hours, followed by intervals of sleep.

Wolverines are nomadic, with daily movements of up to 35+ km per day in the conterminous United States. Home ranges are highly variable, being functions of habitat, topography, food supply, and season. Summer home ranges are larger than in winter, and males have larger home ranges than females. Average annual home ranges in Montana were 422 km^2 for males and 388 km^2 for females. Dispersing wolverines are capable of traveling great distances. For example, the recent North Dakota record would have traveled a straight-line distance of at least 1,600 km from its point of capture in Wyoming.

Population densities are low, to be expected for a solitary predator with a large home range. In the conterminous United States, estimated densities have ranged from 1 wolverine per 48 km^2 to 304 km^2. Wolverine mortality has been largely attributed to human activity (hunting, trapping, habitat alteration). Other factors have included predation (wolves, cougars), infanticide, and starvation. Disease and parasitism are not believed to be significant.

Reproduction

The breeding season can extend from April through August, but usually does not peak until May or later. Mating is polygynous but, consistent with their solitary lifestyle, males and females remain together for only a few days. Wolverines, like other mustelids, exhibit delayed implantation in which the early embryo develops to a blastocyst stage and then remains suspended in the uterus for an extended period prior to implantation. In the case of the wolverine, total gestation lasts an estimated 215–272 days, but active post-implantation development is only about 30-50 days.

Natal dens are dug by females in snow 2–3 m deep, near trees or rocks. Kits are born in late winter or spring. Litter size averages 2–6. Young are born fully furred, with their eyes closed and teeth not erupted. Weaning occurs after 7–8 weeks, after which young start following the mother while foraging. Young attain adult size in early winter, and then disperse. Females attain reproductive maturity at 1.5–2.5 years, and male maturation appears to be about the same.

Status and Conservation

The wolverine is a high-value furbearer because of its scarcity, beauty, durability, and capability to shed frost. However, in the United States wolverines are legally harvested only in Alaska and Montana. Most are taken incidentally while trapping for other furbearers. The species is not listed as a furbearer in North Dakota, South Dakota, or Minnesota.

Wolverine populations and range have undergone drastic declines since the 1800s, resulting in small, isolated enclaves. Overharvest, fragmented habitat, timber harvest, human recreation activities, mineral extraction, and declining snow cover resulting from climate warming, have all been proposed as causative factors.

Selected References

Aubry et al. (2007), Bailey (1926), Copeland and Whitman (2003), Coues (1897), Pasitschniak-Arts and Larivière (1995), Ruggiero et al. (2007)

River Otter

Lontra canadensis

Pta^{n}–Lakota; Petakè–Mandan; Midapóka–Hidatsa; Citapat–Arikara

Deborah Hanson

Description

The body form of the river otter is similar to the weasels and mink, long and cylindrical. The tail is long, muscular, and tapering, making up nearly a third of the body length. The head is flattened, having a broad nose and small eyes and ears. The legs are short and muscular, with webbed toes. The fur is dark brown above and lighter below, with a gray or silvery cast on the throat. The fur is very dense, providing effective water repellence and insulation. It has the highest durability rating of North American furs. Body weights range from 7 to 11 kg, with males averaging 5 to 17% heavier than females. Average standard body measurements are: (Males) Total Length - 1,129 mm, Tail Length - 444 mm, Hind Foot - 129 mm, Ear - 24 mm; (Females) Total Length - 978 mm, Tail Length - 358 mm, Hind Foot - 112 mm, Ear - 23 mm.

Distribution

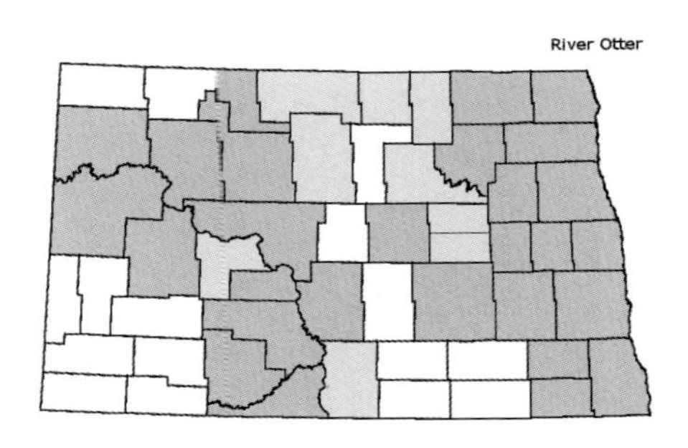

The historical distribution of the river otter in North America included practically all of the United States and Canada, excluding some high arctic areas and portions of the desert Southwest which did not have permanent water. By the 1970s, human settlement, habitat destruction, and overharvest resulted in the loss of about 25% of its historic range. Since then, reintroduction programs and protection have resulted in establishment of new populations throughout most of the states where it had been extirpated. As of 1999, New Mexico was the only state with no verified reports of the species.

In North Dakota, river otters were never abundant, but were present in all of the major streams in the state. By the early 1960s, there had been only two reports of otters in the state, probably wanderers from Minnesota or Canada. Since 1964, there have been numerous reports of river otters from the Red, Sheyenne, Missouri, and Souris rivers and nearby wetlands. Of these, most have been since the year 2000. Today, river otters can be found associated with major lakes and rivers throughout eastern North Dakota and along the Missouri River drainage.

Habitat

River otters live in a variety of aquatic habitats, ranging from marine environments to lakes in montane areas. In interior areas, they prefer slow-moving waters, such as meandering streams and lakes, which include abundant fish and riparian vegetation. They tend to avoid polluted waters and settled areas. During severe winters with thick ice, otherwise suitable habitat may be limited. If water levels in lakes drop below ice levels, however, air pockets may allow otters to continue to use the habitat, traveling and hunting under the ice.

Ecology and Behavior

River otters are active year-round. They are mainly nocturnal or crepuscular during the warmer months, but become more diurnal during winter. They are social animals, forming relatively stable family groups consisting of an adult female, her young of the year, and one or more young of previous litters or occasionally an unrelated individual. They are curious and playful. Play behavior includes sliding, tail-chasing, playing with sticks or captured prey, and wrestling.

River otters do not dig burrows, but enlarge the abandoned dens of other species such as beavers, or use natural shelters such as hollow trees and logs, rock piles, or log jams. Dens are usually close to water. Otters readily change their den sites. In Idaho, one individual used 88 different dens and resting sites in a 16-month period.

The diet of river otters is predominately fish. Frequency of occurrence of fish in otter stomachs from Minnesota was 100%, followed by crustaceans (17%), insects (33%), and amphibians (17%). Of the fish species, suckers constituted 44% by volume of stomach contents, perch and walleye 14%, and sunfish and bass 20%. Other species were less than 10% each. Birds and mammals are relatively unimportant in the diet, but occasionally muskrats become a significant food item.

Home ranges can be very large, depending on resources and severity of seasons. In the United States, home ranges of 8 to 78 km of shoreline appear to be fairly typical. River otters are not territorial, but family groups tend to avoid each other. Home ranges within and between sexes frequently overlap. Daily movements can typically range from 2 to 5 km per day, and have been known to be as high as 42 km in one day. Yearlings disperse in the spring, moving 3 to 5 km per day, ultimately dispersing up to 200 km. Otters released in reintroduction programs dispersed up to 1,138 km from the point of release.

The few estimates of river otter population densities from the interior United States and Canada range from one otter per 3 to 17 km of waterway. Otters are relatively long-lived, and have been estimated to live in the wild for 10 to 15 years. Survival in captivity has been as long as 25 years. The most significant causes of mortality are human-related, including trapping, hunting, road kills, and drowning in nets. Habitat destruction is the most significant cause for the decline of otter populations. Oil spills, water acidification from mining operations, waterway development for human use, destruction of riparian habitat, and decline of water quality due to siltation and pesticide accumulation have all contributed to population declines. As top carnivores, river otters concentrate many contaminants such as mercury and PCBs in their tissues, and are now recognized as indicators of the health of aquatic ecosystems. Although a number of larger predators have been known to take otters, especially the young, this cause of mortality is negligible.

Reproduction

Occasional yearling females may reproduce, but most do not breed until they are 2 years of age. Males become sexually mature at 2 years, but most are not successful breeders until they are 5 to 7 years old. Mating is polygynous, with a single male breeding with more than one female.

Breeding usually takes place during early spring during a 42- to 46-day estrus period following the birth of the previous litter. There is an extended period of delayed implantation of about 274 days, followed by a two-month gestation period when development actually occurs. In the Midwest, birth occurs from late February to late May. Litter sizes range from two to four.

Newborn otters weigh about 132 g, are blind and sparsely furred. Their eyes open at about a month. Young begin taking solid food at 9 to 10 weeks, and are fully weaned at 12 weeks. Adult size is not reached until 3 to 4 years. Juveniles disperse from the family group at about 1 year.

Status and Conservation

Because of its history of population decline on the Great Plains, the river otter has been listed as endangered or threatened in South Dakota, Iowa, Nebraska, Colorado, and Oklahoma. As indicated above, early records indicate that the species has been rare in North Dakota since white settlement, but reports show increases of sightings in recent years. The North Dakota Game and Fish Department lists the river otter as a furbearer with a limited season, and has categorized it as a Level II Species of Conservation Priority.

By 1990, river otters had been successfully reintroduced in 17 states and provinces, and additional feasibility studies were under way. The most successful introductions involved larger numbers of animals (about 160) released into high-quality habitat in core areas of the historic range. No releases of river otters have been made to date in North Dakota. Management strategy in North Dakota has involved protection of existing animals. Its state listing as a Species of Conservation Priority gives it a level of priority and eligibility for conservation funding.

Selected References

Adams (1961), Bailey (1926), Dyke et al. (2004), Hazard (1982), Jones et al. (1983), Knudsen and Hale (1968), Lariviere and Walton (1998), Melquist and Dronkert (1987), Melquist et al. (2003)

American Marten

Martes americana

Wa'anksica–Lakota

Gerald and Buff Corsi, © California Academy of Sciences

Description

The American marten is long and slender bodied, and is arboreal like the fisher. It is about the size of a mink, weighing 500 to 1,400 g. Males average 30% larger than females. The ears are large and rounded, limbs short, and tail bushy. The fur is long, silky, and dense, with the color ranging from reddish to yellowish brown. There is a pale, orangish bib on the throat and chest. Ranges of external body measurements are: (Males) Total Length - 570 to 682 mm, Tail Length - 170 to 240 mm, Hind Foot - 79 to 90 mm; (Females) Total Length - 513 to 560 mm, Tail Length - 165 to 195 mm, Hind Foot - 70 to 78 mm. Martens can be distinguished from fishers by their smaller size, lighter pelage, and the orangish throat and chest patch. They differ from mink by their longer fur and bushier tail.

Distribution

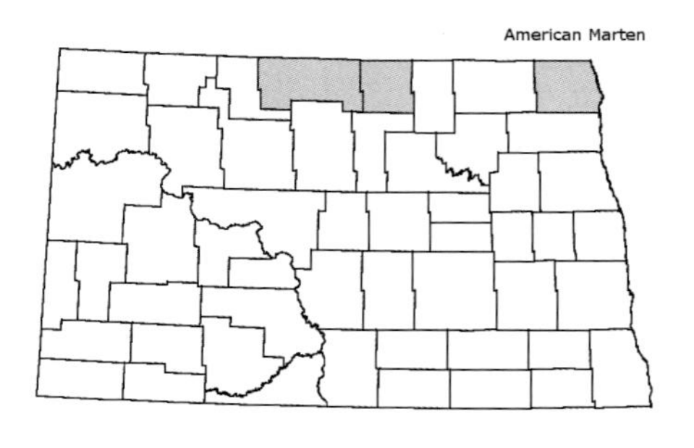

Martens originally occurred widely throughout the boreal forests of the United States and Canada, but were extirpated in many areas by overtrapping. The current distribution includes Alaska, much of Canada, the forested Rocky Mountains, Cascades, Coast Range, and Sierras into central California. An introduced population occurs in the Black Hills of South Dakota, and the species has recovered throughout much of northern Minnesota.

In North Dakota, martens were once fairly common in the forested Turtle Mountains and Pembina Hills of the northeastern corner of the state. However, like the fisher, the species was extirpated by 1900. There were no further reports in the state until two confirmed observations in the Turtle Mountains (Bottineau County) in 2002 and 2004. Since 2004 there have been thirteen additional reports from Rolette and Bottineau counties.

Habitat

Over much of its range, the marten prefers mature coniferous or mixed forest with 30 to 50% crown cover and abundant physical structure near the ground. Although marten and fisher habitats overlap significantly, martens venture into more diverse cover types, and are better adapted for travel through soft, deep snow conditions. Spruce-fir forests having >30% canopy, abundant logs and stumps, and lush ground vegetation are considered to be optimal habitat. Large openings from recent clear cuts or severe burns are avoided. More recently, extensive use of young (<40 years) stands of deciduous forest having complex horizontal and vertical woody structure has been documented. These reports are consistent with the historic North Dakota range, which is dominated by aspen and other deciduous species.

Ecology and Behavior

Martens, like fishers, are opportunistic predators, feeding on rabbits, rodents, birds, reptiles, insects, and some plants. Their staple is probably voles and mice, although they also actively pursue squirrels and take snowshoe hares when available.

Dens and resting sites may be in hollow trees, on large branches, hollow logs or stumps, slash piles, or in ground burrows. Winter dens are frequently under snow, or under stumps and brush piles, especially during severe weather and heavy snow conditions. Mothers with newborn young typically den in tree or log cavities, but occasionally natal dens may be found underground or in slash piles, rock crevices or red squirrel middens. As the young develop, dens are usually shifted to less specialized sites otherwise used for resting.

Overall levels of activity are greatest during summer, when martens are active about 60% of the time, in contrast to 16% during the winter. Daily activity is highly variable, and has been described by different workers as diurnal, crepuscular, or nocturnal. Reported activity patterns, and seasonal shifts in activity, seem to be primarily related to the availability of principal prey species. For example, in Montana, marten activity coincides with that of voles and red squirrels rather than other available species such as snowshoe hares.

Martens, like fishers, are solitary except during a mid-summer mating season, when breeding age adults come together. Home ranges vary greatly, depending on the density of prey, but do not shift in size with season. Males will shift their home ranges with increases in available space, but females display a high level of home range fidelity. Reported home ranges vary from 0.6 to 27 km^2. Home ranges of males may be up to four times those of females. Martens exhibit intrasexual territoriality, but home ranges of males may overlap the home ranges of two to six females.

Populations consist of residents (present >90 days), temporary residents (present seven to 90 days), and transients. Ranges of density run from 0.4 to 1.5/km^2. Martens are relatively long-lived. Longevity records from both captivity and the wild indicate life spans of up to 15 years. Survival from 9 months to 4 years is about 65%. Causes of natural mortality are poorly known. They are believed to be more subject to predation than are fishers, and are known to be taken by a number of larger mammalian and avian predators. While they are well-adapted to severe weather, declines of prey populations are known to reduce carrying capacity, resulting in reductions of marten populations through loss of females, reduced reproductive success, and emigration of juveniles.

Reproduction

Like the fisher, courtship behavior in the marten is polygynous or promiscuous. The female actively attracts males by scent marking, and vocalization. Courtship may last up to two weeks. Breeding takes place in July or August, about four months after the birth of the year's litter and significantly later than the breeding season of the fisher.

Following breeding, the embryo develops to the blastocyst stage and then suspends further development for 190 to 250 days (delayed implantation). Implantation occurs in late February, with parturition in March, about 27 days later. Litters average three to four, ranging from one to five.

Newborn young are altricial, i.e., they are naked, blind, and helpless, and weigh about 28 g. They develop rapidly, with their eyes opening at 39 days and weaning at 6 weeks. The young are able to leave the nest at 46 days, and adult length is reached at 3 months. They are able to make their own kills by 2 to 3 months.

Status and Conservation

The marten is not protected under the Endangered Species Act, and is listed as a furbearer in most states and provinces. With sparse confirmed records in North Dakota in over 100 years, it must be regarded as a rare species in the state.

In the 1980s, about 35,000 martens were harvested annually in the United States, a significant increase over the previous 20+ year period. Seasons were closed in Minnesota until 1985, when trapping reopened. Minnesota currently harvests 2,000 to 3,000 martens annually.

The marten is listed as a furbearer in North Dakota, with a continuing closed season. Recovery of this rare species in the state will be contingent on continued monitoring, protection of mature forests with abundant ground structure, especially in the Turtle Mountains and Pembina Hills areas, and continuing to maintain a closed season.

Selected References

Adams (1961), Bailey (1926), Buskirk and Zielinski (1997), Clark et al. (1987), Foresman and Pearson (1999), Hall (1981), Higgins et al. (2000), Jones et al. (1983), Powell et al. (2003), Raine (1983), Strickland and Douglas (1987)

Fisher

Pekania (Martes) pennant

Skeca–Lakota

L. L. Master, American Society of Mammalogists Mammal Images Library

Description

The fisher is a semiarboreal weasel, spending a significant amount of time in trees. Weighing an average of 4 (2 to 6) kg, it is about the size of a small fox. It has a long body and a bushy, cylindrical tail. The fur is dark brown to nearly black, with white bands on the guard hairs, giving it a grizzled appearance, especially on the anterior half of the body. Average standard body measurements for males are: Total Length - 987 mm, Tail Length - 372 mm, Hind Foot - 116 mm, Ear - 50 mm. Measurements of females are about 15% less than those of males, and their weight is about half that of males.

Distribution

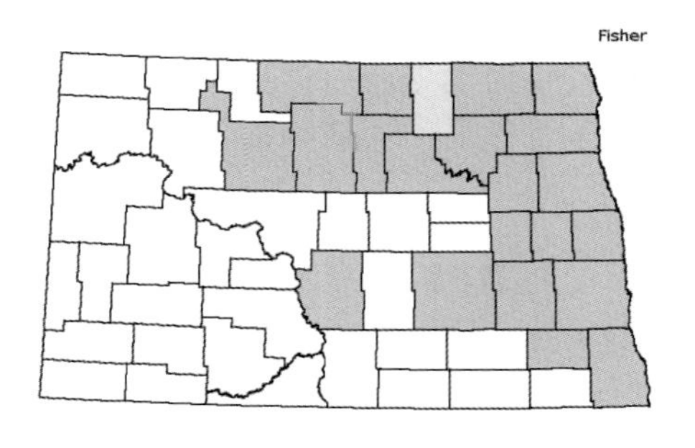

The fisher occurs only in North America, inhabiting northern forests as far north as 60°N latitude. Fishers are found across Canada, and are established in the United States in the Rocky Mountains of Idaho and Montana, Cascades and Sierras of Oregon and California, northern Minnesota, Wisconsin, Michigan, and New England.

Fishers originally ranged into northeastern North Dakota as far south as Grand Forks, but were extirpated from the state by 1900. In 1976, a specimen was trapped in Pembina County. Since that time, there have been additional reports from the eastern half of the state.

Habitat

Most studies indicate that fishers prefer mid- to late successional coniferous forests with extensive overhead cover, and avoid open areas and aspen-birch forests. However, recent North Dakota reports indicate they also readily use deciduous forests, including riparian woodlands adjacent to lakes and rivers. Available den sites such as cavities in logs and trees, and an abundant supply of prey are major factors in habitat selection. Early successional stages, open areas, and forests with little overhead cover are avoided.

Ecology and Behavior

Fishers are generalized, opportunistic predators, taking small to medium-sized mammals, birds, and carrion. Common prey include squirrels, mice, shrews, rabbits, and porcupines. If porcupines are available, fishers are among the few carnivores that can effectively take them by having evolved a behavior of avoiding the quills by repeatedly attacking the head until the porcupine is subdued. The porcupine is then consumed by first eating the head, throat, and underparts, thus largely avoiding the quills. In addition to mammals, fishers also eat birds, reptiles, insects, fungi and fruit.

Fishers are usually crepuscular, active during the twilight hours, but activity can be variable. Overall levels of activity are greatest during summer and least during winter, when they are active about 30% of the day, spending significant time in snow dens. Movement of males increases during the mating season, and activity of females is least during pregnancy but increases with development of their litters

When resting, fishers use hollow logs and tree cavities, brush piles, rock piles, burrows of other species, snow dens, and even the body cavity of big-game carcasses. Maternity dens are usually in hollow logs or trees, and a female with a litter may use from one to three dens during the four-month denning period.

Fishers are solitary animals, except during the breeding season. Home ranges average 15 to 35 km^2, with those of males larger than those of females. Home ranges are smallest during winter, and largest for males during spring and early summer. Females markedly decrease their home ranges during the breeding season. There can be significant overlap of home ranges, at least between the two sexes, but there may be intrasexual territoriality. Fishers routinely travel for long distances within their home ranges, and may average about 5 km per day. They tend to follow irregular circuits of 65 to 160 km, visiting certain areas regularly.

Populations can undergo extreme fluctuations in response to the abundance of their principal prey. Reported densities range from 0.05 to 1.5 animals per km^2 in various parts of the United States and Canada. Fishers have been estimated to live more than 10 years in the wild, but most probably do not survive more than seven years. Causes of mortality are poorly known. However, overharvesting resulted in decimation of populations in the 19th century. They have a low incidence of disease, and there is little evidence that fishers are subject to significant predation. Presumably, overwintering losses, including those resulting from crashes of prey populations, are significant causes of mortality.

Reproduction

Both males and females attain sexual maturity at 1 year, but effective breeding may not take place until the second year. The mating system is polygynous or promiscuous, with males breeding with multiple females. Males and females normally remain together for several days. On occasion, the male may remain with the female throughout the breeding season, but is not allowed near the kits.

Breeding takes place in late March or April, shortly after the birth of the kits. Implantation is delayed, with the blastocyst remaining dormant until late winter. Changing day length apparently induces implantation and development, which continues normally for about 40 days to parturition in late March. Litter size averages two or three, but may range from one to six.

Newborn fishers weigh about 40 g, are blind, helpless, and covered with fine hair. They remain in the maternal den (usually a tree cavity) for about eight weeks, at which time they are mobile and are moved to a den on or under the ground. Their eyes open at 56 days, and solid food is first taken at 62 days, but they continue to nurse until 114 days. They have learned to take prey effectively by 125 days. Females reach adult size at 6 months, but males do not reach adult weight until 1 year. Kits remain in their mother's home range until winter, but tend to avoid her and have dispersed to their own home ranges by 1 year.

Status and Conservation

Early 19th century fur harvest reports indicate that fishers may have been more abundant than mink *(Neovison vison)* in North Dakota, but by the turn of the 20th century, they had been essentially extirpated from the state. Since the 1970s, there have been a number of confirmed reports, primarily from the heavily forested Turtle Mountains and Pembina Hills areas. Reports from elsewhere in the state have been from counties with rivers and associated riparian woodland providing adequate habitat for dispersal far from major forests. Although there have been increasing reports in recent years, the fisher is still considered to be a rare species in the state. The North Dakota Game and Fish Department lists the fisher as a furbearer, with a limited annual harvest.

By the 1980s about 3,000 to 4,000 fishers were harvested annually in the United States, especially in the northeastern states and Minnesota, and 16,000 to 21,000 in Canada. In neighboring Manitoba, the harvest was over 4,000. Overall, United States harvests appear to have increased significantly since the 1940s. Concern about vulnerable populations in danger of extirpation and interest in sustainable fur harvests have driven conservation efforts for fishers. Recoveries have occurred through recolonization of suitable habitat and reintroductions into formerly occupied portions of the species' range.

In North Dakota, the fisher appears to have benefited from protection in the state, combined with dispersal from recovering populations in adjacent Minnesota and Manitoba. Continued monitoring, protection of mature forest habitats, and protection by limited closed seasons will be necessary for full recovery of fisher populations in North Dakota.

Selected References

Adams (1961), Bailey (1926), Berg (1982), Douglas and Strickland (1987), Jones et al. (1983), Leonard (1986), Powell (1981), Powell et al. (2003)

Ermine, Short-tailed Weasel

Mustela erminea

Mach-schipka–Mandan; Itunkasanla–Lakota

P. K. Anderson, American Society of Mammalogists Mammal Images Library

Description

The ermine is intermediate in size between the least weasel *(M. nivalis)* and long-tailed weasel *(M. frenata)*. Like other weasels, it has a long, slender body, long neck, rounded ears, and short limbs. During summer, the upper parts are brown, and the under parts white to yellowish white. The tip of the tail is black. The winter pelage is totally white, except for the black tip on the tail. Males are nearly twice the size of females. Average standard body measurements are: (Males) Total Length - 307 mm, Tail Length - 85 mm, Hind Foot - 40 mm, Ear - 19 mm, Weight - 113 g; (Females) Total Length - 251 mm, Tail Length - 65 mm, Hind Foot - 32 mm, Ear - 16 mm, Weight - 75 g. Hence, the ermine can be distinguished from the least weasel by its larger size and longer, black-tipped tail, and from the long-tailed weasel by its smaller size and proportionately shorter tail. However, some large ermines approach the size of a smaller long-tailed weasel.

Distribution

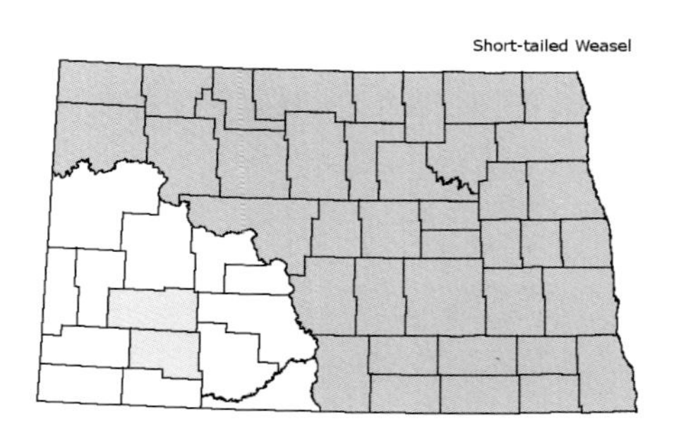

The ermine is circumboreal in distribution throughout much of the Northern Hemisphere, and has the widest range of any member of the family. It has also been introduced into New Zealand. In North America, it occurs throughout Alaska and nearly all of Canada, including the high arctic regions. In the western United States, it occurs as far south as northern California, Nevada, Utah, Colorado, and northern New Mexico. Farther east, it is in eastern North Dakota, the Great Lakes states, and the eastern seaboard as far south as Maryland.

Most reports indicate the North Dakota distribution to be north and east of the Missouri River. However, visual observations from Stark and Hettinger counties in the southwest may indicate a wider distribution in the state.

Habitat

Ermines tend to prefer successional and forest edge habitats, riparian woodlands, hedgerows, meadows, and marshes, wherever small rodents are abundant. In North Dakota, they are mostly associated with woody areas around streams and lakes, prairie marshes, and the aspen parkland.

Ecology and Behavior

Principal foods of ermines are mice and voles, and larger males may even take cottontail rabbits. When their principal prey are scarce, they will switch to alternates such as nestling birds, shrews, reptiles and amphibians, invertebrates, and even berries. For example in one study, shrews constituted the most frequent food category, more than voles and deer mice. Their hunting behavior is characteristic of all weasels, moving rapidly, investigating burrows and crevices, and occasionally standing upright to better see the surroundings. A kill is made with a rapid dash, grasping the prey with the forelegs, and biting into the neck. Food items not immediately consumed are usually cached. Ermines have a high metabolic demand; their daily requirements averaging 19 to 32% of their body weight.

Dens are frequently in rock piles, hollow trees, or in the burrows of previous prey. Side galleries often contain cached prey. The nest may be lined with grass, leaves, and fur of prey.

Ermines are active year-round and throughout the 24-hour cycle, timing their activity with their principal prey. They engage in short bouts of activity, usually less than one hour, alternating with regular rest periods. During winter, much of their activity is spent under the snow cover, where they prey on voles and other small mammals, and use their nests for dens.

Home ranges vary with gender, season, and food supply. Male home range can be as small as 1 ha, and as high as 200 ha, but most vary from 10 to 40 ha. Female home ranges generally run from 4 to 10 ha. When principal prey species are scarce, home ranges may be two to three times the size of when they are abundant. Males increase their home ranges during spring, coinciding with the breeding season. Home ranges of males overlap with those of one or more females, but other males are excluded. Resident females exclude other females from the home range and, except during the breeding season, avoid the resident male. Home range boundaries are marked by scent from the anal glands.

As with home ranges, ermine populations vary seasonally and with food supply. Density estimates have ranged from one per 17 ha to one per 50 ha. Major factors affecting populations have been identified as small mammal abundance, spring flooding, meteorological conditions, catastrophic events such as fires and habitat modification, and epizootics such as tularemia. Of these, abundance of prey is regarded as most significant. Weasels are taken by a variety of snakes and avian and mammalian predators, and predation has been suggested as a factor controlling populations. Estimates of longevity in the wild have ranged from 1.5 to 4 to 6 years. Few ermines have been shown to occupy their home ranges for more than one year.

Reproduction

Like other weasels, mating is promiscuous or polygynous, with males enlarging their home ranges at the beginning of the breeding season and mating with females whose territories overlap with their own. However, unlike most related species, the male ermine may assist in caring for the litter of the previous mating.

Breeding takes place from late May to early summer, shortly after young of the previous year are born. Shortly after fertilization, early embryonic development is suspended prior to implantation in the uterus. The early embryo (blastocyst) remains quiescent until the following spring, when development resumes. This "delayed implantation" results in a total gestation period of 10 months, only three to four weeks of which there is actual embryonic development occurring.

Litter size averages about six, and ranges from four to 13. Newborn young weigh less than 2 g, are blind, and covered with fine white hair. A brownish "mane" soon develops on the back of the neck, probably serving to assist the mother in carrying the young. The young begin taking solid food before their eyes are open. Their eyes open at about 6 weeks, at which time they are fully weaned. They are completely independent and nearly fully grown by autumn. Females become sexually mature during their first summer, but young males delay until the following spring.

Status and Conservation

Harvests of all weasel species in North Dakota peaked in 1940 with a take of over 80,000 pelts. These declined drastically in the 1940s and more gradually in the 1950s to about 6,000 pelts in 1961. Most recently, in 2001-2002 a total of only 56 weasels were taken in the state. While these figures include all species of weasels and may reflect the vagaries of the fur market, drastic habitat losses resulting from conversion of forests, wetlands and grassland to intensive agriculture have also been cited as a major contributing factor leading to overall population declines. Short term local fluctuations reflecting prey abundance are expected, but ermines, are now considered to be uncommon in their North Dakota range.

Through the 1950s, weasels were ranked as important furbearers in North Dakota, but their current economic significance is negligible. Where they occur in rural areas, they are an asset because of their impact on crop-destroying rodents. Weasels occasionally kill poultry and raid game bird nests, but this is negligible compared to their impact on pest species. Population monitoring and habitat protection must be included in any management scheme for weasels in North Dakota.

Selected References

Adams (1961), Bailey (1926), Hall (1981), Hazard (1982), Jones et al. (1983), King (1983), Seabloom et al. (1980), Svendsen (2003)

Long-tailed Weasel

Mustela frenata

Ohsisa – Hidatsa; Mahchpach-piraka – Mandan; Itunkasan – Lakota

C. D. Grondahl, North Dakota Game and Fish Department

George Rohde

Description

The long-tailed weasel is the largest of the three weasels in North Dakota. Like the ermine, its dorsal coat color in summer is brown, and the tail is black-tipped. The underparts are white to buff-colored. Males are significantly larger than females, which might be confused with large male ermines. However, the tail is proportionately longer, about one-third the total length, and the black tip is proportionately shorter than in the ermine. The winter pelage is totally white, except for the black-tipped tail. Average standard body measurements are: (Males) Total Length - 427 mm, Tail Length - 155 mm, Hind Foot - 48 mm, Ear - 24 mm, Weight - 295 g; (Females) Total Length - 362 mm, Tail Length - 127 mm, Hind Foot - 43 mm, Ear - 18 mm, Weight - 155 g.

Distribution

The long-tailed weasel occurs only in the Western Hemisphere. Its northern limits are in southern Canada below the aspen parkland. It is

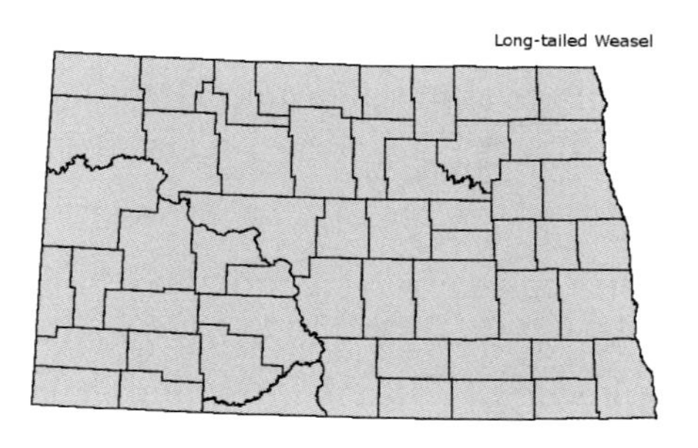

found in all 48 of the contiguous United States, absent only from the Mohave and Sonoran deserts. Its range extends south through most of Mexico and Central America to northern South America.

The long-tailed weasel was once considered to be the most common weasel in North Dakota, occurring throughout the state. There are specimens in the University of North Dakota collection from scattered sites around North Dakota, and surveys indicate that they are fairly common in the southwest.

Habitat

One only needs to note the extensive distribution of this species to realize that physical components of habitat, such as temperature or vegetative composition, are not limiting. The most important habitat component is apparently the availability of abundant and diverse prey species. It appears to favor dense cover, including wetland edges, brush, and rocky areas. It is frequently most abundant in areas near standing water. These physical features can all relate directly to prey abundance. In North Dakota, long-tailed weasels appear to be most common in the Prairie Pothole Region of the eastern half of the state, and in rolling grasslands and roadsides of the southwest.

Ecology and Behavior

Long-tailed weasels are more generalized predators than least weasels or ermines, taking a wide variety of prey and switching prey with varying abundance. They concentrate on small- to medium-sized rodents and rabbits, but also take other small vertebrates. Voles and deer mice are among the most common prey taken, but larger prey include muskrats, pocket gophers, chipmunks, various squirrels, and rabbits and hares. They readily take birds, and raid nests for eggs and nestlings. In captivity, long-tailed weasels require a food intake equivalent to about 1.5 voles per day, or 17 to 33% of their body weight.

Hunting behavior involves active searching of burrows, zigzag investigation of above ground areas and runways, and arboreal searching. They are capable of pursuing arboreal prey, such as red squirrels, through the tree canopy. Prey are typically subdued by biting the nape of the neck until the spinal cord is severed, or grasping the throat and suffocating the animal. Unconsumed food items are usually cached in the burrow, but they have also been known to cache foods in old nests of passerine birds in shrubs during winter.

Nests are typically in old ground squirrel or pocket gopher burrows. The nest may be up to 30 cm in diameter, and lined with dried grasses and fur of previous prey. Other tunnels in the burrow lead away from the nest to sanitation sites and food caches.

Like the other weasels, long-tailed weasels are active throughout the year. They can be active at any time of day or night, but there is some indication of most activity between dusk and dawn. Hot weather seems to result in diminished activity, with the resumption of higher levels during cool mornings and evenings.

Home ranges are larger than those of ermines or least weasels. Summer home ranges average from 16 to 24 ha, but winter home ranges may be more variable, running from 10 to 160 ha, depending on the food supply. Male home ranges do not overlap with home ranges of other males, but normally include the home ranges of one or more females. Daily movements can be highly variable, probably reflecting the abundance of prey. Movements of up to 775 m have been recorded, but a daily radius of 100 to 200 m seems typical.

Long-tailed weasel populations seem to be more stable than populations of least weasels and ermines, possibly because of their ability to use more varied prey resources. However, multi-year fluctuations have been recorded, reflecting variations in prey abundance. Because of sparse populations and movement patterns, density estimates have been difficult to obtain. Estimates usually run six to seven/km^2, but have been as high as 0.1 to 0.3/ha in some habitats. Longevity in the wild may be as high as three years. Populations are primarily limited by prey availability. Other principal causes of mortality include predation (primarily by foxes and raptors), diseases, catastrophic events such as fires, and human-related mortality resulting from trapping, hunting, and road traffic.

Reproduction

As with other weasels, the mating system has been described as polygynous. However, other reports indicate males and females hunting together and sharing the den, suggesting at least short-term monogamy.

The breeding season begins in March or April, and the young are born in April or May. As with many other mustelids, embryonic development is suspended at the blastocyst stage, about two months after fertilization. Implantation takes place around day 251, and all further stages of development occur in the following 27 days, resulting in a total gestation of 278 days.

Litter sizes range from four to nine. Newborn young are blind and naked, weighing 3 g. By 3 weeks, they are furred, and have begun taking solid food. Eyes are open and weaning completed by 5 weeks. Young begin accompanying the adult female at 6 to 7 weeks, and adult size is attained by 10 to 11 weeks. Young females are sexually mature by 3 to 4 months, but males do not attain maturity until 14 to 15 months.

Status and Conservation

Because of its status as a more generalized predator, the long-tailed weasel is probably more common in the state than the other two weasels, and not as subject to drastic population fluctuations. Although it may be fairly common in the grasslands of southwestern North Dakota, recent records from the eastern half of the state have been rare. Harvest records do not consider weasels by species, but, as the account for M. erminea indicates, a total of only 56 pelts were reported in 2001-2002, compared to 6,000 to over 80,000 in the 1940s and 1950s. There has been a great decline in the fur market over the past several decades, but there has also been a vast shift in agricultural practices, resulting in conversion of wild lands to intensive agriculture. This has undoubtedly had a major impact on most wildlife, especially sparsely dispersed carnivores, such as the weasels, depending on an abundant prey resource. Low populations throughout North America have resulted in its classification as endangered, threatened, or of special concern in a number of states and Canadian provinces.

Like the other two weasels, the long-tailed weasel is no longer an important furbearer in North Dakota, and is of negligible economic significance. Where it does occur, it is probably significant locally in control of rodent populations.

Because of the apparent population declines of the three weasel species in North Dakota, combined with its special concern status elsewhere in North America, any management strategy should include continuing monitoring to clarify its statewide status. Existing habitat should be protected. Establishment of corridors connecting large habitat blocks should be investigated in order to provide for dispersal and re-establishment of locally depleted populations.

Selected References

Adams (1961), Bailey (1926), Genoways and Jones (1972), Jones et al. (1983), Seabloom et al. (1978), Sheffield and Thomas (1997), Svendsen (2003)

Black-footed Ferret

Mustela nigripes

Etopta sapa – Yankton Sioux, Lakota;
Nazi – Mandan; Tahu akukahak napish – Hidatsa

Bob Gress

Description

The body form of the black-footed ferret is typical of the weasels, with an elongated neck and body, rounded ears, short legs, and long tail. The body is generally yellowish or buffy, with the face and underparts more whitish. The body may have a sprinkling of black hairs. The face has a black mask, and the feet and tip of the tail are black. Weighing up to 1,100 g, the black-footed ferret is about the size of a mink *(Neovison vison)*. Average standard body measurements are: Total Length - 519 mm, Tail Length - 126 mm, Hind Foot - 63 mm, Ear - 31 mm. Males are about 10% larger than females.

The distinctive coloration of the black-footed ferret makes it relatively easy to distinguish from any other regional mammals. Occasionally, it might be confused with a long-tailed weasel *(M. frenata)*, which is totally brown dorsally and white during winter, or the closely related introduced domestic ferret *(M. putorius)* which occasionally has been released into the wild. Domestic ferrets have more black on the body and shorter black part of the tail.

Distribution

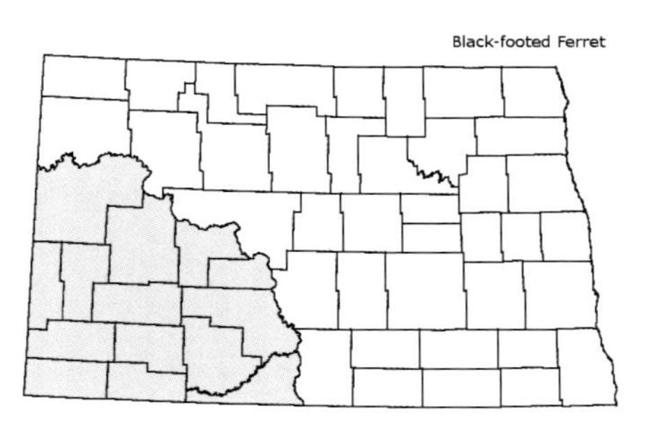

The black-footed ferret originally occurred throughout much of the Great Plains, from southern Alberta and Saskatchewan south to central Texas. Its distribution conformed almost exactly to the three species of prairie dogs *(Cynomys spp.)* occurring in the United States and Canada. By the 1960s, the only known population occurred in southwestern South Dakota. That population died out in the 1970s, and there was concern that the species may have become extinct. Subsequently, an additional population was discovered in Wyoming, brought into captivity, and provided stock for a captive breeding program. Progeny were released into Wyoming, South Dakota, Montana, and Arizona, forming the only known wild black-footed ferret populations in existence today.

In North Dakota, black-footed ferrets originally occurred south and west of the Missouri River, coinciding with the distribution of black-tailed prairie dogs in the state. However, since the early 1900s, there have been only about 30 observations or specimens for the state. There have been no confirmed reports of black-footed ferrets in the state for over 30 years.

Habitat

A highly specialized predator, the black-footed ferret is closely associated with the three prairie dog species on the Great Plains. Thus, they are found only in short-grass prairie.

Minimum habitat requirements are expressed in terms of prairie dog town size, population density, and dispersion. In South Dakota, the mean distance between ferret-occupied towns was 2.7 km. This suggested the minimum habitat requirement for ferrets was eight towns per township of at least 12 ha in size. Of these, at least two towns should be at least 40 ha to support females nursing litters. Further estimates suggest a density of 3.63 prairie dogs or 40 active burrows per ha as the lower limit of good ferret habitat, and that 763 prairie dogs are needed, under typical conditions, to support a family group of ferrets for one year. A viable population of 200 ferrets has been estimated to require 7,923 ha of short-grass habitat in Montana.

Ecology and Behavior

Like all weasels, black-footed ferrets are active year-round. However, during winter, they substantially decrease their amount of activity and travel, spending up to a week at a time in an underground burrow. By contrast, during summer and fall, they may travel up to 6 km in a single 24-hour period. Activity is primarily nocturnal, with peaks at dusk and dawn.

Black-footed ferrets are the most specialized of North American carnivorous mammals, with over 80% of their diet consisting of prairie dogs. The remainder of their diet includes mice and other rodents, rabbits, birds, reptiles, and insects. Daily dietary requirements for females with litters have been estimated at 0.6 to 0.9 prairie dogs per day. Most kills are within the confines of the prairie dog burrow. Killing behavior is similar to that of weasels, involving bites to the neck or base of the skull. Prairie dogs are not easily killed, and usually fight fiercely before being subdued.

Prairie dog burrows are used as den sites. Unlike other weasels, black-footed ferrets are good burrowers, and readily dig out soil plugs constructed by prairie dogs to impede entrance. Prairie dogs also plug the entrances of burrows inhabited by ferrets. Ferrets readily switch dens, at least during summer.

Home ranges average from 40 to 50 ha in size, females having smaller home ranges than males. Male home ranges may overlap with those of females, but adults do not share burrows, except during the breeding season. Adults of the same sex do not seem to share the same town.

Population densities vary from about 0.02 to 0.08 ferrets per ha, depending on season, prey resources, and gender. Disappearance from the population, via emigration and mortality, can range from 53 to 86% per year, primarily involving juveniles. Adult life expectancy is probably less than five years. The primary factor limiting black-footed ferret populations and resulting in their decline has been the near elimination of prairie dogs from their historic range. Diseases, primarily canine distemper and sylvatic plague, can result in catastrophic losses. Other diseases, including tularemia, rabies, and human influenza, are less significant. Black-footed ferrets have many potential predators, including badgers, coyotes, foxes, bobcats, hawks and owls, and snakes.

Reproduction

Much of what is known about black-footed ferret reproduction is through the captive breeding program, but there are few details on reproduction in the wild. The breeding season is probably during March and April, corresponding to closely related species. Three to four young are born in May or June, following a 41- to 45-day gestation. Newborn young are blind and helpless, but develop rapidly. They begin taking solid food at about 1 month, their eyes open shortly after, and they are weaned at 42 days. Litters emerge from the burrow at 6 weeks to 2 months. They begin hunting by August, the female separates them into different burrows, and they are completely independent by September or October. Young ferrets attain sexual maturity at 1 year of age.

Status and Conservation

The black-footed ferret seems to have always been an uncommon species on the Great Plains, probably a result of its specialized habits. The rarity of the species was a likely basis for its veneration by Native American tribes, which used it for ceremonial purposes. Early fur traders never found it abundant enough to be economically important.

With the development of agriculture and ranching, programs of extermination greatly reduced prairie dog colonies throughout their historic range. This destruction, along with habitat loss, poisoning, and sylvatic plague, brought about the near extinction of the black-footed ferret. The black-footed ferret is federally classified as an endangered species.

Although originally a resident of southwestern North Dakota, there have been no confirmed reports of black-footed ferrets in the state in over 30 years. However, significant increases in prairie dog towns over the past two decades, combined with identification of suitable habitat nearby in eastern Montana and northwestern South Dakota have have improved the potential for eventual reestablishment of black-footed ferrets in North Dakota.

A black-footed ferret recovery plan developed by the U. S. Fish and Wildlife Service has outlined the steps for recovery of the species throughout its historic range. A captive breeding program has been established with the goal of maintaining 240 breeding adults in captivity, and continuing to reintroduce ferrets into the wild. The national goal is to establish 10 free-ranging populations over the widest possible area, each with at least 30 breeding adults. An eventual goal is to have 1,500 ferrets in the wild by the year 2010. As stated above, introductions have now been made in South Dakota, Wyoming, Montana, and Arizona.

Although there are currently no known ferrets in North Dakota, a black-footed ferret conservation program has been established by the North Dakota Game and Fish Department, and the species has been designated a Level II Species of Conservation Priority. This plan established goals for determining existing habitat in North Dakota, determining the population, protection of ferret habitat, and a contingency plan for management should ferrets be found in the state.

Selected References

Bailey (1926), Biggins et al. (1989), Clark et al. (1987), Dyke et al. (2004), Grondahl (1973), Hillman and Clark, (1980), Hillman et al. (1979), Jones et al. (1983), Seabloom et al. (1980), Sheets and Linder (1969), Svendsen (2003), U.S. Fish and Wildlife Service (2013)

Least Weasel

Mustela nivalis

Hitu^{n}ka-sa^{n}–Dakota

M. Andera, American Society of Mammalogists Mammal Images Library

Description

The least weasel is the smallest carnivore in North America, and weighs only about 30 to 60 g. The body is long and slender, limbs short, neck long, and ears rounded. During summer, the back and sides are chocolate brown, and the underparts are white, usually with brown spots. During winter, the pelage is totally white. Average standard body measurements are: Total Length - 180 mm, Tail Length - 33 mm, Hind Foot - 21 mm, Ear - 8 mm. Only a small ermine *(M. erminea)* is likely to be confused with this species. It can be distinguished by its smaller size and the absence of a black tip on the tail.

Distribution

The least weasel is circumboreal in distribution, occurring in northern North America Europe, and Asia. Introductions have also been made on islands, including New Zealand. In North America, least weasels occur in Alaska and across Canada. In the conterminous

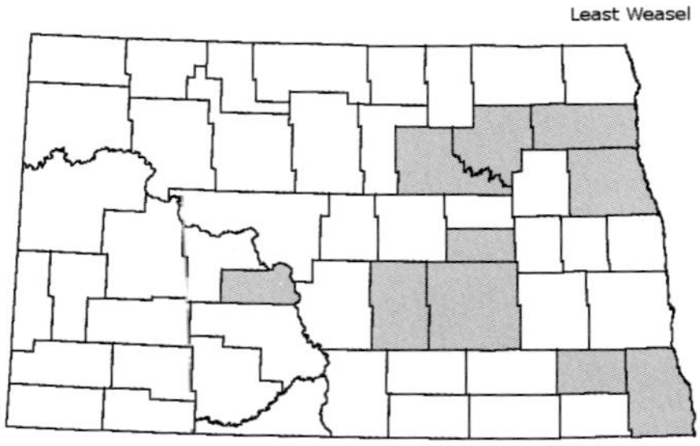

Least Weasel

United States, the western limits are in northern Montana, the Dakotas, Nebraska, Kansas, and Oklahoma. To the east, they are in the lake states, Iowa, and the Appalachian states.

There are scattered reports of least weasels in North Dakota as far west as Fort Clark, on the Missouri River. Other reports are from Ramsey, Benson, Barnes, Kidder, Stutsman, Foster, Walsh, Grand Forks, Ransom, and Richland counties.

Habitat

Least weasels can be found in a wide variety of habitats, their occurrence determined by the abundance of small rodents and available cover. They do well in the harsh climates of the tundra and tree line in the far north, in woodlands, hedge rows, and agricultural areas. In farm-land, they concentrate their activity along fence lines, shelterbelts, and under brush piles. The only habitats that are avoided are sandy deserts, dense forests with little ground cover, and other open areas where prey abundance is low and they are more vulnerable to predation. In North Dakota, they have been found in tall- and mixed-grass prairie, heavy cover around wetlands, the forested Devils Lake area, and riparian woodland.

Ecology and Behavior

Least weasels are specialized predators, feeding primarily on voles and mice. Along with other weasels, their long bodies and short fur result in a high rate of heat loss and subsequent high metabolic demand resulting in a daily consumption of prey equivalent to 40 to 60% of their body weight. When there is a scarcity of their primary prey, they switch to shrews, larger rodents, eggs and nestling birds, other small vertebrates, and even insects. Their hunting behavior involves random searching, using sight, sound, and odor to locate potential hiding places of their prey. Prey capture involves dashing to the prey, biting the base of the neck, and grasping the body with the feet until it is dead. Killing behavior is innate; 2-month-old young separated from their mothers since infancy have successfully killed mice, with no prior experience.

The entire body may be consumed, or the weasel may select only certain parts, such as viscera, muscle, and brain. When food is abundant, portions may be cached in a side tunnel off the main burrow, or near the kill site.

Nests are in shallow burrows constructed by other species, often animals they have killed. The nest is about 10 cm below the surface, and constructed of plant materials brought in by the previous occupant. It may be further lined with fur brought in from previous prey items.

Least weasels are active year-round, with activity during winter concentrated under the insulating layer of snow. They are active day and night, timing their foraging to the activity patterns and vulnerability of their principal prey. Activity periods are dominated by hunting and, during the breeding season, seeking of mates.

Male home ranges vary from <1 to 26 ha, and are defended against other males. Female home ranges are smaller (<1 to 7 ha), and may be included within the home ranges of males. Transients regularly pass through occupied home ranges, and are available to move in should one be vacated. Least weasels are solitary, and tend to avoid each other except during the breeding season. Even when male and female home ranges overlap, they tend to use portions that are unoccupied at a given time.

Least weasel populations fluctuate with food supply, seasonally, and annually. Under favorable conditions, they may reach 25/km^2, but in times of prey scarcity, they may completely disappear from a local area. One adaptation to low prey density is adjustment of reproduction. Young females may delay reproduction until the next season, and adults may not breed at all. A minimum prey density of 10 to 15 voles/ha has been suggested as necessary for reproductive success. Hence, least weasel populations are closely attuned to the three- to five-year vole and lemming cycles of the north. Populations can undergo irruptions of nine- to ten-fold in a given year. In the north-central states, irruptions have been documented in Nebraska, Minnesota, and in 1969-1970 in North Dakota.

Least weasels are taken by most larger predators, including snakes, hawks and owls, foxes, larger weasels, and cats and dogs. The impact of predation on least weasel populations is unclear. Average longevity is about 11 months in the wild, but a few individuals survive into their third year. Annual mortality averages 75 to 90% for all ages.

Reproduction

Mating is promiscuous in least weasels, and males do not participate in any care of young. Breeding may occur throughout the year, but is concentrated in spring and late summer. Females may produce two or three litters per year, but one or two is probably more common.

The litter size averages four to five, but may range from one to 10. The total gestation is 34 to 37 days. Least weasels do not exhibit delayed implantation, as seen in the ermine *(M. erminea)* and long-tailed weasel *(M. frenata)*. Newborn young weigh less than 2 g, are blind and naked. Their brown pelage appears at 3 weeks, and they begin to take solid food. Their eyes open between 26 and 30 days, and after a month they are fully weaned and begin to leave the den. At this time, the female begins teaching them to hunt. The young attain adult size at about 14 weeks and disperse. They are sexually mature at 3 to 4 months of age.

Status and Conservation

Least weasels are uncommon, living in sparse populations in eastern North Dakota. However, populations may erupt to fairly high numbers when food resources are abundant. In neighboring Minnesota, the least weasel is listed as a "species of special concern."

Although categorized along with other weasels as furbearers, least weasel pelts have essentially no economic value. As efficient predators with a high demand for small prey, they kill large numbers of rodents considered agricultural pests. Because of sparse records in North Dakota and its "special concern" status in neighboring Minnesota, continued monitoring of least weasels in the state should be included in its management strategy.

Selected References

Bailey (1926), Hazard (1982), Hibbard (1972), Jones et al. (1983), Lokemoen and Higgins (1972), Sheffield and King (1994), Svendsen (2003)

American Mink

Neovison (Mustela) vison

Dokshi^{n}ca–Dakota; Ikusan–Lakota; Daktsuà, Naksua–Hidatsa; Monika suntike, Mini-gasundek–Mandan; Eruch–Eruch—Arikara, Zhon-gwayzh'–Ojibway

George Rohde

Description

The mink is typical of members of the weasel family, having a long body and short legs. It is somewhat stockier than other weasels. Most wild mink are dark brown except for a white chin and a tail that is tipped with black. Some mink have one or more irregularly shaped white patches on their throats or chests. The ears are short, barely extending above the fur, and the tail is about two-fifths the length of their body. The fur is dense and luxurious, an adaptation to semiaquatic life. Standard measurements are: Total Length - 490 to 720 mm, Tail Length - 160 to 200 mm, Hind Foot - 60 to 80 mm, and Ear - 22 to 27 mm. Mink are sexual dimorphic, with females significantly smaller than males. In North Dakota, spring weights for males and females average 1,508 and 866 grams, respectively.

Distribution

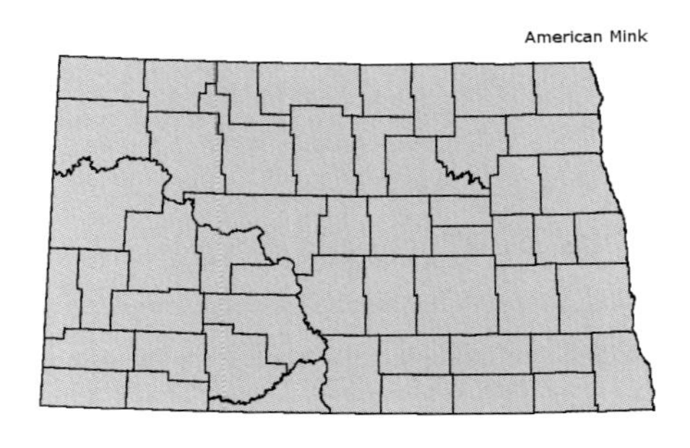

The mink can be found throughout North America except for northern Canada and areas of the southwestern United States. Mink abundance varies widely across the range, but they are most common in areas with ample wetlands and waterways. Mink are found throughout North Dakota and are fairly common along most streams in the state and throughout the Prairie Pothole Region when drought is not affecting the wetland conditions.

Habitat

Mink depend on wetland habitats for foraging, including rivers, streams, lakes, and semipermanent and permanent wetlands. These habitats contain prey such as fish, crayfish, and muskrats. Semipermanent or permanent wetlands with irregular shorelines, high water levels, and large areas of open water are preferred. In the prairies, mink also occupy shallow prairie wetlands, which provide abundant food resources. Muskrat burrows in shorelines often are used for dens. In the Prairie Pothole Region of North Dakota, hundreds of wetlands may fall within the home range of a mink, but foraging activity is focused on a small subset of these (between 10 and 30).

Sites selected by mink for denning also are closely associated with water. Throughout their range, mink are known to make use of available shelter, including ground squirrel burrows, rock piles, brush piles, culverts, dense stands of emergent vegetation, beaver lodges, and natural crevices.

Ecology and Behavior

Mink are mostly nocturnal; during the winter months; however, they become active during the twilight hours and even in midday. They are inactive at low temperatures immediately after a snowfall. They are active 5 to 20% of a 24-hour period; their inactive time is spent resting in a den. Mink usually spend each day in a different den, except for females with litters that use only a few dens until their young are mobile. Mink dens are mostly those of other animals, mainly muskrats, which have been abandoned by their original occupant. They will also use natural cavities in trees or logs, beaver dams, log jams, and rock piles as dens. In North Dakota, den location seems to be related to habitat suitability for muskrats and grazing intensity. In areas of heavy grazing, the mink need a fairly active muskrat population to keep the dens in good repair. Dens are used most often in spring, when the females are rearing young. The average distance moved by a mink family between consecutive den sites is <0.5 km.

Mink inhabiting the prairie are vulnerable to annual and seasonal fluctuations in water levels that affect food abundance and availability of den sites. Frequent and often widespread droughts, characteristic of the Prairie Pothole Region, greatly affect availability of prey, particularly in winter (e.g., winter freeze-out of shallow wetlands) and thus can lower the reproductive performance of mink. During droughts, mink often are forced to move to river systems that maintain water flow. In North Dakota, they benefit from increasing numbers of man-made permanent wetlands, which buffer populations from the effects of droughts.

Mink are opportunistic and will readily exploit any available food supply. Prey is often cached, and mink have been reported to kill more prey than they can use. Mammals are the most important part of the diet in spring-summer and fall-winter. Most common mammal resources include lagomorphs, muskrats, and small rodents. Diet also includes waterbirds, aquatic insects, fish, and tiger salamanders. During the summer, waterfowl and marsh birds are important foods.

Prairie mink tend to have larger home ranges than mink in riverine or coastal systems, averaging about 8 km^2. Typically there is no overlap between ranges of males or between ranges of females; however, the range of a male often overlaps the ranges of several females. Range size is a factor of habitat quality; in good habitat, ranges are smaller than those in poor habitat.

Drought often has severe effects on mink populations, necessitating long dispersal to find suitable water. Populations in the upper Midwest frequently vary from one to five per km^2, depending on water conditions and prey availability (usually muskrats). Human activities, mostly trapping, are the most significant causes of mortality in the mink population. Mink are preyed upon by coyotes, foxes, large cats, large hawks, and great-horned owls. While mink have lived up to eight years in captivity, few survive past their third year in the wild.

Reproduction

In North Dakota, mating typically occurs in March. Males often mate with several different females, and the males often bite the females in the back and neck during copulation. An average of 51 days passes between mating and parturition; however, this period can be as short as 40 days or as long as 75 days because implantation is delayed. Delayed implantation is common among members of the weasel family, and involves a dormant period between the time the egg is fertilized and the time it implants in the uterus to begin development. In general, the earlier mating takes place, the longer the delay before embryos begin to develop. Young females and those with smaller litters tend to have longer gestation periods. Mink are capable of breeding at 10 to 12 months of age. A single litter of two to seven young (average four) is born around the first of May. Newborns are altricial (born blind and lack hair), and weigh 6 to 10 grams. In two weeks, they grow soft, reddish-gray hair. Their eyes open at about 3 to 4 weeks, they begin to eat solid food at 5 weeks, and by 7 weeks the young are fully homeothermic. The young start learning how to hunt at 6 to 8 weeks of age, and can care for themselves by late summer. They will disperse in late August through September.

Status and Conservation

Mink are common throughout North Dakota in wetland habitats. They are most common around large permanent/semipermanent bodies of water. Their distribution is limited in the southwest part of the state due to little wetland habitat. Human development along shorelines has caused mink declines in certain areas. During periods of drought, their distribution recedes to remaining available water habitats (e.g., rivers), but when wetland conditions improve, they can quickly re-establish in wetland habitats.

The mink is managed as a furbearer in North Dakota and harvest is regulated. It is also an important predator and can have significant impact on water bird populations. Habitat protection and enhancement are important management practices for sustaining healthy mink populations. Some of these include (1) wetland protection and restoration;

(2) management of grazing by restricting livestock access along streams and rivers, which helps to maintain natural vegetation and dens by reducing grazing and trampling; this provides habitat for mink and other wildlife, and reduces erosion of stream banks; (3) maintaining structural diversity (e.g., snags, rocks) along rivers and streams and discourage development of shorelines; and (4) maintaining vegetation strips in areas next to wetlands not only provides good areas for mink (and other predators) to hunt mice and other small rodents, but also improves water quality.

Selected References

Arnold and Fritzell (1987, 1989, 1990), Eagle (1989), Eagle and Whitman (1987), Eberhardt and Sargeant (1977), Lariviere (1999, 2003), Marshall (1935)

American Badger

Taxidea taxus

Choka–Dakota; Hoka–Lakota; Maté–Mandan; Amaka, Awaga–Hidatsa; Sunuh-katuh–Arikara

C. D. Grondahl, North Dakota Game and Fish Department

Description

The American badger is a stocky, robust member of the weasel family. Badgers have shaggy fur that is grizzled grayish, with a buff cast. Their hairs are longer on the sides than the back, which adds to their broad appearance. A narrow, white medial stripe extends from the black nose to the back. Their distinctive black cheek patches are surrounded by white. The ears are upright, rounded and black with broad white edges on the front. The tail is relatively short. Well adapted for digging, they have short stout legs and relatively wide front feet with long curved front claws. Front toes are angled slightly inward causing a 'pigeon-toed' gate. Average standard measurements (male, female, respectively) are: Total Length - 732, 716 mm; Tail Length -131, 125 mm; Hind Foot - 113, 109 mm; Ear - 53, 53.

Distribution

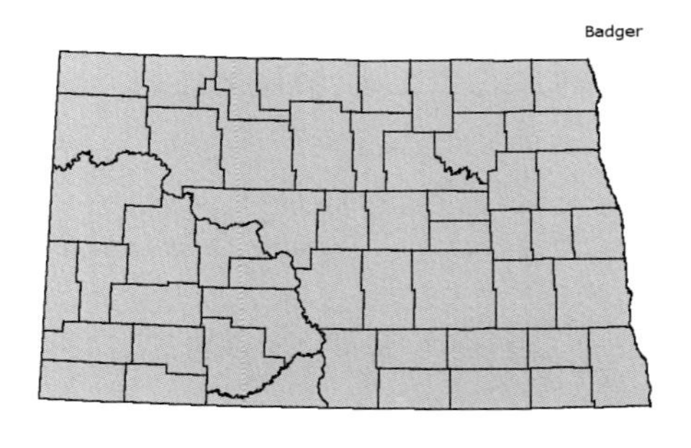

Badgers are found from central British Columbia across the Canadian provinces to southern Ontario, southeastward through Wisconsin, Michigan, and northern Ohio. From Ohio, the eastern limits extend southwesterly through Missouri, Oklahoma, eastern Texas, and to central Mexico. Western limits reach the Pacific coast. Badgers occur throughout North Dakota.

Habitat

Badgers are particularly suited to open, grassland habitat. Locally, they tend to avoid cultivated areas, but they easily occupy agriculturally dominated landscapes if human-caused mortality is contained. In such landscapes, they use scattered isolated pastures or hayland, road rights-of- way, fencerows, field borders, woodlots, cemeteries, and idle crop fields.

Habitat use generally reflects the fossorial habits of badgers, and they are drawn to areas with an abundance of ground-dwelling prey and loose soils. The conversion of cultivated land to grassland, under the Conservation Reserve Program, benefited badgers by increasing the amount of suitable habitat.

Ecology and Behavior

Badgers are solitary animals, but home ranges will overlap. Adults have established home ranges that expand and contract seasonally, but persist year after year. Male badgers tend to use larger areas than females. There is much regional variation in home range size (2 to 17 km^2), likely related to local abundance of resources. In North Dakota, ranges average about 10 km^2. Juveniles do not have permanent home ranges; they move erratically and sometimes travel long distances as they disperse from their natal areas.

One of the most conspicuous signs of badger presence is the den, the badger's daytime refuge. Sites of overnight dens used by badgers change daily in summer, but old den holes are frequently used. Denning habits and activity levels change with the onset of cold weather. Extended use of a single den becomes more frequent as activity declines in the fall, and badgers may stay underground for several days or weeks in winter. Active badger dens have a distinctive furrow at the entrance.

Badgers are largely nocturnal but occasionally are active during the day. Individuals gradually become less active during late fall and are inactive for prolonged periods during winter. An increase in hunting and caching of food in late fall coincides with badger weight gain, in preparation for periods of reduced activity in the winter. The activities of females with litters differ from those of other resident badgers. They use a shifting sequence of maternal dens and their movements remain within 200 ha of the active maternal den.

Badgers are opportunistic foragers and consume a wide variety of small mammals, birds, eggs, reptiles, amphibians, fishes, arthropods, and plant material. They are adapted morphologically to digging and their diet, consisting largely of fossorial mammals, reflects this ability. Small mammals, primarily mice and voles, but also pocket gophers and ground squirrels, are the most common food items during spring and summer. Most prey is eaten as soon as it is caught, but occasionally prey items are cached.

Little is known about the density of badger populations, but they seem to vary between 0.4 and 5/ km^2. Humans cause most badger mortality, either intentionally or by vehicle collisions. Although young badgers occasionally fall victim to predators, such as golden eagles, coyotes, or dogs, adult badgers have few natural predators. Little is known about the role of disease, parasitism, and starvation in badger populations because mortality caused by these factors is not easily documented. Badgers have a low reproductive rate, resulting in slow recovery of depressed populations. However, they are relatively long-lived, probably four to five years in the wild. Captive animals are known to survive 15 years.

Reproduction

Badgers breed in July and August, but delay implantation of the blastocyst until between December and February. Young females may breed when they are only 4 to 5 months old and give birth their first year, but more commonly they will not breed until they are 1 year old. Males, however, do not become sexually active until after their first year.

One to four young are born in March or April, but typically females give birth to only two young. At birth, young badgers have fur but are blind and ears are closed. Juveniles remain with their mother for about four months before dispersing in late summer. Dispersal of young badgers is characterized by erratic movement and sometimes long-distance travel.

Status and Conservation

Badgers are relatively common in North Dakota and are managed as furbearers. This also allows removal of badgers that are causing damage to croplands, roads, or cemeteries. It also allows for the use the badger fur, a natural renewable resource, without endangering the health of the overall population.

Factors that influence badger abundance and distribution in the region are unclear; however, variability in availability of grassland habitat due to agricultural practices, as well as human-inflicted mortality (e.g., trapping, hunting), likely are important factors. Distribution of local populations might depend on the occurrence of fossorial prey. Management actions that enhance or create grassland habitats are beneficial for badgers. Badger populations in northern grasslands seem to be increasing since the onset of the Conservation Reserve Program.

Selected References

Apps et al. (2002), Errington (1937), Lampe (1982), Lampe and Sovada (1981), Lindzey (2003), Long (1973), Long and Killingley (1983), Messick (1987), Messick and Hornocker (1981), Michener (2000), Sargeant and Warner (1972), Snead and Hendrickson (1942), Sovada et al. (1999)

Family Mephitidae
Skunks

Striped Skunk
Mephitis mephitis

Ma^{n}ka – Dakota; Maka – Lakota; Su^{n}kte – Mandan; Hohga – Hidatsa; Nichwit – Arikara

C. D. Grondahl, North Dakota Game and Fish Department

Description

The striped skunk is a mid-sized carnivore about the size of a domestic cat. Striped skunks have a triangular-shaped head; small, rounded ears; black eyes; and a rounded, black nose. Most of the skunk's head and body is glossy black, with a narrow white stripe between the eyes. The back of the head and neck has a white patch that splits into two stripes that continue down the back. The width of the stripe across the back varies from pencil thin to wide, nearly covering the whole back. Other markings include some white hairs mixed in with the relatively long black hairs of the tail and occasionally a small white patch on the chest. Females, on average, are smaller than males. Both sexes gain substantial weight in late summer and fall as they accumulate fat reserves in preparation for the winter denning period. Range of measurements are: Total Length - 560 to 795 mm, Tail Length - 225 to 295 mm, Hind Foot - 68 to 85 mm, Ear - 26 to 35 mm. The weight ranges from 1.8 to 4.2 kg.

Distribution

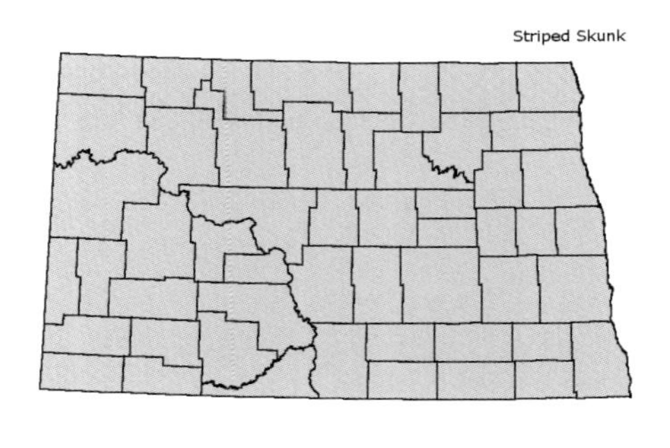

The striped skunk is found throughout most of North America from the southern border of the Northwest Territories in Canada to northern Mexico, with the exception of areas in the desert Southwest. They are seldom found at high elevations (>1,800 m). Striped skunks are found throughout North Dakota.

Habitat

Striped skunks live in a variety of habitats, including woods, prairies, and desert areas. They are abundant on agricultural landscapes where there is ample food and cover. They also have adapted to urban areas, living under houses and garages. They are less common in unfragmented, large forested areas where the food supply is low.

In the Prairie Pothole Region of North Dakota, skunks forage most often in wet meadows and upland grasslands, but they also venture into cropland, road rights of way, and tree plantings. Edges of wetlands and farmsteads are common resting sites. Buildings, brush piles, and culverts are often used as diurnal resting sites or as natal dens. Burrows excavated by skunks or other mammals, such as badgers, red foxes, or muskrats, may also be used as natal dens.

Ecology and behavior

Striped skunks are largely solitary except for brief encounters for mating and communal denning in winter by multiple females or females and a single male. Spring populations comprise loose aggregations of skunks that include an adult male with one or several adult females living in smaller areas within the male's home range. Home ranges of males and females overlap, but females with young are often aggressive toward males. In mid- to late spring, some adult males disperse long distances into new areas, sometimes >100 km away. The average home range of adult skunks in the Prairie Pothole Region of North Dakota is about 2 km^2 for females and 3 km^2 for males. Female home ranges overlap extensively and skunks will concurrently forage in the same area.

Skunks are largely nocturnal, and travel extensively each night (except on extremely cold nights). Limited daytime movement may occur near rest sites. Skunks are most active in summer and fall, particularly in September as they gain weight to prepare for colder weather. In warm months, skunks move daily to a different daytime retreat, which may be a simple aboveground bed. Females with kits, however, are restricted to the area of a natal den. Some female skunks prepare and use just one rearing den, but many females use multiple dens.

Adults start using winter dens by November and early December, but juveniles enter dens slightly later. Skunks are not true hibernators and depend on stored fat, inactivity, warm underground dens, and slightly reduced body temperatures to survive. Typically, the winter denning period extends from December through March, although the actual length is dependent on the severity of winter. During this period, skunks are largely inactive and may lose over half of their body weight. In late winter and early spring (including the mating period), activity outside of the den increases, and by May, the usual pattern of nocturnal activity has resumed.

Although primarily insectivorous, skunks are very opportunistic foragers; a variety of plant and animal foods are included in the diet, particularly when insects are not abundant. Focal areas of foraging correspond to the availability of invertebrate larvae, and the diet reflects abundances of different prey types over time. In North Dakota, large quantities of sunflower seeds are consumed in early spring. As animal prey becomes more abundant, skunks focus on beetles, caterpillars, grasshoppers and crickets, and other insects. Bird eggs (primarily of songbirds and ducks) and voles, mice, and other mammals are important vertebrate prey.

The striped skunk, like other skunks, is widely known for its ability to spray a foul smelling fluid when threatened. The fluid is stored in paired anal glands and can be discharged up to 6m, but with accuracy only up to 3m. The fluid is a strong eye irritant and can cause severe discomfort or death if ingested by mammals.

Densities of skunk populations vary by geographic area. Reports of >5 skunks/km^2 are common throughout the geographic range, but in the Prairie Pothole Region, densities of 0.5 to one skunk/km^2 are typical. Many factors contribute to variable geographic densities, including increased stress on vulnerable young animals by severe winter conditions in northern latitudes.

Diseases, particularly rabies and distemper, can limit local populations. Epizootics such as rabies and distemper likely are among the greatest mortality factors for skunks. Other major causes of mortality are human-inflicted deaths (hunting, trapping, poisoning, road kills, and encounters with farm machinery). Skunks are vulnerable to trapping, and at times of high fur prices and higher trapping activity, harvest for fur can significantly contribute to mortality. Due to their unique defense system, skunks have few natural predators. Several predators have been observed killing skunks, such as badgers, coyotes, dogs, red foxes, fishers, great horned owls, and golden eagles.

Reproduction

Skunk populations are characterized by high recruitment rates and rapid turnover. Males are polygamous, mating with several females. Both males and females are likely to breed their first year, and nearly all females become pregnant. Breeding can occur between mid-February and late April, but peaks in February to early March. Mating most often occurs in the winter dens; however, males will travel from winter dens as soon as conditions allow, seeking other receptive females. Rutting activities of males may last as long as 36 days. Gestation ranges from 59 to 77 days, and delayed implantation may be involved. Parturition usually occurs in early May. Litter size averages seven kits in the prairies of North Dakota and Minnesota.

Young skunks are nursed for six to seven weeks, and then are gradually introduced to foraging with their mother. Males do not participate in raising offspring. Most juvenile skunks will disperse from their natal areas during late July through November.

Status and Conservation

The striped skunk is common throughout North Dakota. The striped skunk is classified as an unprotected species in North Dakota, thus state residents may kill skunks at any time without a license. Few striped skunks are harvested for their fur.

Management of skunks can be directed at fostering or reducing populations depending on management goals. Wildlife managers' primary interest in skunks is as an egg predator of grassland nesting birds, and intensive removal programs in local areas have sometimes succeeded in improving duck nest success. High reproductive rates and superb dispersal abilities limit the reasonable application of removal as a broad-scale tool for enhancing nest success. Conversely, skunks consume large quantities of insects that can damage crops and some rodents, which is beneficial to farmers.

No management activities occur specifically for striped skunks; however, habitat management practices aimed at improving conditions for other wildlife species, such as breeding ducks, are good for skunks as well. Striped skunks are the principal vector of rabies in North Dakota. Although, rabies is common in local populations of skunks, humans occupying those areas are seldom aware that an epizootic of rabies is occurring. Therefore it is crucial, particularly in rural areas, for the public to be aware of the dangers of rabies and the importance of vaccinating pets.

Selected References

Bjorge et al. (1981), Greenwood and Sargeant (1994), Greenwood et al. (1997, 1999), Phillips et al. (2003), Rosatte (1987), Rosatte and Lariviere (2003), Verts (1967), Wade-Smith and Verts (1982)

Eastern Spotted Skunk

Spilogale putorius

Bob Gress

Description

Spotted skunks, sometimes called civet cats, have striking coloration; they are black with four to six white or yellowish-white broad stripes (dorsal and lateral) that are increasingly broken into large spots toward the posterior. The striping pattern and amount of black and white color varies among individuals. They have a white triangular patch on the forehead and white in front of the ears. The long, bushy tail may be tipped with white. Spotted skunks are smaller, more slender, and more weasel-like than the striped skunk, and their hair is finer and shorter. Males are larger than females (approximately 600 g and 425 g, respectively. Ranges of standard measurements are: Total Length - 459 to 539 mm, Tail Length - 175 to 228 mm, Hind Foot - 44 to 51 mm, Ear - 25 to30 mm.

Distribution

The eastern spotted skunk occurs from the border between Canada and Minnesota, continuing southwestward to southern North

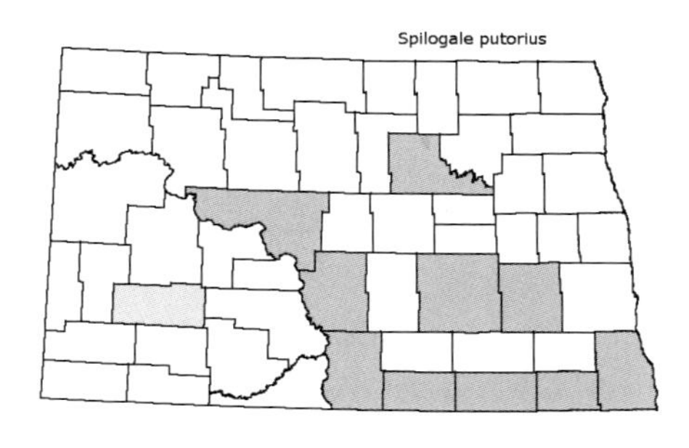

Dakota, through the Great Plains states down to the northeastern corner of Mexico. The eastern limits include a small portion of western Wisconsin southward along the Mississippi River and generally south of the Ohio River in the southeastern United States.

The spotted skunk appears to be a recent arrival to North Dakota, with no reports in the state until 1924. Since that time, there have been a few reports from the southern tier of counties east of the Missouri River and in Barnes, Benson, Burleigh, Cass, McLean, and Stutsman counties. There is a single report from west of the Missouri River (Stark County).

Habitat

The spotted skunk occurs in a wide range of habitats, but seems to prefer brushy areas, forest edge, riparian areas, and upland prairie grassland, especially if rock outcrops or shrubs are present. It tolerates the presence of humans. Spotted skunks den almost anywhere protected from light, wind, rain and predators, such as underground burrows (abandoned by other species), abandoned buildings, woodpiles, rock piles, or in hollow logs or trees.

Ecology and Behavior

Spotted skunks are largely solitary except for brief encounters for mating and communal denning in cold conditions. They are more nocturnal than striped skunks, and are seldom seen during daylight hours. When foraging, they tend to stay close to shelter such as dense brush and fence rows. Spotted skunks are opportunistic foragers, eating a wide variety of animals and plants, including small rodents, birds, bird eggs, insects, fruits, seeds, agricultural grains (e.g., corn) and carrion. In summer and fall, insects are their primary foods; in winter and spring small mammals are a more important food source.

Home range estimates have varied from <1 to about 2 km^2 during the non-breeding seasons. Males may wander over much larger areas when breeding. Nightly movements can be as much as 5 km. Spotted skunks are not known to defend territories.

Within their home range, spotted skunks usually have more than one den site. The use of particular dens seems to be related to the local abundance of food. In natal dens, leaves and grasses are sometimes gathered for nesting material. Most often winter dens are dug by the skunk or modified from existing burrows, and extend below the frost line. In fall, they gain weight to store energy for the winter, but they are not true hibernators.

Similar to other skunks, spotted skunks are able spray a foul-smelling fluid as a defense mechanism. The fluid is stored in paired anal glands and is considered to be stronger and more pungent than striped skunk musk. When threatened, spotted skunks display warning behaviors; spraying seems to be the final defense. First, skunks may stomp their front feet on the ground, and if further threatened, they will roll back on their hind feet and lunge forward, stomping their front feet down and then draw them back. Spotted skunks also display "handstanding behavior" consisting of the skunk standing on its front feet, with the tail bending forward over its back, and walk toward the source of the threat. The musk is then expelled by returning to all four feet, assuming a U-shaped stance with back arched, and contracting the muscles surrounding the scent glands.

Reported densities range from 2 to 40 skunks/ km^2. Humans are the main cause of mortality in spotted skunks, largely from automobile collisions, trapping and poisoning. Great horned owls, bobcats, coyotes, and domestic dogs are known predators. Although striped skunks are the more important vector for rabies, spotted skunks are also susceptible to the disease. Because of their low numbers, however, they probably do not represent a major health problem. Spotted skunks may live six years in captivity, but probably live only two or three years in the wild.

Reproduction

Eastern spotted skunks breed in March and April. Following conception, implantation of the fertilized egg in the uterus may be delayed briefly (up to 14 days) or not at all. They have one litter per year, typically five young (range two to nine) in a litter. The young are born in May or June following a 50- to 65-day gestation period. Their eyes and ears are closed and they weigh about 10 grams at birth. Their eyes and ears open about one month after birth, and their teeth begin to erupt after five weeks. Young are born with a thin coat of fine fur, the black and white markings already apparent. At about 5 or 6 weeks, the young begin to emerge from the den, but they remain nearby. They are weaned by about eight weeks. Dispersal occurs at about 16 weeks of age. Skunks reach adult size by mid-October of their first year. Skunks are capable of breeding the spring after their birth.

Status and Conservation

The spotted skunk is rare in North Dakota, and has been listed by the North Dakota Game and Fish Department as a Species of Conservation Priority. However, it is not protected, and may be trapped by North Dakota residents without a license. Because they are rare, captures or observations should be reported to the North Dakota Game and Fish Department. Like other skunk species, these animals can be a nuisance if they select denning sites under occupied buildings.

Selected References

Bailey (1926), Crabb (1948), Kinlaw (1995), Mead (1968), Rosatte and Lariviere (2003), Van Gelder (1959)

Family Procyonidae
Raccoons, Ringtails, Coatis

Raccoon
Procyon lotor

Wica–Dakota; Wiciteglega–Lakota; Miká–Omaha; Isat–Arikara; Shunte-pusa–Mandan; Sida-buzhe–Hidatsa; Asebun–Ojibway

George Rohde

Description

The raccoon is a thickset medium-sized carnivore. Markings include a long, bushy tail alternately ringed with five or six black and yellowish-gray bands and tipped with black. A black mask covers the eyes and cheeks, surrounded mostly by white across the nose and forehead. The grizzled pelage varies in color from yellow-brown to dark gray-black. Ears are upright, medium length, and rounded. The legs are medium length and toes are equipped with long, non-retractable claws. Front toes are relatively long and flexible, adapted to manipulate food. Weights and measurements are variable, but males are typically larger than females. Range of measurements are Total Length - 655 to 960 mm, Tail Length - 200 to 288 mm, Hind Foot - 105 to 132 mm, Ear - 48 to 58 mm, Weight - typically 5 to 10 kg in summer, but individuals over 25 kg are encountered in fall.

Distribution

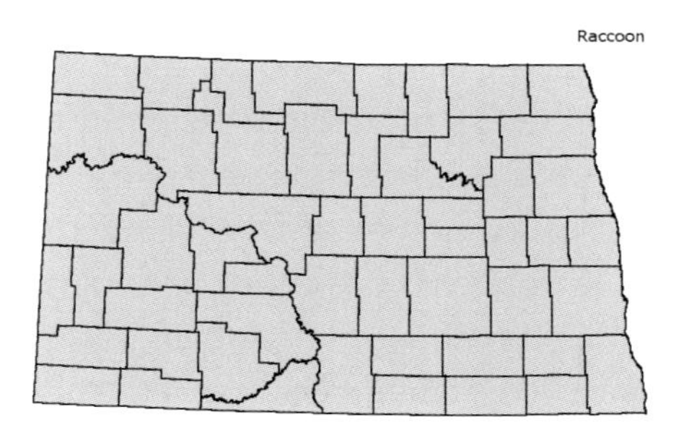

Raccoons range from southern Canada, throughout much of the United States, Mexico and Central America. They occur primarily in wooded areas, especially near water. During the early days of white settlement, raccoons were common in wooded areas of the east, such as the Red River and tributaries, but with expanding agriculture they became more widely distributed throughout North Dakota.

Habitat

Raccoons are well-suited to coexisting with humans, and they benefited from agricultural development in North Dakota. During the day, raccoons bed in thick vegetation of pond edges, abandoned buildings, trees, or other places of shelter. Rearing dens are often found in farmyards; haylofts, attics, and buildings are commonly used. Females with young restrict their activities to areas around their natal den. Nightly foraging by raccoons may include trips to farmyards and wetlands, often several kilometers apart.

Ecology and behavior

Adult raccoons are primarily solitary but may be found together around concentrated food sources, during the mating season, and while denning during winter. Home ranges vary greatly depending on habitat, population density, and food supply. Raccoons in the prairies have larger home ranges than elsewhere. Adult males occupy the largest home ranges, averaging >25 km^2 in North Dakota and may encompass the entire range of multiple females. Home ranges of adult females and yearlings are considerably smaller (16 km^2) and may overlap considerably with those of conspecifics. Raccoons have been reported to occur at relatively low densities of 0.5 to 1/km^2 in North Dakota, compared to estimates of up to 31/km^2 elsewhere.

Raccoons are omnivores, feeding on carrion, insects, amphibians, mammals, crayfish, mussels, birds, and a wide variety of grains, fruits, and other plant materials. In early spring, grain and livestock feed obtained at farmyards is an important supplement to the limited naturally occurring food sources. Later, they exploit increased production and availability of wetland-related foods such as invertebrates, amphibians, and nesting waterbirds.

Raccoons are largely nocturnal, but occasionally may be active during the day. Nightly foraging activity in central North Dakota is characterized by extended periods of localized foraging interspersed with long-distance travel. Diurnal resting sites change daily, except for those of females with young, whose foraging areas and denning locations are restricted to areas around the rearing den.

Raccoons remain active well into fall. Weight gain during this time is critical for overwintering success because it is common for raccoons to lose 30 to 50% of their body weight over the winter. Severe winter weather limits raccoon activity, eventually causing them to retreat to winter dens, commonly located in or under buildings, scrub piles, cellars, burrows, or trees. Although raccoons do not hibernate, they remain inactive for extended periods during cold weather.

Hunting, trapping, and road kills, account for the majority of raccoon mortalities in North Dakota. Additional causes of mortality include disease (especially canine distemper), overwinter starvation, and extreme parasite loads. Overwintering and dispersal may represent periods of greatest vulnerability for raccoons. Predation is not a significant cause of mortality, although coyotes and fishers are known to kill raccoons. When threatened, raccoons retreat into wetlands or climb trees for protection.

Reproduction

Studies across the breeding range of raccoons have revealed substantial variability in reproductive biology. Breeding ages, whelping dates, and litter sizes vary with latitude, habitat, and density. Reproduction in northern latitudes is characterized by delayed maturity and large litter size. Fewer yearling females (38 to 77%) reproduce than adult females (84 to 100%). In southeastern North Dakota, only two of 14 yearlings had bred, but 26 of 28 adults had reproduced. Breeding typically occurs from February through April in North Dakota, followed by a 63-day gestation. Average whelping date is around May 8 (range mid-April to mid-June), but some litters may be born in late August or early September.

Litter sizes of adult females typically range from three to seven kits, although an exceptionally large litter of 10 was recorded near Jamestown. Males do not assist with raising young. Females and young remain together through the summer until weaning, and may travel together thereafter. The family gradually breaks up but may den communally in winter. Most young males disperse the following spring (May-June), while most females remain in their natal area.

Status and Conservation

Raccoon distribution and abundance have generally increased in North Dakota over the past 100 years. They are highly suited to coexist with humans and benefit from agricultural activities in the state. Because of high reproductive rates, populations are relatively resilient to localized disease outbreaks and typical trapping and hunting pressures.

Raccoons are valued as furbearers, providing one of the largest revenues for trappers in North America. But raccoons are also recognized as pests and they may damage crops, buildings, campgrounds and gardens. They are recognized as important predators of breeding birds. They are one of the most frequent nuisance animals reported in urban and suburban areas, and are often difficult to deter.

Much of North Dakota is well-suited to raccoons and little management beyond regulated trapping is needed to maintain populations.

In North Dakota, raccoons are managed as a furbearer and they may be harvested by trapping or shooting.

Selected References

Bailey (1926), Clark et al. (1989), Cowan (1973), Fritzell (1977), Fritzell (1978a,b,c), Fritzell (1987), Fritzell, (1989), Fritzell and Greenwood (1984), Fritzell and Matthews (1975), Gehrt (2003), Greenwood (1981, 1982), (Jones et al. 1983), Mech and Turkowski (1966), Sanderson (1987), Sargeant et al. (1993), Yeager (1937)

Order Rodentia (Rodents)

The rodents comprise the largest order of mammals, with 28 living families and over 2,000 species. More than 43% of all mammal species are rodents. They are the most widespread terrestrial mammals, occurring on every continent except Antarctica, and have filled nearly every ecological niche. Some species live in trees (tree squirrels and tree mice), some glide from tree to tree (flying squirrels), some spend most of their lives underground (pocket gophers), and some are partially aquatic (beaver, nutria, and muskrats). They range in size from only about 5 g (the pygmy mice) to nearly 80 kg (capybara of Central and South America).

Rodents are easily recognizable, although their body forms are highly variable, reflecting their diverse ecological niches. Superficially, rodent skulls resemble those of lagomorphs (hares and rabbits), their most diagnostic features being the lack of fenestrations on the side of the rostrum and only a single incisor on each side of the upper jaw. There are four chisel-edged incisors, which are continuously growing. There is a large space (diastema) between the incisors and the cheek teeth. Canines and most of the premolars are absent. Hence, the maximum tooth formula is 1/1, 0/0, 2/1, 3/3. Many species have fewer teeth.

There are about 218 species of rodents in North America north of Mexico. Of these, 34 species in eight families occur in North Dakota. This is a fairly high species diversity, considering the state's lack of significant altitudinal gradient and relatively harsh climate. By comparison, Minnesota to the east with its deciduous and coniferous forests has 29 reported species, while Montana to the west with its mountainous regions reports 45 species of rodents. North Dakota rodent families include the Sciuridae (squirrels), Castoridae (beavers), Geomyidae (pocket gophers), Heteromyidae (pocket mice and kangaroo rats), Dipodidae (jumping mice), Cricetidae (New World rats and mice), Muridae (Old World rats and mice), and Erethizontidae (porcupines).

Family Sciuridae
Squirrels; Hetkala—Lakota

Black-tailed Prairie Dog

Cynomys ludovicianus

Achks–Arikara; Shopka–Mandan;
Si^{n}hpa,Tsipá–Hidatsa, Pispiza–Lakota

Robert Seabloom

Description

The black-tailed prairie dog is a robust ground squirrel averaging about 884 g in weight. Its dorsal coloration has been described as "pinkish cinnamon grizzled with buff and black." Its underparts are whitish, and the tail is tipped with black. The head is broad, with moderately large eyes and small ears. Average standard body measurements are: Total Length - 364 mm, Tail Length - 83 mm, Hind Foot - 60 mm, Ear - 10 mm. Males are about 10% larger than females.

Distribution

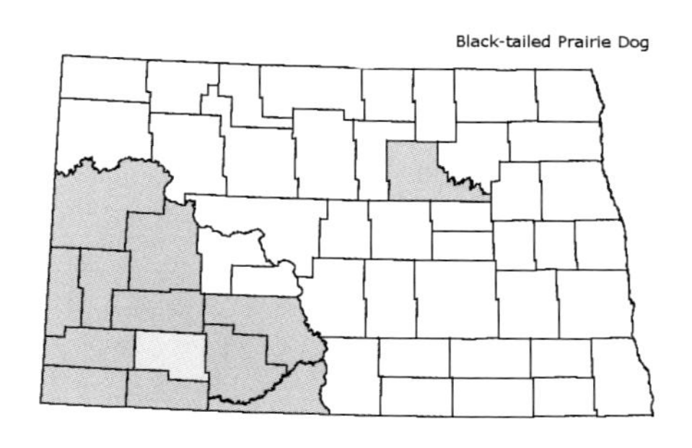

This species is distributed from extreme southern Saskatchewan to northern Mexico. Its western limits are the eastern foothills of the Rocky Mountains, while to the east it extends to the ecotone between the short-grass and tall-grass prairies in the Dakotas, Nebraska, Kansas, Oklahoma, and Texas. While prairie dogs originally occupied 40 million ha in North America, large scale control efforts, disease, and land use changes have resulted in a scattering of small surviving colonies.

In North Dakota, black-tailed prairie dog colonies occur naturally south and west of the Missouri River, especially in the drainages of the Little Missouri and Cannonball rivers. Attempts to introduce prairie dogs elsewhere in the state have been largely unsuccessful, but two such colonies remain viable at Fort Stevenson State Park (McLean County) and Sully's Hill National Game Preserve near Fort Totten (Benson County).

Habitat

Black-tailed prairie dogs show a distinct preference for short-grass prairie. Locally, they are most abundant on overgrazed pastures; their occurrence in such situations is more symptomatic than causal. Indeed, they tend to avoid tall-grass habitats. Where colonies occur, they engage in deliberate clipping as well as foraging of grasses. This is believed to be an adaptation providing an unobstructed view of approaching predators.

Soil texture, slope, and exposure may be additional factors influencing the occurrence of black-tailed prairie dogs. One study suggested that colony distribution in Roosevelt National Park was correlated with fine- to medium-textured alluvial soils, and dry south-facing slopes of less than 30%. These situations probably enhance burrowing activity and favorable vegetative community structure. In high-density colonies, however, burrows may also occur in a variety of soil types, including clay and soft lignite, suggesting that prairie dogs may not be limited by soil type as much as soil texture, moisture, and vegetation. Habitat models have been proposed based on vegetation, slope, landownership, and prairie dog dispersal capabilities. Based on these parameters, it was concluded that 31% of the Little Missouri National Grasslands contained highly suitable habitat for prairie dogs.

Ecology and Behavior

Black-tailed prairie dogs are highly social animals, living in well-organized colonies. Colony sizes are highly variable, in North Dakota ranging from 9 to 95 ha, with densities of seven to 108 prairie dogs per ha. When undisturbed, colonies can include thousands of individuals and extend for kilometers in all directions, the record being a Texas colony estimated at 400 million! Larger colonies may be divided by topographic features into "wards," but the functional social unit is the "coterie," involving two to 35 related prairie dogs living on an area averaging one-third ha. Coteries usually involve an adult male, several breeding females, and several non-breeding yearlings and juveniles. Relations are friendly among all members of the coterie, including use of burrows, except during the breeding season. All activity, including foraging, is conducted within the boundaries of the coterie, and it is vigorously defended against members of neighboring coteries. As juveniles (especially males) mature, they may disperse to the periphery of the colony or to new areas in an attempt to establish new coteries.

The gregarious nature of the black-tailed prairie dog has resulted in a complex repertoire of communication, functioning in territorial defense, cohesiveness within the coterie, and warning of the presence of predators. There are at least 12 different vocalizations, including "jump yip" barks (territorial calls) and antipredator calls warning the colony of potential danger. In addition, a variety of postures, including play, allogrooming, "kissing," bluff charges, flairing the tail, staring, tooth chattering, and chases have been identified.

Permanent burrows descend vertically from 1 to 3 m, and then angle horizontally for 5 to 10 m. There are blind extensions off of the main burrow, including one for the nest chamber. Burrows typically have multiple entrances with associated mounds of excavated soil. Dome-shaped mounds have a diameter of 2 to 3 m, and a height of about 0.3 m, while crater-shaped mounds, resembling small volcanoes, may be as high as 1 m. It is believed that entrance mounds provide some protection against flooding, facilitate air circulation in the burrow, and offer vantage points for detection of approaching predators.

Prairie dogs are strictly diurnal, and are active from dawn to dusk. Unlike other ground squirrels, they do not hibernate, and may be observed on the surface throughout the year. However, during severe winter weather, they may remain underground for several days. Recently, it has been shown that during these periods, black-tailed prairie dogs may enter into a "facultative hibernation," depending on fat reserves for survival.

Black-tailed prairie dogs are selectively herbivorous, and their foraging and clipping activity affects the vegetative composition and structure of the colony. Primary components of the diet are grasses (87%) and forbs (12%), with the remaining 1% consisting of shrubs, seeds and insects. Favored foods in this region include sand dropseed, sun sedge, blue grama and wheatgrass. Selective grazing of perennial grasses on prairie dog towns, combined with low forb consumption, results in altered plant species composition compared with the surrounding range land. Occasionally, insects are taken, including cutworms, grasshoppers, and beetles.

Dispersal of juveniles occurs during late May and June upon emergence from the natal burrow, and dispersal of yearling males before their first copulation. Yearling males may disperse within the colony, but females usually remain in the natal coterie. Juvenile dispersal is male-biased, but may include females, and may involve movement to other colonies or attempts to establish new colonies. These movements may be as far as 5 km from the natal colony.

Maximum longevity of male black-tailed prairie dogs may be as high as five years, while females have been reported to live as long as eight years. In North Dakota, a spring sample of 94 animals was 46% yearlings, 35% 2 years old, and 19% 3 years or older. Only two individuals (2.1%) were older than 3 years of age.

An important cause of pre-emergent juvenile mortality is infanticide and cannibalism by related lactating females. This infanticide can involve as much as 39% of a year's litter production. Badgers, black-footed ferrets, and snakes also account for losses of pre-emergent juveniles. Upon emergence, a variety of mammalian and avian predators account for additional losses, but the colonial nature of the species and its vigilance behavior ameliorate some of this mortality. Prairie dogs are highly susceptible to sylvatic plague, which may eliminate entire colonies within weeks. Human-caused mortality by hunting, trapping, poisoning, and agricultural and development practices has accounted for the precipitous decline of the species in historic time.

Reproduction

Breeding in North Dakota begins in mid-March, soon after resumption of expanded spring activity. This timing of reproduction is later than in other portions of the species' range, but comparable to other ground squirrels at similar latitudes and elevations. Severe winters and chronic disturbance, such as persistent shooting, can affect a number of parameters of reproduction in the colony. In Montana, breeding was delayed following a winter of unusually low temperatures and persistent snow cover. Severe winters did not seem to influence the breeding season in North Dakota, but may have affected the reproductive performance of age classes. Following a severe winter in Roosevelt National Park, only half of the yearling females reproduced, and probably none of the yearling males. By contrast, in an earlier study of nearby undisturbed colonies, 90% of the yearling females reproduced following a mild winter. Other towns which had been subjected to heavy hunting pressure exhibited little (32%) yearling reproduction. Hence, reproduction by young females may be contingent on their capability to build up requisite energy reserves for the coming breeding season.

A single litter is produced, following a 35-day gestation. Litter sizes in North Dakota range from one to 11, but average about four. Intrauterine losses of embryos can be variable. North Dakota studies indicated no embryo resorptions, and similar results have been observed elsewhere. However, in Montana, resorptions occurred in nearly 25% of pregnant females. Yearling females produce the same-sized litters as the adults.

As with other squirrels, newborn young are blind, pink, hairless, and weigh less than 20 g. Their fur appears at about 3 weeks of age, and eyes open at 5 weeks. Juveniles emerge from the natal burrow at about 40 days of age. At this time, they begin foraging on their own, but may continue nursing as well. By October, both males and females average over 500 g.

Status and Conservation

Prior to white settlement, the black-tailed prairie dog may have been the most abundant mammalian herbivore on the Great Plains, numbering as high as 5 billion. Habitat destruction, combined with intensive control efforts via shooting, poisoning, trapping and drowning, have reduced colonies to less than 2% of the historic range.

In North Dakota, black-tailed prairie dogs originally occurred throughout suitable habitat south and west of the Missouri River. However, extensive habitat destruction and control operations (primarily poisoning) drastically reduced the species to <1% of its former range in the state. During the 31-year period from 1939 to 1972, there was a reduction of 89% in number and 93% in acreage of towns in the Little Missouri National Grasslands. Subsequent surveys have revealed a significant increase in North Dakota prairie dog populations over the past several decades, probably related to the cessation of poisoning programs on public lands. Active colonies in North Dakota currently occupy about 140 km^2, according to recent estimates. However, in view of the species' historic decline to a small fraction of its original levels, the U. S. Fish and Wildlife Service listed the black-tailed prairie dog as a candidate species for threatened status.

The decline of the black-tailed prairie dog has generated extensive concern. As a keystone species, it affects vegetative structure and species composition; alters soil texture, moisture, and aeration; and influences overall biological diversity because of the large numbers of invertebrates and vertebrates attracted to or dependent on the colony. Vertebrate species that are especially dependent on prairie dog colonies include the black-footed ferret, mountain plover, and burrowing owl. These have undergone marked declines, especially the black-footed ferret, which is on the verge of extinction.

For many years, the black-tailed prairie dog has been perceived by ranching interests to be a competitor with livestock for forage. This resulted in extensive governmental and private efforts to destroy colonies, primarily by poisoning. However, despite their huge numbers, prairie dogs had coexisted for thousands of years with thriving herds of bison and other ungulates numbering in the tens of millions. While some competition may exist, recent studies have indicated that prairie dogs have differing preferences for forage species. It has been further demonstrated that continuing control operations are not economically feasible, with the costs greater than the value of the added forage gained.

When control is warranted, alternative approaches to poisoning may aid in holding populations to more acceptable levels while still maintaining the integrity of the ecosystem. Prairie dogs are most abundant under heavy cattle grazing. Periodic exclusion of cattle or reduced stocking rates, combined with more direct control, may help regulate colony expansion. Recreational hunting of prairie dogs may provide an additional tool in such situations. In Montana, hunted colonies were reduced by 35% over a two-year period, compared with a 15% reduction in non-hunted colonies. Spring hunting may be especially effective, for it would target the overwintering survivors and also disrupt reproductive activity, especially among yearling females. Furthermore, habitat treatment adjacent to existing colonies, such as mechanical brush removal and prescribed burns, may influence the direction of colony expansion to areas where there is lower likelihood of conflict with human activity.

Conservation issues around this potentially threatened species have resulted in significant activity at private, state, and federal levels. Large-scale aerial surveys have been conducted, resulting in habitat models for the Little Missouri National Grasslands and Theodore Roosevelt National Park. Conservation strategies have been proposed which involve setting aside core reserves on federal and state lands, buffer areas around core reserves which would allow compatible human activities, and connecting linkages which would provide for dispersal between core areas.

The state of North Dakota lists the black-tailed prairie dog as a non-game wildlife species, with no restrictions on hunting by residents. It is also listed by the State Game and Fish Department as a Level I Species of Conservation Priority. However, state law lists it as a pest species, with extermination authorized on private lands. A working group has been developed to formulate a state management plan that may balance these diverse goals.

Selected References

Bailey (1926), Bishop and Culbertson (1976), Collins et al. (1984), Dyke et al. (2004), Grondahl (1973), Hoogland (1995, 1996, 2003), Knowles (1985, 1987), Koford (1958), Jones et al. (1983), Milne (2004), Reading and Matchett (1997), Reid (1954), Roe and Roe (2003), Seabloom and Theisen (1990), Seabloom et al. (1978), Sidle et al. (2001), Stockrahm (1979), Stockrahm and Seabloom (1988), Theisen (1981), United States Fish and Wildlife Service (2000), Uresk (1984), Vermeire et al. (2004), Vosburgh and Irby (1998)

Northern Flying Squirrel

Glaucomys sabrinus

Psinca–Lakota

Dr. Lloyd Glenn Ingles, © California Academy of Sciences

Description

Flying squirrels are the only North American scansorial rodents, those that use gliding locomotion to escape predators and quickly cover distance. They have a loose fold of skin, the patagium, along each side, and attached to the wrist and ankle. The spread patagium, used in conjunction with the tail for maneuverability, enables the squirrel to "hang glide" over significant distances up to 30 m or more.

The northern flying squirrel is a relatively small squirrel weighing about 90 g, slightly smaller than an eastern chipmunk. Its dense, silky fur is brownish gray on the back, and the underparts are creamy white. The base of the white belly hairs is lead-colored which, in addition to size, distinguishes the species from the southern flying squirrel. Average standard body measurements are: Total Length - 273 mm, Tail Length - 128 mm, Hind Foot - 37 mm, Ear - 22 mm.

Distribution

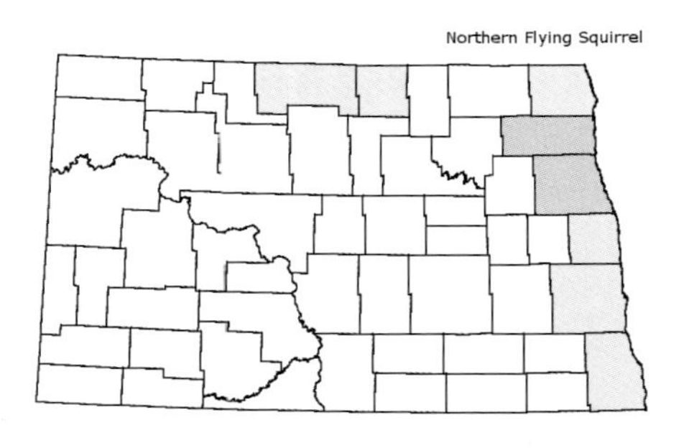

Northern flying squirrels occur mainly in coniferous forests of North America, in the north from Alaska to eastern Canada. In the west, it extends down the Cascades and Coast ranges to the Sierra Nevada of California. It occurs in the Rocky Mountains of Montana, Idaho, Wyoming, and Utah. There is an isolated population in the Black Hills of South Dakota. In eastern North America, they occur in North Dakota, the lake states, New England, and isolated populations in the Appalachians.

In North Dakota, northern flying squirrels seem to be restricted to mature forests along the Red River and its major tributaries. It is thought that they may occur in the Turtle Mountains (Rolette, Bottineau counties), but there are no known records from that area.

Habitat

The primary habitat of the northern flying squirrel is mature coniferous forest. In the Midwest and eastern states, however, they are found in a variety of deciduous and coniferous forests. In North Dakota, flying squirrels inhabit mature riparian forests along the Red River and its tributaries. These forests are dominated by American elm, green ash, basswood, box elder and cottonwood. Mature forests with tree cavities are required for denning. Lichens and fungi accommodating their unique food habits are especially important for the survival of this species.

Ecology and Behavior

Northern flying squirrels inhabit tree cavities, such as abandoned woodpecker holes, and outside nests. Cavity nests are lined with shredded bark, moss, lichens, or other fine material. Artificial structures, such as wood duck houses, are also used. Outside nests are located on a supporting structure, such as a branch or abandoned bird nest. Twigs, bark, roots, grass and lichens are used to encase the nest, which is lined with fine material. Cavity nests are probably the only ones used during severe North Dakota winters, and the availability of suitable trees may be a limiting factor. During winter, cavity nests may be shared by several squirrels.

The diet of the northern flying squirrel is dominated by various species of fungi and lichens, and at times these may be the only food items taken. They have also been known to feed on acorns, other hardwood seeds, buds, staminate cones, tree sap, and catkins, as well as eggs and nestling birds. Unlike other tree squirrels, northern flying squirrels are not known to hoard food items.

Northern flying squirrels are nocturnal, beginning their activity shortly after sundown. There are two peaks of activity, for about two hours after sunset, and again for several hours before sunrise. Inclement weather delays, but does not prevent, activity. Flying squirrels remain active throughout the winter, even in temperatures as low as -24°C. During winter, their tracks may be seen in the snow, as well as evidence of tunneling under the snow while foraging for food.

Little is known about the social structure, territoriality and home range of northern flying squirrels. Adults may dominate younger squirrels around food sources, but there is no evidence of territorial behavior. Groups of squirrels may occupy the same nest during winter, especially young of the year with their mother. Home ranges tend to have a radius of 100 to 200 m.

The population ecology of northern flying squirrels is also poorly known. Population density estimates range from one squirrel per 2 ha to 10 per ha. Juvenile survival is probably less than 30%, while adult survival is about 50% per year. Maximum longevity in the wild is about four to six years. Northern flying squirrels succumb to accidents and predation by a variety of hawks, owls, and predatory mammals. Populations appear to be primarily limited by the availability of food and den sites (tree cavities).

Reproduction

Mating is later than in other tree squirrels, ranging from March to May. Litters are born from May to July in the northern part of the range, following a gestation period of 37 to 42 days. Litter size typically averages from two to four, but may be as high as six. A second litter may be produced in the southern part of the range, but probably only one is born in North Dakota.

Newborn young are blind, hairless, and weigh 5 to 6 g. They begin to acquire fine hair after a week. They are furred and their eyes open by 1 month. Young begin to leave the den at about 40 days, and begin to take solid food. They are fully weaned at 2 months of age, and approach adult size by 4 months. They do not breed until their second year.

Status and Conservation

Because of their nocturnal habits, few people encounter northern flying squirrels. But they are probably fairly common in mature forests of the Red River valley. Their status in the forested Turtle Mountains is unknown.

The northern flying squirrel is of no economic significance, but confers an aesthetic benefit to those who appreciate its presence and fascinating life history. Its continued occurrence in marginal parts of its range, such as eastern North Dakota, will depend on mature forested areas, especially older trees, containing cavities for dens. Large-scale sanitation cutting of older and diseased trees, such as for Dutch elm disease control, may jeopardize the occurrence of this squirrel in the state.

Selected References

Bailey (1926), Clambey (1986), Hazard (1982), Jones et al. (1983), Wells-Gosling and Heaney (1984)

Woodchuck (Ground Hog)

Marmota monax

G. L. Tweist, American Society of Mammalogists Mammal Images Library

Description

Weighing up to 4,000 g, the woodchuck is this region's largest squirrel. It is a stocky animal, with short ears and legs, and a relatively short, bushy tail. Its dorsal coloration is dark brown, with whitish hairs, giving it a grizzled appearance. The feet are black. Average standard body measurements are: Total Length - 536 mm, Tail Length - 117 mm, Hind Foot - 78 mm, Ear - 27 mm.

Distribution

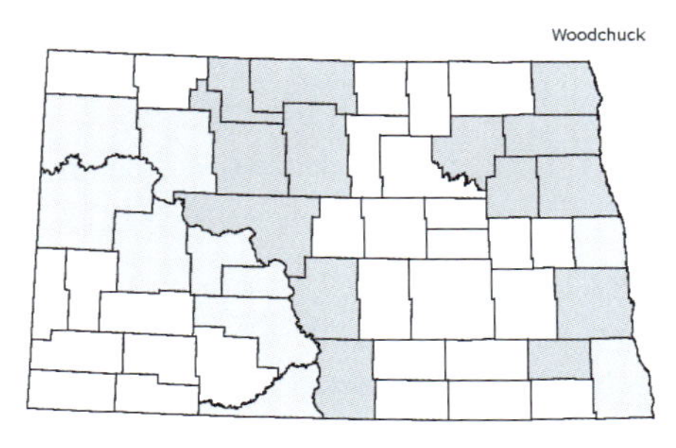

The woodchuck ranges from eastern Alaska across the forested regions of Canada, south into much of the eastern United States. Its southern limits include Georgia, Alabama, northwestern Louisiana, and Arkansas. Except for a small extension into northern Idaho, its western limits are the eastern portions of the Plains states.

Early North Dakota records indicate a distribution restricted to forested areas of the eastern counties (Ransom, Cass, Grand Forks, Walsh, Pembina, Nelson, Ramsey). Recent reports indicate a range extension to include the Mouse (Souris) River valley and Missouri River flood plain. This appears to be consistent with reports of westward range extension in the southern Plains as well.

Habitat

Woodchuck habitat primarily consists of forest edges and openings, where heavy woody cover provides suitable den sites and open areas used for breeding and foraging. Fence rows, rock piles, stream or lake banks, and human structures provide additional sites for burrows. Consequently, forest openings created by clearing for agriculture may have contributed to range expansion of the species, as have the planting of trees in the Plains states.

Ecology and Behavior

Woodchucks are relatively solitary, asocial animals. However, home ranges may overlap, and animals may communicate recognition, dominance, and territoriality with a fairly broad repertoire of communication, including various postures, scent marking, and vocalizations ("whistles").

Burrows are fairly extensive, with a main entrance, several concealed "plunge holes," and an enlarged, leaf- and grass-lined nest chamber. There are several blind pockets in the burrow, used as sanitation sites for depositing feces. The burrow is quite large; it may be >2 m deep and up to 13 m in length.

Woodchucks are deep hibernators, and put on extensive fat during late summer and fall to prepare for winter dormancy. They retreat to their dens in October, enter hibernation, and do not emerge until March or April. Hibernation is not continuous; arousals occur every 4 to 10 days, possibly to rid the body of metabolic wastes. Unlike light hibernators such as chipmunks, there is no food intake during these arousal periods.

Daily activity during the warm months is typically diurnal, but animals may spend only 1 to 5 hours per day above ground. Activity following emergence in spring, prior to immergence in the fall, and during bad weather, peaks at mid-day, while early morning and late afternoon peaks occur during mid-summer.

Woodchucks eat a wide variety of plants, and can be serious garden pests. They may readily switch plant foods, depending on location, seasonal availability, and individual preference. Some animal foods are taken as well, including beetles, grasshoppers, and snails.

Home range size can be highly variable, but is frequently less than 2 ha. Factors influencing home range size include locality, sex, population density, reproductive activity, and dispersal status. Male-female and female-female home ranges may overlap, but male-male home ranges do not. Population densities generally average less than four woodchucks per ha.

Woodchucks are long-lived for squirrels. Longevities of four to six years have been recorded in the wild, and nearly 10 years in captivity. Causes of mortality include flooding of burrows, disease, predation, and shooting. In addition to carrying a variety of parasites, they are susceptible to hepatitis, tularemia, Rocky Mountain spotted fever, and rabies. Predators include foxes, coyotes, mink, weasels, hawks, and owls.

Reproduction

Breeding takes place shortly after emergence from hibernation. Both adult and some yearling females are bred at this time. Yearling males do not breed during the year following their birth. Gestation is 31 to 32 days; hence in this area, most births can be expected by mid-May. The average litter size is about four, but may run from one to eight. As with most other rodents, the young are altricial, i.e., hairless, eyes and ears closed, and have poor temperature control. The eyes are open at about 1 month, and the young begin taking solid food. Young are weaned about two weeks later, and begin emerging from the burrow. Growth is not completed until the second or third year even though females achieve sexual maturity as yearlings.

Status and Conservation

Woodchucks occur in scattered wooded areas over eastern North Dakota. Being at the western extreme of their range, and subjected to a harsh climate, they are uncommon in the state.

Although woodchucks have appeared to have expanded their range in recent years, North Dakota, with its limited forest land and harsh climate, probably provides marginal conditions for their survival.

Although woodchucks do occasional damage to garden crops, large-scale control is not warranted. Large-scale land clearing associated with mature forest removal, together with intensive agriculture, are probably the most significant threats to the species in this region.

Selected References

Armitage (2003), Bailey (1926), Hazard (1982), Hibbard (1972), Kwiecinski (1998), Michener (1984)

Eastern Gray Squirrel

Sciurus carolinensis

Zicahota–Lakota

G. L. Tweist, American Society of Mammalogists Mammal Images Library

Description

This well-known tree squirrel is highly visible and relatively easily recognized in the field. It is a medium-sized squirrel, averaging about 600 g. The back and sides are gray, sometimes with a brownish cast, and the underparts are usually white. Melanistic (black) gray squirrels are fairly common in some areas. The tail is long and very bushy. Some may confuse the gray squirrel with the fox squirrel, which is larger and has more orange coloration, or the Franklin's ground squirrel, which is smaller and has a much less bushy tail. Average standard body measurements are: Total Length - 479 mm, Tail Length - 211 mm, Hind Foot - 65 mm, Ear - 31 mm.

Distribution

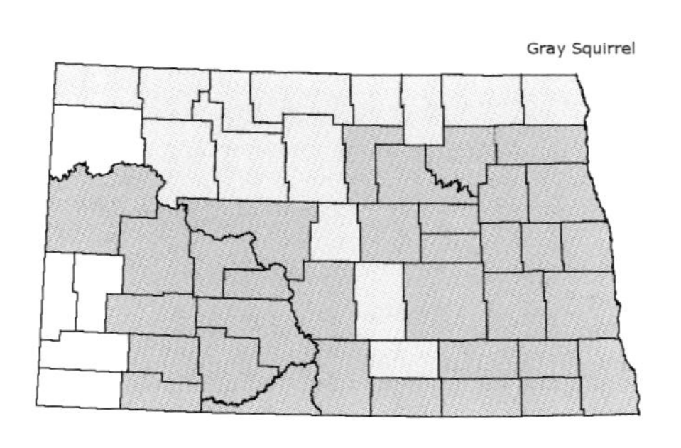

Gray squirrels occur throughout much of eastern North America from southeastern Saskatchewan and southern Manitoba, south to the Gulf of Mexico. There are scattered introduced populations in a number of western states, British Columbia, and Ontario.

In the early 20th century, gray squirrels apparently only occurred in wooded areas of southeastern North Dakota. Since that time, they have expanded their range naturally and by introduction throughout much of the timbered areas east and north of the Missouri River. This expansion was aided by introductions in a number of communities, including Valley City, Jamestown, Bismarck, and Minot.

Habitat

Preferred habitat of gray squirrels is mature hardwood forest with a diverse but sparse understory. Large woodlots (>40 ha) having a high proportion of trees with den cavities and producing winter-storable foods (e.g., oaks) sustain the most stable populations. In North Dakota, populations are essentially confined to forested areas having bur oak. This includes areas along rivers, timbered fringes around lakes, some shelterbelts, the Pembina Hills (Pembina, Cavalier counties), and Turtle Mountains (Rolette, Bottineau counties). They are very common in urban areas where adequate den sites and mast-producing trees are available.

Gray squirrels are often seen sharing habitat with fox squirrels and red squirrels. While they have very similar food habits to fox squirrels, gray squirrels seem to be more efficient foragers, tend to prefer larger tracts of woodland, and generally occupy broader ecological niches.

Ecology and Behavior

Occurrence of the gray squirrel in North Dakota is contingent on large, mature forest tracts containing trees with den cavities and winter-storable foods. Cavities are probably essential for survival during the severe winters of this region. Natural cavities take up to 30 years to form, have entrances 5 to 20 cm in diameter, are up to 50 cm deep, and are typically situated in the trunk. Nest boxes, such as those placed for wood ducks, are commonly used when available. Leaf nests are also used, especially during the warmer months. These consist of a platform of twigs on a tree limb, a base of decaying material, and an outer shell of twigs and leaves.

While gray squirrels are known to consume 97 plant and 14 animal items, only a small number can be regarded as available in the northern part of the species' range. Acorns show up as staples throughout the year, and in North Dakota, these are produced by just one species, the bur oak. Other regional plant foods taken, when available, include oak buds and flowers; elm buds, flowers, and seeds; maple buds, flowers, and seeds; fungi; corn; wheat; and soybeans. During summer, animal foods include insects, bones, and bird eggs and nestlings. Storable foods are often cached by "scatterhoarding." Rather than storing large numbers of seeds in one location, individual nuts are buried less than 2 cm below the soil surface. These are subsequently located by memory or olfaction.

Gray squirrels are active throughout the year. Like other tree squirrels, they do not hibernate, but may remain in their dens for days during unusually severe winter weather. Their activity patterns are diurnal. During the warmer months, activity peaks during the hours after sunrise and before sunset. During winter, activity is unimodal, peaking at mid-day, about four hours before sunset. Overall level of activity is higher during the warm months, compared to winter.

Home ranges vary from 0.5 to >5 ha. There can be considerable overlap between adjacent home ranges. Gray squirrels are not known to have territories, although core areas may be defended during the fall. Dispersal, especially among juvenile and adult males, may be over as much as 100 km. Mass emigrations have been recorded, usually occurring during the fall, probably in response to food shortages. These involved long distances and even crossing lakes and rivers. Such movements have been uncommon in the 20th century, but mass emigrations were recorded in 1969 and 1985.

Communication among gray squirrels includes visual, olfactory, auditory, and tactile interactions. Piloerection, posturing, and various tail movements all provide visual signals. Social grooming establishes bonds between individuals, especially siblings. Various growls, screams, "chucks," and tooth chattering are common. Scent marking at traditional sites is via gnawing followed by cheek/chin rubbing or urination, and probably signifies social status or reproductive condition.

Population densities tend to range from <3 to 16 per ha, but in urban settings can be higher. Fluctuations in density are correlated with the crop of tree seeds, especially winter-storable nuts. In North Dakota, the acorn crop produced by bur oaks, the only significant nut producer in the state, can be expected to be a significant limiting factor regulating density. Bur oaks produce abundant acorn crops every two to three years, and without other significant mast-producing trees, periodic gray squirrel starvation losses can be expected. Mortality during the first year of life averages 75%, and ranges from 42% to 57% annually for adults. Maximum longevity is 12 years for females, and nine years for males. Causes of mortality, in addition to starvation, include predation by owls, hawks, foxes, coyotes, bobcats, dogs, and cats. Hunting, accidents, and diseases such as tularemia, tetanus, and leptospirosis result in additional losses.

Reproduction

The beginning of the breeding season is indicated by mating chases during January. A number of males may follow a single female for several days prior to her coming into estrus. When in estrus, a female typically copulates with more than one male. The first litter of two to four young is born in March, following a 44- to 45-day gestation. A second littering peak occurs during July or August, and the first litter is driven out at that time.

Newborn young, like most rodents, are altricial, i.e., are naked, eyes and ears are closed, and have poor temperature control. Their eyes and ears open at about 1 month, and weaning occurs between the seventh and 10th week. Adult body mass is achieved at about 9 months. Early litters may breed during the following winter, but late litters probably delay until the next year's second breeding peak.

Status and Conservation

The gray squirrel is common in North Dakota wherever there are large forested areas with nut-producing trees (bur oak). It can also be abundant in urban settings with adequate food resources.

The gray squirrel is a popular game animal in many parts of its range. Some eastern and southern states harvest over a million squirrels annually. It is a game species in North Dakota, but of relatively minor importance. Its status in the state is dependent on the maintenance of large blocks of mature forest dominated by bur oak, North Dakota's only significant mast-producing species. These forests provide storable foods and den sites required for surviving our severe winters.

Selected References

Bailey (1926), Edwards et al. (2003), Hazard (1982), Hibbard (1956, 1972), Koprowski (1994a), Steele (1998)

Fox Squirrel
Sciurus niger

C. D. Grondahl, North Dakota Game and Fish Department

Description

The fox squirrel is very similar in appearance to the gray squirrel, except for its larger size and orange cast to its fur. Its back and sides are gray mixed with orange, and the tips of the ears, the feet, the underparts and top of the tail are all orange. Weighing between 800 and 900 g, the fox squirrel is about 20% heavier than the gray squirrel. Average standard body measurements are: Total Length - 550 mm, Tail Length - 246 mm, Hind Foot - 73 mm, Ear - 28 mm.

Distribution

Fox squirrels occur over much of the eastern United States, and have extended into southern Manitoba and southeastern Saskatchewan. They have expanded westward along river systems through much of eastern Montana, and occur throughout the Plains states.

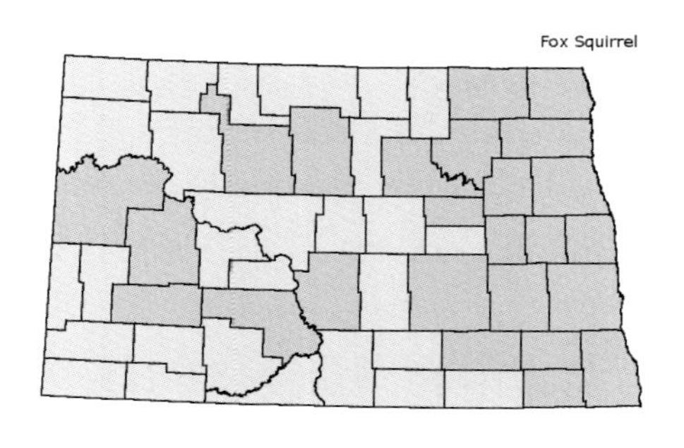

The fox squirrel is a recent arrival in North Dakota. Early biological surveys of North Dakota did not record it as present, or even its possible occurrence. It is believed to have arrived in southeastern North Dakota from Minnesota and South Dakota in the 1940s, and became well-established along wooded streams and in tree plantings and shelterbelts. It also spread up the Missouri River in the 1940s and'50s, extending into its tributaries, including the Yellowstone River. Its extension up the Yellowstone River probably accounts for its expansion into Montana. Human introductions have also been made along the Yellowstone River in McKenzie County, in the Killdeer Mountains of Dunn County, Jamestown (Stutsman County), and Washburn (McLean County). The species can now be regarded as occurring in suitable wooded habitats throughout the state.

Habitat

Like the gray squirrel, the fox squirrel is an animal of diverse mature forests which supply sufficient winter-storable foods and adequate den cavities. However, this species tends to prefer smaller stands of <40 ha and with an open understory. Originally a species of savannah-type woody patches in the Midwest, fox squirrels spread with white settlement and the breakup of continuous forests, and corn production. They spread into the prairie states with the development of tree plantings and shelterbelts, which simulated the patchy woodlands of their original range. Although aggressive interactions between fox squirrels and red squirrels *(Tamiasciurus hudsonicus)* and gray squirrels *(S. carolinen-*

sis) occur, habitat sharing, especially in urban settings, is common in North Dakota.

Ecology and Behavior

Except for its preference for smaller forest tracts, the life history and ecology of the fox squirrel is very similar to that of the gray squirrel. It also seems to prefer more open, drier conditions, and relies more heavily on corn as a food source. Hence, this species tends to be more associated with agricultural lands.

Fox squirrels depend heavily on dens in tree cavities, especially during the winter months. Development of suitable cavities may require eight to 30 years, hence the dependence of squirrels on mature forests. Sites where the tree has been injured, such as broken limbs, may encourage decay and the beginnings of cavities which are further opened by woodpeckers and squirrels. Leaf nests may also be used, especially in the summer months. These consist of a platform of twigs on a tree branch, with a shell of twigs and leaves, and a lining of shredded plant material.

While fox squirrels are known to consume over 100 plant foods, they depend on a limited number of species. They feed heavily on tree seeds throughout the year, but also take tree buds and flowers seasonally. Important natural foods in North Dakota include bur oak acorns, buds and flowers, and elm and maple buds, flowers and seeds. The tendency for bur oaks to produce heavy acorn crops every two to three years, followed by light production, can be a significant limiting factor for fox squirrel populations in North Dakota, a state which does not have other native mast-producing species. Fungi and agricultural crops, especially corn and wheat, are also taken when available, and may be especially important where mast-producing trees do not occur. Animal foods, including insects and bird eggs and nestlings, are taken in small amounts. Like gray squirrels, fox squirrels store foods by scatterhoarding. Individual nuts are buried under the leaf litter and subsequently located by memory or olfaction.

During spring, summer, and fall, activity peaks occur during the early morning and early evening hours, while during winter, there is a single mid-day peak. These seasonal shifts in activity patterns probably reflect avoidance of the heat of the day during summer and extreme low temperatures during winter. Overall, activity is suppressed by high winds, temperature extremes, and heavy precipitation. Mass migrations, such as recorded for gray squirrels, have not been observed.

Home ranges can vary from <1 ha to >40 ha, with male home ranges about double those of females. These are larger than those of the gray squirrel, and most likely reflect the more open nature of fox squirrel habitat. There is considerable home range overlap, and territories are not defended, except for certain core areas. Juveniles and subadults disperse from their natal areas from spring through fall. These movements may be over long distances, ranging from 5 to 64 km.

Fox squirrels utilize similar modes of communication as gray squirrels. These include body signals, scent marking, tactile communication, and vocalization. Body signals involve piloerection, posturing, and tail movements, and may indicate aggressive interactions. Social grooming reinforces individual relationships. Scent marking may indicate resident and social status. Vocalizations, including barks and tooth chatters, frequently involve mating, agonistic behavior, and alarm calls.

Population densities are highly variable, reflecting the quality of the habitat. The highest densities are in the midwestern states, averaging 1 to 3.5 squirrels per ha. North Dakota, because of its severe winters and less diverse food resources, probably harbors smaller populations than other midwestern states. Mortality rates are thought to be similar to those for gray squirrels. During the first year, mortality is believed to be about 75%, while adults experience 45 to 55% mortality annually. Maximum longevity is 12.6 years for females and 8.3 years for males. Causes of mortality are also essentially the same as for gray squirrels. Limited food resources leading to starvation are undoubtedly a major mortality factor in North Dakota. Avian and mammalian predators, hunting, accidents, and diseases account for additional losses.

Reproduction

The breeding season begins in January with mating chases involving multiple squirrels. During her one day of estrus, a female copulates with several males. A litter of two to four young is born in March following a 45-day gestation. A second littering peak may occur later in the summer.

Newborn young weigh about 15 g, are naked, and the eyes and ears are closed. At 3 weeks, they begin to acquire fur and the ears open. Their eyes open during week five. Weaning occurs between weeks eight and 12. Young squirrels are not fully grown until over 1 year of age. Some females may bear young at 8 months of age, but most do not reproduce until over 1 year.

Status and Conservation

The fox squirrel, a relatively recent arrival, is now fairly common in smaller woodlots and urban settings throughout North Dakota. Its expansion into North Dakota has undoubtedly resulted from development of tree planting, shelterbelts, and agricultural crops.

The fox squirrel is a popular game species over much of its range. Some states report a harvest of over 1 million taken annually. It is a game animal, categorized under "tree squirrels" in North Dakota, but is not as important as other upland game. Its continued success in North Dakota will be contingent on maintaining relatively small, mature woodlots containing a high proportion of bur oak, the state's only significant mast-producing tree.

Selected References

Bailey (1926), Edwards et al. (2003), Foresman (2001), Hazard (1982), Hibbard (1956, 1972), Knapp (1986), Koprowski (1994b), Nelson (1981), Seabloom et al. (1978)

Franklin's Ground Squirrel

Poliocitellus (Spermophilus) franklinii

C. D. Grondahl, North Dakota Game and Fish Department

Description

The Franklin's ground squirrel, with its gray coat and rather bushy tail, somewhat resembles a small gray squirrel. However, its back and sides are speckled with light and dark flecks, and it has shorter ears and a shorter, less bushy tail. The fur on its sides is paler, and the underparts are yellowish-white. Its body weight in North Dakota ranges from 326 to 477 g. Average standard body measurements are: Total Length - 374 mm, Tail Length - 127 mm, Hind Foot - 51 mm, Ear - 14 mm.

Distribution

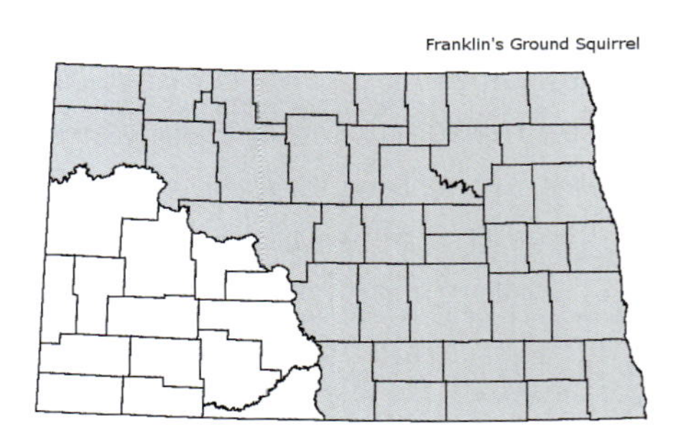

This species occurs from central Alberta, southern Saskatchewan, and southern Manitoba, south through much of the eastern Dakotas, Nebraska, and northern Kansas, and east through much of Minnesota, Iowa, northern Missouri, southern Wisconsin, northern Illinois, and northeast Indiana. In North Dakota, Franklin's ground squirrels occur in all of the counties east of the Missouri River.

Habitat

The Franklin's ground squirrel prefers dense grassy vegetation, frequently bordering forest edges. Unmowed roadside and railroad rights of way are also utilized. Areas which have been burned, mowed, or grazed tend to be avoided. In some areas, it has expanded its range as a result of forest clearing for agriculture. Although secretive, it has a relatively high tolerance for humans, and is commonly seen in parks and campgrounds.

Ecology and Behavior

Franklin's ground squirrels are the least social members of their genus. They are secretive, and live alone or in pairs in small colonies. Except in park situations, they are seldom seen because of their secretive habits and use of heavy cover. Unlike other ground squirrels, they do not stand upright to detect danger but rather retreat directly to their burrows. Even though they cannot be seen, their presence may be readily detected by their characteristic vocalizations

Burrows are constructed in well-drained soil in heavy cover, frequently at the edge of woody areas or marshes. Burrow entrances generally have conspicuous spoil piles of excavated soil. The burrow has one or two entrances, and slopes down 1 or 2 m to a set of tunnels and cavities for the nest and food storage.

Like other members of the genus, Franklin's ground squirrels are deep hibernators. In North Dakota, adults retreat to their burrows between the end of July and the first of September, males preceding females by up to two weeks. Young-of-the-year do not enter hibernation before September or October, requiring the extra time for development and fat deposition. Emergence from hibernation is from mid-April to early May, with males emerging one or two weeks prior to the females.

During the active months, Franklin's ground squirrels are diurnal. In North Dakota, surface activity may run between 7 a.m. and 9 p.m. They are more arboreal than their close relatives, frequently climbing low shrubs and small trees in their activities.

Upon spring emergence, about 75% of the diet consists of vegetation, including succulent roots and shoots of grasses and forbs. As the season progresses, leaves, stems, blossoms, seeds and fruits are added. Consumption of animal material makes up the balance and increases in importance by mid-summer. A wide variety of animal prey is taken, including insects, fish, amphibians, birds and their eggs, and small mammals. Franklin's ground squirrels have been demonstrated to be significant predators on waterfowl nests. In one study of predation on artificial waterfowl nests, 97% of blue-winged teal and mallard eggs were removed by Franklin's ground squirrels, and clutches were completely depredated within five days.

Home ranges are relatively large, averaging about 25 ha for males and 9 ha for females, with daily movements of 213 m and 153 m for males and females, respectively. Daily movements are more restricted for females during gestation and for both sexes prior to hibernation.

While relatively uncommon over most of their range, Franklin's ground squirrels may be locally abundant, with population densities of 1.2 to 2.5 adults per ha. Overwinter survival is 20 to 60%, and average total longevity is 0.74 year. However, total life expectancy may be as high as five years for females and two years for males. Causes of mortality include overwinter losses, predation by hawks and mammals (especially badgers), and human factors (e.g. road kill, control efforts).

Reproduction

The breeding season begins shortly after emergence from hibernation. Only full adults participate in breeding. One litter, averaging seven to nine young, is produced following a 28-day gestation. Newborn young are naked and blind; they are fully furred by 16 days and eyes are open at 18 to 20 days. Weaning occurs at about a month of age.

Status and Conservation

While locally common in heavy cover in much of eastern North Dakota, the Franklin's ground squirrel is rare to uncommon over most of its geographic range. Its area of occupancy has declined more than 30% over the past 10 years, and the decline is projected to continue for the next decade as well. It has listed status in six of the 14 states and provinces of its historic range, and has been categorized as vulnerable in the IUCN Red List of Threatened Species.

Franklin's ground squirrels can be significant nest predators in North Dakota, especially on wildlife refuges and waterfowl production areas, which also provide optimal Franklin's ground squirrel habitat. However, in view of the species' decline and vulnerable listing over most of its range, its status in North Dakota is in need of further study. Any efforts at control should be local in scope and approached with caution.

Selected References

Bailey (1926), Choromanski-Norris et al. (1986, 1989), Erlien and Tester (1984), Hazard (1982), Iverson and Turner (1972a), Jones et al. (1983), Ostroff and Finck (2003), Sargeant et al. (1987b), Turner et al. (1976)

Richardson's Ground Squirrel (Flickertail)

Urocitellus (Spermophilus) richardsonii

Honkóta - Arikara; Pinsa - Dakota;
Shopka-sop - Mandan; Tsipá sopa - Hidatsa

C. D. Grondahl, North Dakota Game and Fish Department

Description

The Richardson's ground squirrel (Flickertail) is the North Dakota state mammal. Its alternate common name comes from its tail-flicking behavior. It is a heavy-bodied ground squirrel, superficially resembling a small prairie dog. Its coat is cinnamon-buff; the back is shaded with brown hairs. The tail is brown above, mixed with blackish hairs, and lighter below. There is a light eye ring. The head is broader and the tail shorter than other area ground squirrels. Adult body weights in North Dakota average 233 g (females) to 371 g (males) upon emergence from hibernation, and 372 g (females) to 396 g (males) prior to immergence. Average standard body measurements are: Total Length - 285 mm, Tail Length - 74 mm, Hind Foot - 45 mm, Ear - 12 mm.

Distribution

The Richardson's ground squirrel is truly an animal of the northern Great Plains. Its northern range includes southern Alberta, southern Saskatchewan, and southwestern Manitoba. In Montana, it occurs north of the Missouri River west to the foothills of the Rocky Mountains. More recently, it has extended its range in Montana from the foothills eastward to areas between the Missouri and Yellowstone rivers. Its North Dakota and South Dakota range is north and east of the Missouri River, although there is one North Dakota observation from McKenzie County, west of the Missouri River. Therefore, the Missouri and Yellowstone rivers appear to have functioned as effective barriers to range expansion of the species. The eastern range limits are extreme western and southwestern Minnesota and northwestern Iowa.

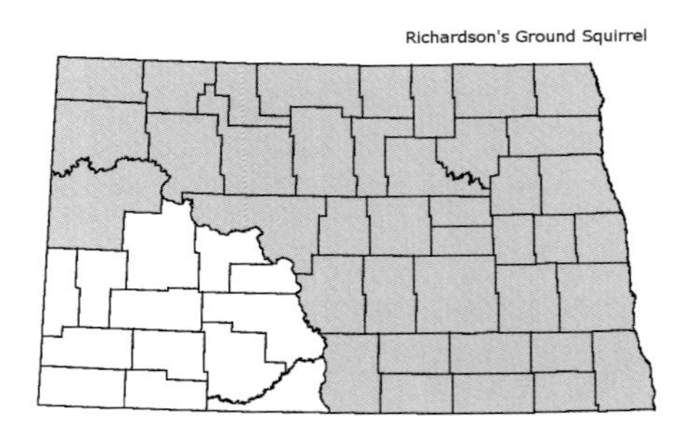

Habitat

This is a species of open grasslands and croplands. Richardson's ground squirrels seem to be especially common in moderate to heavily grazed pastures, which afford good visibility, possibly for predator defense. Heavy cover, including shrubs and woodland, tend to be avoided. Like other common ground squirrels, they readily adapt to urban situations, establishing colonies in parks, playgrounds, golf courses, etc.

Ecology and Behavior

It has been suggested that the Richardson's ground squirrel is an eastern ecological equivalent of the prairie dog, being morphologically similar, preferring short grassland habitats, and colonial. Such comparison is superficial, for it also has other distinctive attributes such as deep hibernation and specific behaviors. Although burrows are frequently close together, cohesive behavior is not high and only female kin tend to live in close proximity to each other. In spring, males emerge from peripheral burrows and move into the rest of the colony to establish breeding territories. Following the breeding season, males tend to return to the periphery of the colony or disperse.

Burrows are as deep as 2 m, and extend 4 to 15 m from the entrance. They include both grass-lined nesting chambers and food storage chambers. Burrows are well-sealed for the winter, and can withstand significant spring flooding.

In North Dakota, emergence from hibernation occurs in late March to early April. Most reports indicate emergence of males preceding females by one or two weeks, but simultaneous emergence of the two sexes has also been reported. Adult males are active until the end of May, when they again enter hibernation. Adult females are active longer, returning to hibernation between late June and late July. Juveniles enter hibernation in September or early October. The hibernating state is not continuous. Squirrels enter the hibernating season by allowing their metabolic rates and body temperatures to drop in a stepwise fashion followed by periods of arousal. Hence, body temperatures are highest and arousal periods more frequent and torpor periods relatively short (two to eight days) early in the season. As the season progresses, body temperatures may drop to a few degrees above freezing, with periods of torpor lasting as long as 19 days. This process is reversed during the weeks prior to spring emergence.

Like other ground squirrels, Richardson's ground squirrels are diurnal. At the beginning and end of the active season, they are active throughout the day, while during the heat of mid-summer, they have two activity peaks, early morning and late afternoon.

Richardson's ground squirrels are vegetarians, 90% of the diet being plant material. Flowers, leaves and seeds of native plants and agricultural crops are taken. Crops subjected to frequent cultivation or irrigation receive less damage, as these practices tend to destroy or flood out burrow systems. In addition to plants, about 10% of the diet may consist of insects or other animal material.

Average home range size is small, estimated as 0.32 ha, but varies by season, sex, and reproductive state. For example, female home ranges are larger following emergence of the litter than before, while those of males decrease following the breeding season. Both sexes defend very small territories. Male territories average about 0.05 ha, and after the breeding season only the immediate proximity of the burrow is defended. Female territories are half the size of male territories, and are maintained throughout the summer. Dispersal from home ranges primarily involves juvenile or yearling males. Average dispersal distance is unknown, but individual distances of up to 10 km have been recorded.

Adult population densities usually average one to five per ha, but this increases dramatically with the emergence of juveniles when densities may reach over 50 per ha. Annual mortality is high: juvenile females - 75%; juvenile males - 88%; adult females - 50%; adult males - 70%. Maximum longevity has been estimated at three to four years.

A wide variety of avian and mammalian predators take Richardson's ground squirrels, especially long-tailed weasels, badgers, and buteo hawks, much of this predation occurring during dispersal. They are also known to carry the tick and flea vectors of Rocky Mountain spotted fever, tularemia, plague, and equine encephalomyelitis. Overwintering losses during hibernation undoubtedly account for significant mortality as well.

Reproduction

Females normally mate within a few days of emergence from hibernation. A single litter is produced annually, following a 22-day gestation. Average litter size is eight, ranging from four to 13. Young are altricial at birth, i.e., naked, eyes closed, and limbs not fully developed. They are weaned at about 25 days, fully furred at 28 days, and emerge from the burrow at about 1 month. Juveniles weigh about 110 g upon emergence from the burrow, early June in North Dakota, and have achieved adult size by early September. Both sexes are sexually active upon emergence from hibernation the following spring.

Status and Conservation

Prior to white settlement, Richardson's ground squirrels occurred in vast numbers in their historic range. Early reports cited as many as 4,000 taken from a single section of land. With settlement and the advent of agriculture in the Plains, agricultural practices, poisoning, and shooting greatly curtailed populations. This population decline has resulted in the North Dakota Game and Fish Department designating it as a Level II Species of Conservation Priority. However, the species remains common in pastures, field edges, rights of way, and other grassland habitats.

The Richardson's ground squirrel has a reputation as a significant agricultural pest, especially around hayland and grain crops, and efforts at control may be warranted in these situations. Bounties have proven ineffective, but judicious use of poisons may provide some control. Any control efforts would be most effective if targeted in early spring to the overwintering survivors prior to breeding. However, conservation of the species is also important because of its limited geographic range, fascinating life history, and contributions to the richness of the regional fauna. It is a species of great importance to the prairie ecosystem, providing an important source of prey for many predators, and its burrows provide refuge for other species.

Selected References

Bailey (1926), Dyke et al. (2004), Eisenberg (1981), Foresman (2001), Hazard (1982), Jones et al. (1983), Michener and Koeppel (1985), Pigage (1975), Quanstrom (1966, 1968), Swenson (1981)

Thirteen-lined Ground Squirrel

Ictidomys (Spermophilus) tridecemlineatus

Tashnáheca–Dakota, Lakota; Tshíkarani–Arikara; Naksátshi–Hidatsa; Mashedónikcha–Mandan; Miniwakao–Cheyenne

M. D. Carleton, American Society of Mammalogists Mammal Images Library

Description

This small- to medium-sized ground squirrel is easily identified by its 13 longitudinal stripes running from the nape of the neck to the base of the tail. There are seven pale yellow stripes alternating with six dark brown stripes. Each dark stripe has a series of pale yellow or white spots running down the middle. The ears and legs are very short, the body elongate, and the tail about a third of the total length. Post-emergence weights of adult males range from 140 to 200 g, and increase to 160 to 240 g prior to immergence into hibernation. Post-partum weights of adult females average about 160 g, and increase to about 250 g prior to immergence. Average standard body measurements are: Total Length - 282 mm, Tail Length - 93 mm, Hind Foot - 37 mm, Ear - 9 mm.

There are two subspecies of thirteen-lined ground squirrels in North Dakota. The darker, slightly larger *I. t. tridecemlineatus* occurs east and north of the Missouri River, while the smaller, paler *I. t. pallidus* occurs in the southwestern quarter of the state.

Distribution

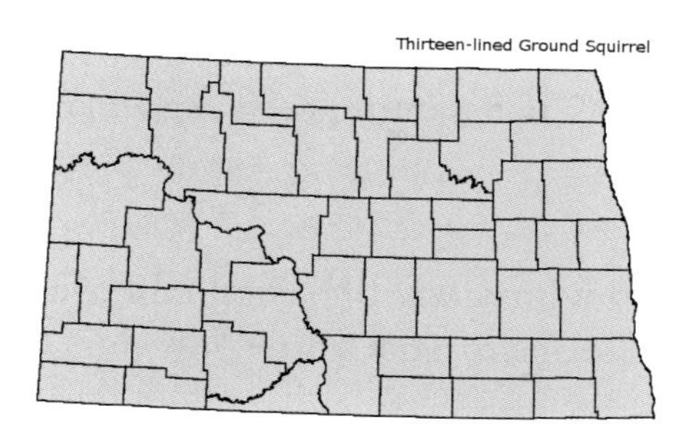

The thirteen-lined ground squirrel occurs from the southern Prairie Provinces of Canada, south through the Great Plains states generally east of the Rocky Mountains, to the gulf coast of Texas. It is found throughout most of the Great Lakes states as far east as Ohio. It is found in grassy habitats throughout North Dakota.

Habitat

This is a grassland species, having habitat preferences similar to those of the Richardson's ground squirrel. It is common from the semi-arid badlands of the Southwest to the more mesic tall-grass prairies of the East. It is especially common where grasses are shorter, such as pastures, mowed roadsides, parks, and golf courses. Numbers are also greater on high ground with well-drained soils, lessening the likelihood of flooded burrows. The clearing of forests in the Midwest may have enabled the species to expand its range eastward since white settlement.

Ecology and Behavior

Thirteen-lined ground squirrels live in loose colonies, but their behavior is generally solitary, except during the breeding season. They are regarded as non-territorial, although specific sites may be defended. They are strictly diurnal during the active season. Following spring emergence, they are most active during the warmer hours of the day, from about noon to early afternoon. As the warm season progresses, the activity cycle lengthens until late summer, when it becomes bimodal, with peaks in mid-morning and late afternoon.

Two types of burrows are constructed. Shallow burrows, less than 0.5 m in length, are used for temporary refuge. Permanent burrows may be over 1 m in depth and over 5 m long, have multiple entrances, and nest and sanitation chambers. At the surface, runways or trails may be sometimes seen, which lead to foraging areas or burrow entrances.

Like other area ground squirrels, emergence from hibernation is in late March or early April, with males preceding females. During late summer, these squirrels put on a heavy layer of fat in preparation for hibernation. Adult males typically enter hibernation during July, followed by adult females in early August, and juveniles by early September. The hibernating state is similar to that described for Richardson's and Franklin's ground squirrels. As the squirrel enters hibernation, it allows its body temperature to drop by stages, followed by fairly short periods of arousal. Arousal periods become less frequent as the cold season progresses, and again become more frequent with the approach of spring.

Thirteen-lined ground squirrels are omnivorous, and 40 to 50% of their dietary volume may consist of animal matter. Seeds, clover leaves, and insect larvae (especially moths and butterflies) frequently make up the bulk of the diet. Other insects, especially grasshoppers, are also readily taken.

Average home range size for males is about 5 ha, and for females less than 2 ha. Male home ranges tend to increase during the breeding season, and then become more restricted prior to hibernation. Females expand their home ranges during pregnancy and lactation, but then restrict their movements upon emergence of their litters. These squirrels are not territorial, defending only the immediate vicinity of their burrows. However, encounters between individuals often result in agonistic behavior.

Population density may run from two to five per ha prior to emergence of the juveniles, when it may increase to about 25 per ha. Mortality is high, with annual survival averaging about 19%. Males have higher mortality than females, probably related to a tendency for juvenile males to disperse into marginal habitat. The bulk of the mortality in a colony occurs among juveniles prior to hibernation. Ground squirrels are vulnerable to predation by a variety of diurnal predators, including snakes, raptors, and predatory mammals. Overwintering mortality during hibernation could be high during severe winters when the frost line may penetrate below the hibernaculum.

Reproduction

Males are in breeding condition upon emergence from hibernation, preceding the females. This species is an "induced ovulator," requiring the stimulus of mating before the female will ovulate. Mating takes place between mid-April and the first week of May. In the northern part of the species range, there is only one litter of eight to nine young per year. Newborn young are altricial, naked and with their eyes closed. They are furred by 3 weeks of age and weaned by 4 weeks, when they emerge from the burrow. Although juveniles gain weight steadily and continue above ground activity into September, they apparently do not attain adult size by the time they enter hibernation.

Status and Conservation

Thirteen-lined ground squirrels are common in grassy habitats throughout North Dakota. Although sometimes blamed for crop damage and burrowing activity in parks and golf courses, this ground squirrel also consumes large numbers of insect pests and is an important prey species in the prairie ecosystem. It has continued to thrive and, in some cases, expanded its range following white settlement.

Selected References

Beer (1962), Criddle (1939), Hazard (1982), Hohn and Marshall (1966), Jones et al. (1983), Seabloom et al. (1978), Streubel and Fitzgerald (1978)

Least Chipmunk

Neotamias (Tamias) minimus

Sachho–Arikara; Hetkada^{n}–Mandan; Kokokshi–Hidatsa

C. D. Grondahl, North Dakota Game and Fish Department

Description

The least chipmunk is the smaller of the two chipmunks in North Dakota. Weighing an average of 45 g, it is about half the size of the eastern chipmunk. Dorsally, there are five dark and four light stripes running from the nape to the base of the tail. The face has two dark and two light stripes from the nose to the ear. There are two distinct subspecies in North Dakota, the darker *N. m. borealis* in the northeast corner, and the pale *N. m. pallidus* characteristic of the southwest. Average standard body measurements are: Total Length - 211 mm, Tail Length - 99 mm, Hind Foot - 32 mm, Ear - 16 mm. From a distance, running least chipmunks can be distinguished from eastern chipmunks by their vertically held tails; eastern chipmunks usually extend the tail straight back. Like other chipmunks, the least chipmunk has internal cheek pouches.

Distribution

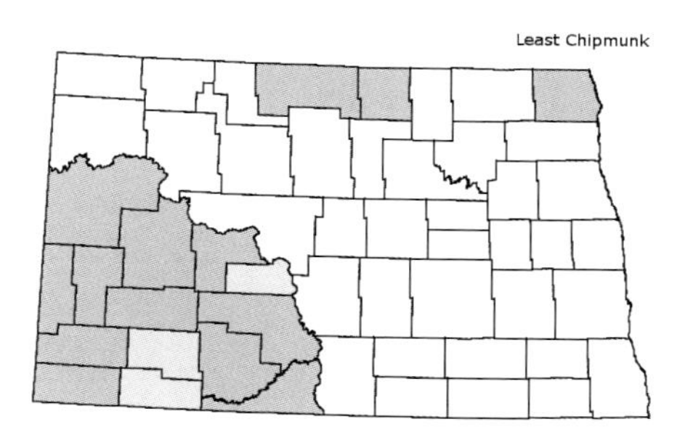

This is a widely distributed chipmunk. In Canada, it occurs primarily in boreal forested areas from Yukon eastward to James Bay and southeastern Ontario. In the United States, it occurs in the Upper Peninsula of Michigan, northern Wisconsin, northern Minnesota, northeastern and southwestern North Dakota, eastern Montana, western South Dakota, southern Idaho and eastern Washington south to eastern California, Nevada, and northern Arizona and New Mexico. Its North Dakota distribution is disjunct, with *N. m. borealis* occurring in the Turtle Mountains (Bottineau, Rolette counties) and the Pembina Hills (Pembina County), while *N. m. pallidus* occurs in most counties south and west of the Missouri River.

Habitat

A wide variety of habitats have been reported for the least chipmunk, but it tends to prefer openings and rocky areas in mixed woodland and conifers. In northeastern North Dakota, the boreal subspecies occurs in aspen-dominated forests. In the southwestern part of the state, these chipmunks are common in upland breaks, rockland, hardwood draws, and forested areas. In Theodore Roosevelt National Park, they are associated with habitat types dominated by sagebrush, juniper, and green ash.

Ecology and Behavior

The least chipmunk, though primarily a ground-dwelling squirrel, is a good climber and is frequently seen perched in shrubs and small trees. When disturbed, however, it rapidly retreats to its burrow under brush piles, stumps, or fallen trees. In the Badlands of southwestern North Dakota, they also burrow under old buildings, and in rocky outcrops and sides of buttes.

While occasionally using tree cavities or leaf nests, least chipmunks usually construct burrows. Generally, burrows are up to 1 m in length, terminating in a nest chamber lined with fine materials, and a food cache. Least chipmunks do not leave spoil piles around burrow entrances, as seen with many other burrowing rodents.

Least chipmunks are diurnal, and during the warm months of the year are most active around mid-day. When temperatures are very high, however, they frequently retreat to their burrows. While solitary, they seem to be less aggressive and territorial than eastern chipmunks. During the colder months (October to April), they hibernate, arousing periodically to subsist on cached seeds. Foods include a wide variety of seeds, grains, nuts, fruits, and other plant materials. Insects, especially grasshoppers, caterpillars, and beetles, and their eggs are also frequently consumed.

Estimates of home ranges range from 0.1 to 5.5 ha. This variability may be a function of habitat structure, food supply, and sex. Home ranges of males may be twice those of females. Furthermore, home ranges of individuals have been observed to change seasonally and from year to year.

As with home ranges, population densities may vary with habitat. Late summer population densities have been estimated as high as 15 to 37 per ha. While potential longevity may run up to six years, average longevity in the wild is about 0.7 year. Least chipmunks are preyed upon by a variety of hawks and mammalian predators. Tularemia has been found in some populations, and they are known carriers of the tick hosts of Colorado tick fever and Lyme disease.

Reproduction

The breeding season occurs from April to mid-May, with females coming into estrus within a week of emergence from hibernation. Most litters are born by early June following a gestation period of 28 to 30 days. Normally only one litter is produced per year, unless the female loses the litter shortly after birth. The average litter size is about six (three to seven). Newborn young are altricial, with eyes and ears open at 27 days and weaning at 36 days. Adult size is attained by September, but not all yearlings reach sexual maturity early enough to breed in their first year.

Status and Conservation

Least chipmunks are common in northeastern and southwestern North Dakota, and are popular residents of areas frequented by the public, including camp grounds and picnic areas. They provide an important prey for diurnal predators, and probably consume significant numbers of pest insects. Occasionally, they may damage stored foods or garden crops. The public should be advised of their potential as carriers of certain human pathogens.

Selected References

Bailey (1926), Erlien and Tester (1984), Genoways and Jones (1972), Hazard (1982), Hopkins (1983), Jones et al. (1983), Seabloom et al. (1978), Svihovec (1967), Verts and Carraway (2001)

Eastern Chipmunk

Tamias striatus

Hetkala–Lakota

Paul Smiths, American Society of Mammalogists Mammal Images Library

Description

The eastern chipmunk is the larger of the two chipmunks in North Dakota. Averaging over 100 g, it is more than twice the mass of the least chipmunk. Like the least chipmunk, there are five dark and four light stripes on the back and sides, but they do not extend to the base of the tail, and the rump area is reddish brown in color. The facial stripes are also less distinct than those of the least chipmunk. Average standard body measurements are: Total Length - 260 mm, Tail Length - 101 mm, Hind Foot - 36 mm, Ear - 18 mm.

Distribution

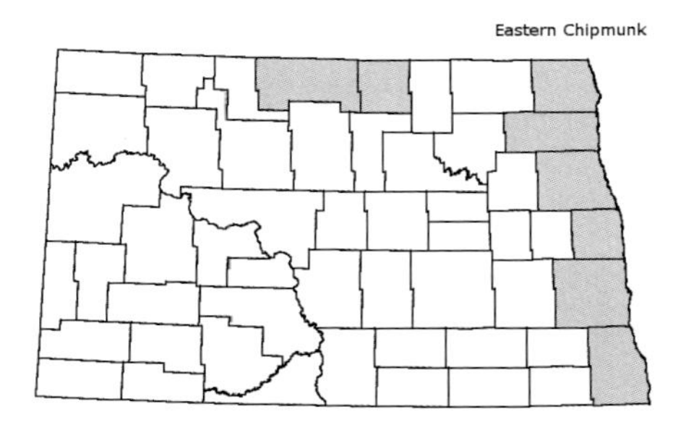

The eastern chipmunk occurs from southern Manitoba across southeastern Canada on the north, and south over much of the eastern United States. Its western limits in the United States are the extreme eastern Dakotas, Kansas, Nebraska, and Oklahoma. Its North Dakota distribution is restricted to timbered areas along the Red River and its tributaries, the Pembina Hills (Pembina County), and Turtle Mountains (Bottineau, Rolette counties).

Habitat

This is primarily a species of the deciduous forest, especially areas with significant brush, logs, and stumps for protective cover. Significant numbers of mast (acorn) producing trees appear to be especially important, as well as close proximity to continuous forest.

Ecology and Behavior

Eastern chipmunks require significant woody or brushy cover, but are conspicuous, tolerant of human activity, and relatively easy to observe. Unlike the least chipmunk, it rarely climbs trees. Eastern chipmunks are aggressive and solitary, except for adult females with newly emerged young. They produce a variety of vocalizations ("chips" and "cuks"), serving as alarm signals and denoting territory.

Like the least chipmunk, the eastern chipmunk is a true hibernator, going into torpor in October, and emerging in March. Unlike other ground squirrels, however, it is a relatively "light" hibernator, and may emerge from its burrow and engage in foraging activity on warm winter days.

Activity is strictly diurnal, with mid-morning and mid afternoon peaks being common. However, timing of peak activity may shift to decrease conflict with red squirrels competing for the same resources. Highest levels of overall activity occur on bright, warm, quiet days.

Burrows are often constructed under rocks, tree roots, or buildings. They tend to be longer and more complex than those of least chipmunks, extending up to 10 m and having a complex system of storage chambers and smaller galleries. Eastern chipmunks may store prodigious amounts of food in the burrow chambers; up to a bushel has been recorded. One of the larger chambers contains a leaf-lined nest.

Foods stored over the winter consist of seeds, nuts, and acorns. These foods are important during the growing season as well, at which time the diet also expands to include fungi, fruits, and animal material (insects, frogs, snakes, birds, and small mammals).

Home ranges are quite variable, ranging from under 100 m^2 to over a hectare. Chipmunks may also make long excursions out of the home range, usually to exploit some food resource. Eastern chipmunks also have significant homing ability. Animals displaced up to 500 m or more commonly return to the home range within a day or two of release.

Population density varies with season and location, and may vary from 0.3 to 37.6 per ha. Population levels may fluctuate by as much as five-fold on an annual basis. In some habitats, populations may fluctuate on a cyclic basis, correlated with the periodicity of mast (acorn) production. Average longevity is slightly over one year, but individuals have been known to survive as long as eight years. Eastern chipmunks are prey for hawks, snakes, and a number of carnivorous mammals. Insecticide application to control forest insect pests can significantly reduce eastern chipmunk populations. The eastern chipmunk is a known viral host for the LaCrosse strain of California encephalitis, and can also carry the tick host for Lyme disease.

Reproduction

The breeding season begins upon emergence from hibernation in March. A second breeding peak may occur in late June or July. Females may breed either once or twice during any given season. Four to five young are born following a gestation of 31 days. Newborn young are altricial, i.e., hairless, eyes closed, and with poor temperature control. Weaning takes place at about 1 month of age, when their eyes are open, they are well-furred, and weigh about 30 g. Emergence from the burrow is at about 6 weeks. Adult size is achieved at 3 months, but sexual maturity is not reached until after the first winter. Some yearling females will breed following their first winter, but most males will not.

Status and Conservation

Eastern chipmunks are fairly common in the mature forests of the Red River valley, Pembina Hills, and Turtle Mountains. They are absent from the rest of the state.

As with the least chipmunk, the eastern chipmunk is a popular subject for wildlife viewing. It consumes harmful insects, and is an important prey species. On the negative side, it can damage garden crops and is a carrier of some human diseases.

Considering its dependence on mature forest and its limited range in North Dakota, further monitoring of its status and the extent of its principal habitat may be warranted. Furthermore, the sensitivity of chipmunk populations to widespread forest insect control efforts should be a significant consideration by management agencies.

Selected References

Bailey (1926), Erlien and Tester (1984), Forbes (1966), Hazard (1982), Iverson et al. (1967), Jones et al. (1983), Snyder (1982)

Red Squirrel

Tamiasciurus hudsonicus

Ahjiduhmo–Ojibway; zica–Lakota

C. D. Grondahl, North Dakota Game and Fish Department

Description

Weighing between 200 and 300 g, the red squirrel is less than half the size of the gray squirrel. The back, sides, and tail have a reddish coloration, and the underparts are white. There is a white ring around the eye, and a dark band separating the back and sides from the white belly. In winter, there are tufts of fur on the ears. Average standard body measurements are: Total Length - 314 mm, Tail Length - 122 mm, Hind Foot - 46 mm, Ear - 23 mm. Two similar subspecies of red squirrels occur in North Dakota, *T. h. pallescens* centered in the Turtle Mountains, and *T. h. minnesota* in the Red River valley.

Distribution

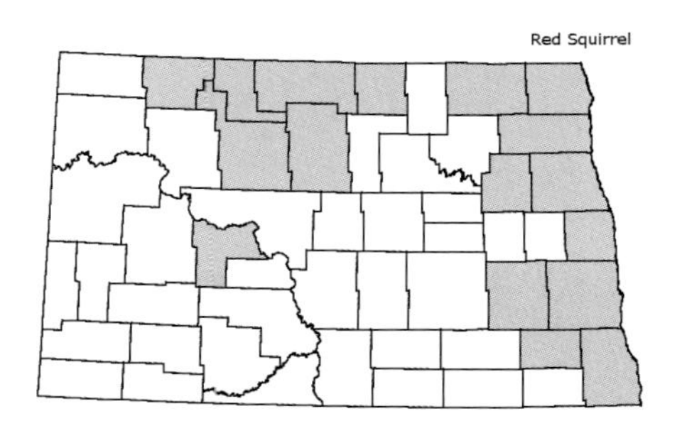

The red squirrel occurs in the boreal forests of North America, ranging in the north from tree line in Alaska to eastern Canada. Its western limits include the Cascades, and the Rocky Mountains as far south as Texas, New Mexico, and Arizona. Red squirrels are found in eastern North Dakota, the Great Lakes states, Iowa, New England, and the Appalachian Mountains as far south as Georgia. They are also found in portions of southeastern Montana, northeastern Wyoming, and the Black Hills of South Dakota.

In North Dakota, red squirrels occur in the riparian forests of the Red River and its tributaries, the Pembina Hills (Pembina and Cavalier counties), the Turtle Mountains (Rolette and Bottineau counties), and the Mouse (Souris) River and aspen parkland of the northwestern portion of the state. Its occurrence in southwestern North Dakota is uncertain. In spite of several mammal surveys of the southwest, there has been only one recorded observation, from along the Knife River (Mercer County). Reports of red squirrels from ponderosa pine habitats in adjacent northwestern South Dakota and southeastern Montana suggest the possibility of the species extending into the pine stands of the North Dakota Badlands.

Habitat

Over most of its range, the red squirrel is regarded as a species of the boreal forest, preferring mature stands of conifers (pine, cedar, spruce, fir), which provide seeds and fungi for foraging. The interlocking branches of such stands also provide nest sites and escape cover. In marginal portions of their range, which lack significant conifers, such as in North Dakota, red squirrels will utilize mature deciduous forests, including riparian woodland, shelterbelts, urban areas, and the aspen-dominated forests of the Pembina Hills and Turtle Mountains. In North Dakota, red squirrels commonly share habitat with gray and fox squirrels.

Ecology and Behavior

Red squirrels prefer tree cavities for their nests, but will also construct leaf nests or utilize abandoned burrows of other species, tree roots, or rock shelters. Factors in nest site selection include proximity to food resources, thermoregulation, and protection from predators. Most nests are located in close proximity to food caches. Nests are lined with any insulating material, including grasses, moss, shredded bark, leaves, feathers, and fur.

Over much of its range, conifer seeds provide the bulk of the red squirrel's diet. During late summer and fall, whole cones are cached in one or more middens near the nest site (larderhoarding) for later use during winter. A single midden may hold up to 3,000 cones! Except in urban situations and plantings such as shelterbelts, conifers are uncommon in North Dakota; thus red squirrels rely on other food sources. These include nuts and seeds of deciduous trees and shrubs, including bur oak, maple, basswood, and elm. Single food items may be stored separately as well (scatterhoarding), and may comprise up to 43% of stored foods. Squirrels also feed heavily on fungi, when available, as well as tree buds and flowers, sap, and bark. Animal foods include insects, bird eggs and nestlings, and young small mammals. In agricultural areas, crops such as apples, corn, and sunflower seeds may be taken.

Like other tree squirrels (except flying squirrels), red squirrels are active during the daylight hours. During the warm months, there are two activity peaks, during early morning and late afternoon, thus avoiding the heat of the day. Winter activity is typically limited to midday. During severe weather, squirrels may remain inactive in their nests for one or two days at a time. They may also avoid extreme low temperatures by tunneling under snow cover during their foraging. Severe weather of all types (temperature extremes, wind, heavy rain, snow) tends to suppress overall activity.

Red squirrels are highly territorial, defending the entire home range against competitors, primarily conspecifics, throughout the year. Territorial behavior, centered around their food middens, is especially pronounced during fall, when foods are being stockpiled. Territory size varies with food supply and other habitat parameters, but is usually less than 1 ha. Territorial behavior involves a variety of vocalizations (rattle calls, screech calls, growl calls), chases, and fights. They may defend against other species, such as gray and fox squirrels, and chases are sometimes observed.

In addition to territorial communication, red squirrels emit alarm calls indicating the presence of potential predators. By varying loudness and band width, they are able to communicate whether it is avian or terrestrial predatory danger.

Population densities have been estimated from <one to as high as seven per ha, depending on the habitat and possibly cyclic fluctuations. Mortality is about 60% during the first year of life. Average overall longevity is only 0.8 year, but some individuals have been known to survive in the wild for seven to 10 years. Food appears to be the primary limiting factor on red squirrel populations. In addition to starvation losses, red squirrels are taken by a variety of avian and mammalian predators, including red foxes, coyotes, bobcats, hawks, and owls.

Reproduction

The reproductive pattern of the red squirrel is similar to that seen in gray and fox squirrels. Females come into estrus during February and March, and pregnancies peak in March and April. The mean gestation period is 33 days. A second pregnancy peak may occur in late summer, but in the northern part of the range, only one litter per year is typical.

Average litter sizes vary from 3.2 to 5.4. Newborn young, typical of most rodents, are naked and blind. There is a covering of dorsal hair by 3 weeks, eyes are open at about 1 month, and young are mostly furred by 2 months of age, at which time they are weaned. Sexual maturity may be attained by yearlings but, in most cases, reproduction is delayed until females are 2 years of age. Age at first reproduction appears to be a function of temperature and food supply which function as constraints on reproductive development.

Young red squirrels tend to disperse from their mothers' home ranges a few weeks after weaning, and establish new feeding territories nearby. In some cases, however, the breeding female may disperse, leaving her territory to one of her offspring, probably enhancing juvenile survival.

Status and Conservation

The red squirrel is a common inhabitant of mature forests and wooded urban settings in eastern North Dakota. At present, it cannot be considered to be resident south and west of the Missouri River, but there is a possibility of future expansion via the Little Missouri River and its tributaries out of South Dakota and Montana.

The red squirrel, along with other tree squirrels, is categorized as a game species in North Dakota. However, very few are taken other than incidental to other hunting activities. Occasionally, red squirrels damage tree plantings through bark stripping, and get into attics and cabins. However, this damage is generally minimal in North Dakota. Red squirrels are dependent on mature forests, and contribute an important boreal element to the faunal diversity of the state.

Selected References

Bailey (1926), Erlien and Tester (1984), Hazard (1982), Hibbard (1972), Higgins et al. (2000), Jones et al. (1983), Lampe et al. (1974), Pruitt (1960), Seabloom et al. (1980), Steele (1998), Yahner (2003)

Family Castoridae
Beavers

North American Beaver
Castor canadensis

Ah-mik´ – Ojibway; Ah-misk´ – Cree; Capa – Lakota

Portland Zoo, American Society of Mammalogists Mammal Images Library

Description

The North American beaver is the only member of its family (Castoridae) on the continent, and closely resembles the Eurasian beaver *(Castor fiber)* of the Old World. With an adult weight of 20 to 27 kg, it is the largest rodent in North America. It is a stout animal, well-adapted for life in the water. It has thick, brown fur overlain with long coarse guard hairs. The hind feet are webbed, and the tail is scaly and flattened—both adaptations for swimming. The small eyes are protected by a nictitating membrane, and the ears and nostrils are valvular, allowing them to be closed when under water. The lips close behind the nostrils, allowing the animal to gnaw under water. Paired anal scent glands ("castors") are present in both sexes and function in marking territory. Average standard body measurements are: Total Length - 1,098 mm, Tail Length - 328 mm, Hind Foot - 182 mm, Ear - 35 mm. Two subspecies of beaver are believed to occur in North Dakota, the larger, darker *C. c. canadensis* in the northeastern half of the state and the smaller, lighter *C. c. missouriensis* in the southwest. Boundaries of subspecies are uncertain because of former large-scale restocking efforts throughout North America.

Distribution

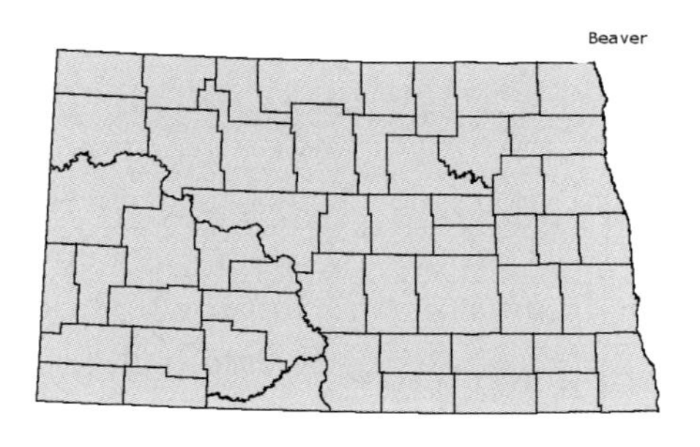

The beaver occurs naturally throughout most of North America from tree line in the far north, south to northern Mexico, wherever there are water bodies and surrounding woody vegetation. It is absent from peninsular Florida and the Southwestern deserts.

Beaver were originally abundant throughout North Dakota, wherever there were suitable lakes, ponds, and streams and associated woody vegetation. Beaver were so heavily harvested in the 19th century that they became practically extinct in the state by 1900. For example, during the first years of the 20th century, there was only one known colony in the Turtle Mountains, and beaver were practically unknown in the Pembina Hills. Subsequent protection and restocking efforts have resulted in beaver again being common in permanent water areas throughout the state.

Habitat

Suitable habitat for beaver includes a stable aquatic habitat, a channel gradient of less than 15%, and a sufficient supply of quality foods. Beaver are capable of creating suitable water conditions on small streams through dam construction, providing channel gradients are less than 15%. Larger streams and rivers may already have suitable water depths to preclude the necessity for dams. Larger lakes of at least 8 ha with irregular shorelines also provide optimum water conditions. Semipermanent and permanent wetlands also provide habitat for at least seasonal occupancy. Beaver tend to avoid adjacent grazed areas, proximity to rowcrop agriculture, and adjacent livestock feed lots.

Aquatic plants and adjacent woody vegetation provide a food supply for beaver. Woody vegetation, preferably with a high percentage of willow and aspen, should be available within 100 m, the average foraging distance from the water body.

Because of their activities in tree cutting and dam construction, beaver have been regarded as "ecosystem engineers" or "habitat-modifying keystone species." Dam construction dramatically alters riparian habitats by creation of impoundments, which affect stream flow, water chemistry, and water temperature. This further creates new habitat for fishes, water birds, mammals, and other aquatic organisms. Their foraging activity also alters species composition, growth form, density, and distribution of the surrounding woody vegetation.

Ecology and Behavior

The fundamental social unit of a beaver population is the colony, consisting of a single, usually monogamous, adult pair, the young of the year (kits), and the yearling young of the previous year. As yearlings approach sexual maturity, they are driven from the colony and disperse. Hence, a typical colony consists of 4 to 10 individuals.

Beaver dwell in constructed lodges or bank dens, depending on water depth and bank conditions. Lodges are typically conical in shape, and constructed of cut branches and mud on a firm underwater substrate. There are usually several underwater entrances, and a single nest chamber several inches above the water level. Some lodges may be up to 5m in diameter at the base, and over 2 m high.

Bank dens are excavated when beaver live in rivers or lakes having steep banks. These are usually constructed under overhanging trees, shrubs, or ledges. They have an underwater entrance leading to a nest chamber above the water level. Small holes from the ceiling to the surface provide ventilation. In North Dakota, with few natural lakes, bank dens in streams and rivers tend to predominate.

Dam construction frequently stabilizes water conditions for the beaver colony. Like lodges, dams are constructed of cut logs and interwoven branches sealed with mud. The dam is anchored by forcing the butt ends into the substrate, and anchoring them further with stones and mud. Dam size varies with topography, from just a few feet in length to nearly a kilometer. They may hold back water less than 1 m in depth, or a sizable fairly deep pond. A dam on the University of North Dakota's Forest River biology area was over 2 m high, and when it broke, the wall of water uprooted sizable trees.

Beaver are strict herbivores, feeding on a wide variety of herbaceous and woody plants, yet they show definite preferences and are referred to as "choosy generalists." During the warmer months, the diet may be dominated by herbaceous plants, such as rhizomes of water lilies, grasses, sedges, rushes, and cattails, along with leaves and twigs of willows. At other times of the year, a variety of woody plants are eaten, based on preference and availability. In the Dakotas, green ash and willow are preferred, along with box elder and cottonwood. Preferred winter foods in North Dakota are red-osier dogwood, green ash, and willow, followed by box elder, American elm, and bur oak. Aspen is also heavily utilized, when available.

During spring, summer, and fall, activity is crepuscular and nocturnal, beginning in late afternoon and continuing until around sunrise. Under heavy snow and ice and in the far north, the activity cycle tends to "free run" and beaver may be active at any time. Much of the active time is spent in dam and lodge maintenance, and foraging. Beaver are "central place foragers," i.e. they cut terrestrial woody vegetation and bring it back to a central location such as a dam, lodge, or food cache. The food cache is established prior to freeze-up, consists of a submerged pile of green branches, and provides a winter food source for the colony. The upper portion of the cache is exposed, and winter cache counts provide a useful tool for wildlife managers counting colonies in areas where beaver are living in bank dens rather than lodges.

Home ranges are variable in size, a function of shoreline configuration, season, social organization of the colony, and food supply. A small pond may have only one colony with a somewhat circular home range, while in larger lakes and rivers, the home range may be more linear, following the shoreline. Home ranges are smaller during winter when ice conditions restrict movements. In Minnesota, tagged adult beaver moved less than 2.5 k, while in North Dakota, most transplanted adults moved less than 1 k during the year subsequent to their release. Young beaver may disperse approximately 16 k to establish new colonies, with the record movement being 110 k. Beaver tend to forage within 100 m of water, but distances of up to 200 m have been reported. Greater distances may be observed during times of food shortage. Beaver-gnawed bur oaks have been observed on the Forest River biology area (Grand Forks County), at least 500 m from the river.

Beaver are territorial, and establish large numbers of scent mounds within the home range. These are piles of mud and debris deposited with secretions of their anal scent glands (castoreum). From the scent mounds, beaver are able to distinguish family members, holders of neighboring territories, and non-neighbors. Distance between colonies averages about 1 k, but may vary contingent on habitat suitability.

In addition to scent marking, beaver communicate by postures, vocalization, and tail slapping. Grunts, hisses, and tooth chatters communicate aggression, and kits engage in whining when begging for food. The well-known tail slap serves to warn other members of the colony of potential danger, and may serve to frighten predators as well.

Recorded colony densities have varied from none to 4.6/k^2, and up to 1.9 per kilometer of stream. Annual mortality has been estimated at about 30% for all age classes. Beaver frequently live up to 10 years, with a known record of 21 years. Principal causes of mortality include trapping, predation, disease, and catastrophic weather events. Heavy trapping pressure during the 19th century nearly resulted in the species' extinction throughout North America. For example, fur trader Alexander Henry recorded 7,233 beaver pelts taken from the Red River valley through his trading post at Pembina between 1801 and 1808. While beaver are relatively well-protected from predation when in their aquatic habitat, on land they may be taken by wolves, coyotes, dogs, and bobcats. Occasionally, mink will enter the lodge and prey on the kits. Waterborne tularemia occasionally reaches epizootic status, resulting in mass mortality, especially when colonies are stressed by other factors such as food shortages. Winter and early spring mortality may result from starvation due to insufficient food stores, or destruction of the colony due to sudden snowmelt and violent ice breakup.

Reproduction

The breeding season in beaver begins in January, but peaks in mid-February. Young are born in May and June, following a gestation period of about 100 days. Average litter size, based on embryo counts, in North Dakota was 4.2 (two to six), and in Minnesota 5.3 (two to nine).

Newborn beaver are precocial, i.e., they are born fully furred, eyes are open, incisors erupted, and weigh about 500 g. They are weaned at 6 weeks to 2 months, but begin taking solid food earlier. They achieve sexual maturity at 1.5 to 3 years, at which time they are driven from the colony.

Status and Conservation

Harvest regulations, reintroductions, and setting back plant succession through fires and logging have led to recovery of beaver populations. Such conservation measures have resulted in the highest continenental harvests ever recorded (over 1 million pelts in the 1980s), this in spite of erratic prices and reduced demand for furs. North Dakota populations have recovered to the point where it is now a common species in suitable habitat throughout the state. During the 20th century, annual beaver harvests in North Dakota have ranged from a low of 556 in 1937 to a high of 7,000 in 1954. In 2001-2002, fur buyers purchased a total of 2,000 pelts in the state. Current recommended harvest rates which still sustain populations are 20 to30% of the population, or one to 1.5 beaver per colony per year. These must be carefully monitored, however, because beaver are very subject to overharvesting.

Because of its role as a keystone species, the beaver has had an enormous impact in improving riparian ecosystems and associated species. It has been suggested that benefits provided by beaver in creation of wetlands, flood control, improved water quality, sedimentation control and recreation outweigh its direct value as a furbearer and offset the cost of any damage.

Beaver can inflict significant economic damage. Damage can involve flooding of roads and croplands by plugging culverts, cutting and flooding of timber, consumption of crops, cutting of ornamental plants, and burrowing activity in roadbanks, dikes, and dams. During the 1980s, annual damage ran from $75 million to 100 million nationwide. Beaver are carriers of the gastrointestinal parasite *Giardia lamblia*, a concern for recreational users of lakes and streams.

Control of damage by beaver can involve killing or transplanting of problem animals, use of repellants or fencing to protect plants or crops, installation of beaver-proof culverts or flow control devices in beaver ponds, or dynamiting dams. Trapping, poisoning, or shooting of problem animals has been the most common approach to control. It can be effective, but may also attract new occupants to the vacated habitat and may be unacceptable to the public, especially in urban areas. Live trapping and transplanting of problem beaver may have more public acceptability but can be expensive and have the same problems as lethal approaches.

Use of animal repellants, fencing, and placement of wire sleeves on ornamental plants can be effective in protecting ornamental trees and shrubs. Construction of beaver-proof culverts and flow-control devices can reduce damage to roads and maintain ponds at acceptable levels.

Selected References

Adams (1961), Bailey (1926), Baker and Hill (2003), Beer (1955), Dieter (1987), Dieter and McCabe (1989), Hammond (1943), Hazard (1982), Hibbard (1958), Jenkins and Busher (1979), Jones et al. (1983), Longley and Moyle (1963), Swenson and Knapp (1980), Van Ballenberghe et al. (1975), Warren (1927), Wheatley (1997a, b, c), Wobeser (1985)

Family Geomyidae
Pocket Gophers

Plains Pocket Gopher
Geomys bursarius

Bob Gress

Description

The plains pocket gopher is similar to the northern pocket gopher, but is significantly larger (weight averaging about 300 g), and has more robust forelimbs and longer claws. Average standard body measurements are: Total Length - 260 mm, Tail Length - 66 mm, Hind Foot - 33 mm, Ear - 7 mm. The pelage is brownish dorsally, with lighter underparts. Like the northern pocket gopher, the eyes and ears are small, and the short tail is nearly naked. Fur-lined cheek pouches extend nearly to the shoulders. Similar to other pocket gophers, the incisors protrude in front of the closed lips, assisting the animal in digging and gnawing without ingesting soil.

Distribution

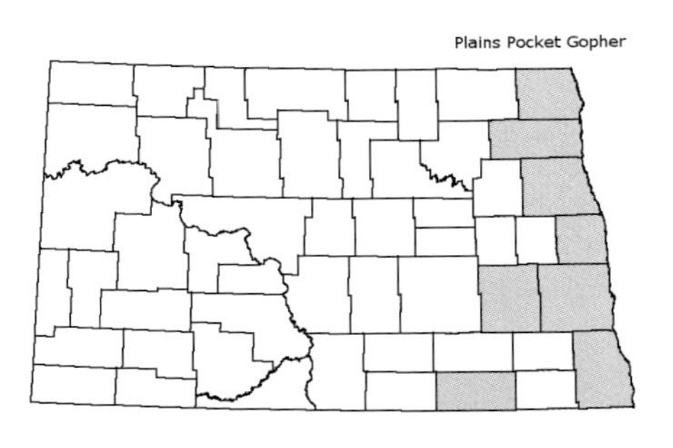

The plains pocket gopher occurs in central North America from southeastern Manitoba to Texas and from northwestern Indiana to eastern Wyoming, Colorado, and New Mexico. In North Dakota, it has been reported as far west as along the James River near Ellendale and along the Sheyenne River near Valley City, but it is generally restricted to the Red River valley, where it abuts the eastern limits of the northern pocket gopher.

Habitat

This species is more of a habitat specialist than the northern pocket gopher. While generally preferring loose, sandy soils, the plains pocket gopher has also invaded the heavier soil types of the Red River valley, where it seems to have prevented occupation by the more generalized northern pocket gopher. While less tolerant of flooding, and generally avoiding wooded habitats, the species' size and relatively aggressive behavior, coupled with its ability to occupy well-drained sites such as road banks, has resulted in its sole establishment over much of the Red River valley. This occupation may have been a relatively recent event, which replaced the more mobile, cold-, and flood-tolerant northern pocket gopher following the last glacial retreat.

Ecology and Behavior

Like other pocket gophers, the plains pocket gopher is a solitary animal, except during the breeding season and when nursing young are present. Territories (the burrow) are vigorously defended at all times.

The plains pocket gopher is a powerful burrower, and its size, limb structure, and heavy front claws may have contributed to its ability to negotiate the heavy soils of the Red River valley. Burrows are similar to those of the northern pocket gopher, but may be shallower and more extensive. Foraging tunnels may be less than 25 cm below the surface, while the nest, food storage, and sanitation sites are frequently less than 1 m in depth. A single burrow system (likely occupied by a solitary individual) was measured to a total length of 157 m, and involved 105 separate surface mounds. Unlike the northern pocket gopher, it rarely emerges from the burrow except when mound-building or when forced out by a catastrophic event such as flooding. Some foraging may occur immediately around the burrow entrance. Plains pocket gopher activity can have significant effects on ecosystems, impacting vegetative composition, soil aeration, and altering the physical characteristics of soil.

As with other pocket gophers, the plains pocket gopher is a strict vegetarian. Roots and underground stems of forbs make up the bulk of the diet, but leaves of grasses and other plants immediately adjacent to burrow entrances may be taken as well. Although succulent plants are preferred, this species seems better able to subsist on grasses than the northern pocket gopher. During winter, cached foods are heavily relied on, as this species does not engage in under-snow foraging as does the northern pocket gopher.

Little information is available on population density for this species. Mound counts have been suggested as an indicator of density, but they are highly variable and, as indicated above, may greatly exaggerate the number of gophers in a given area. Estimates of upper limits of density have run from16.8 to 37.1 gophers per ha. Natural enemies of the species are the same as for the northern pocket gopher, namely snakes, small to medium predatory mammals, and owls. Badgers are probably the main predators on this species.

Reproduction

In North Dakota, the breeding season is thought to begin in early spring, and run through June. The University of North Dakota collection has specimens reported as pregnant as late at July 27. Placental scar counts from collection specimens, an indicator of litter size, averaged 4.6 (two to seven). There is only one litter per year. The eyes are open at about 5 weeks, and young are weaned a few weeks later. Dispersal begins in mid-summer, after the young are more than half grown. Both males and females become sexually mature during their first year. Average longevity is probably two to three years.

Status and Conservation

The plains pocket gopher is common and abundant over much of the Red River valley, and probably along the southern reaches of the Sheyenne and James rivers in the state.

As with other pocket gophers, this species is considered an agricultural pest, resulting in crop losses and damage to agricultural machinery. Its burrowing activity may also damage underground cables, road beds, and dikes. A variety of approaches have been tried to control the species, including trapping, various poisons inserted into the burrows, and repellants.

Pocket gophers play an important role in ecosystem maintenance. Their activities affect soil aeration, texture, and nutrients, as well as the composition of plant communities. They are also a significant prey base for a variety of predators.

Selected References

Bailey (1926), Baker et al. (2003), Hazard (1982), Jones et al. (1983), Kalin (1964), Wrigley and Dubois (1973)

Northern Pocket Gopher

Thomomys talpoides

Machtóhpka–Mandan; Mánica–Dakota;
Cipans–Arikara; Kípapudè–Hidatsa

C. D. Grondahl, North Dakota Game and Fish Department

Description

Pocket gophers are powerfully built rodents, well-adapted to fossorial life. Weighing an average of 160 g, the northern pocket gopher is the smaller of the two pocket gophers in North Dakota. The head and shoulders are heavily muscled, without an apparent neck. Other external features include narrow hips, short legs, and small eyes and pinnae. The front legs, though short, are robust and equipped with three very long claws. Like other pocket gophers, the incisors lie outside the mouth cavity. The pelage is brownish to brownish gray dorsally, relatively long and silky in winter, and short and harsh during summer. There are external fur-lined cheek pouches on either side of the mouth. The tail is naked. Average standard body measurements are: Total Length - 235 mm, Tail Length - 64 mm, Hind Foot - 30 mm, Ear - 8 mm.

Distribution

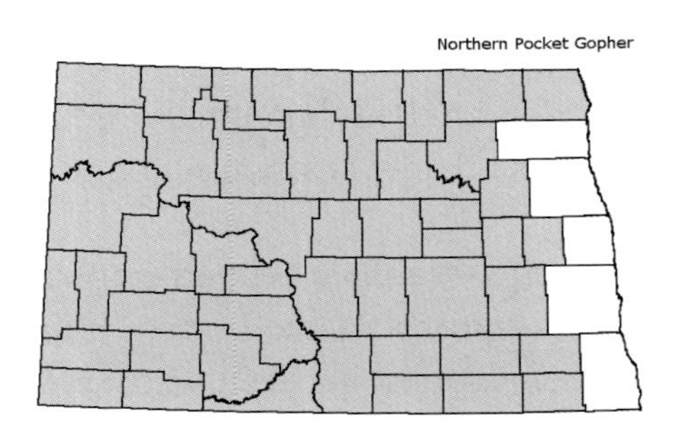

The northern pocket gopher is one of the most widely distributed members of the family. Its northern limits include southeastern British Columbia and the southern Prairie Provinces of Canada. Its western limits are the eastern sides of the Cascade and Sierra ranges of Washington, Oregon, and California, while southern limits include northern Nevada, eastern Utah, most of Colorado, and isolated populations in New Mexico and Arizona. Its eastern limits are eastern North and South Dakota and extreme northwestern Minnesota. In North Dakota, its eastern limits are essentially the western edge of the Red River valley, except in the far north, where it has crossed into extreme northwestern Minnesota and southeastern Manitoba.

Habitat

This pocket gopher is regarded to be a generalist, occupying a greater variation in habitat than any other member of the family. It can occur nearly anywhere lacking significant canopy cover, including alpine meadows, grasslands, sagebrush steppe, and cultivated lands. It seems to prefer soft, easily worked soils, but can also occur in other soils from heavy clays to relatively thin, rocky soil types. In southwestern North Dakota, it has been collected in upland and rolling grasslands, agricultural lands, and sandy flood plains, while in the eastern part of the state it can also occur in forested and brushy areas and lands subject to flooding. Its current habitat occupancy and distribution in North Dakota, Minnesota, and southeastern Manitoba, compared to that of the plains pocket gopher, may be functions of its wider habitat and climatic tolerance, competitive exclusion by the plains pocket gopher in its preferred habitat, and its relatively high mobility to reoccupy the Great Plains following the last ice age.

Ecology and Behavior

Pocket gophers are solitary animals, except during the breeding season and when caring for young. Territories are vigorously defended against all intruders. Hence, during most of the year, any given burrow can be expected to have only one occupant.

Burrows may average about 50 m in length, and occur at two levels. The upper level, about 30 cm below the surface, is used for foraging, while the lower level, which may be as deep as 2 m, contains the nest, food storage, and sanitation sites. In constructing the burrow, excavated soil is deposited in mounds at the surface. Normally burrow entrances are plugged, except when soil is carried to the surface. Burrow excavation and mound building is lowest during summer, and peaks during the fall. Daily peaks of activity are primarily at dawn and dusk. Burrowing activity by pocket gophers can have profound effects on ecosystems. Soil displacement may total over 11 metric tons per hectare, affecting nutrient redistribution, soil texture and aeration, and vegetative diversity.

The northern pocket gopher is more mobile and may engage in more surface activity than the plains pocket gopher. During winter, gophers may emerge from their burrows and tunnel under the snow while foraging and depositing excavated soil. Serpentine casts of their surface activity may be seen following snow melt.

Northern pocket gophers feed solely on vegetation. A wide variety of roots, underground stems, bulbs, and tubers of forbs, grasses, and cacti are taken. Above-ground leaves may be taken as well, but only adjacent to a burrow entrance. Agricultural crops are readily taken, with roots and tubers pulled directly into the burrow.

Recorded population densities have approached 50 per ha. Survival through the summer months is about 63%, and drops to about 17% over the winter. Northern pocket gophers are not thought to live more than two years. Important predators of pocket gophers include bull snakes, rattle snakes, most small to medium predatory mammals (especially badgers), and owls.

Reproduction

The breeding season is estimated to run from March to June. In North Dakota, females captured between June and August carried placental scars, indicating prior pregnancy. The gestation period is estimated to be 18 to 19 days, and litter sizes range from three to seven. Three females captured in North Dakota had five to seven placental scars. There is no indication of more than one litter per year.

Newborn young are naked, blind, and toothless. While they begin to eat solid food by day 17, their eyes are not open until day 26. They are actively foraging at a little over a month of age. They remain with the mother for six to eight weeks, then disperse, and become sexually mature at 1 year.

Status and Conservation

Northern pocket gophers are common and abundant throughout most of North Dakota. They are absent from the Red River valley, except near the Canadian border.

Pocket gophers are regarded by many as agricultural pests. They are also known to cut through buried cables and damage agricultural machinery, road beds and dikes through their burrowing and mound-building activities. Various control measures have been utilized, including trapping, gassing, insertion of poisoned pellets in artificial intersecting burrows, and use of repellants such as capsaicin. Control issues may be important when refuges, national grasslands, or other protected sites abut agricultural lands.

The role of pocket gophers in ecosystem maintenance must also be recognized. As stated above, they play an important role in affecting soil aeration and texture, nutrient redistribution, and maintenance of certain plant communities. They also provide a significant prey base for a variety of predators, and their burrow systems provide overwinter habitat for other species such as salamanders.

Selected References

Bailey (1926), Baker et al. (2003), Genoways and Jones (1972), Hazard (1982), Jones et al. (1983), Kalin (1964), Seabloom et al. (1978), Verts and Carraway (1999), Wrigley and Dubois (1973)

Family Heteromyidae

Pocket Mice and Kangaroo Rats

Hispid Pocket Mouse

Chaetodipus hispidus

Bob Gress

Description

The hispid pocket mouse is distinguished from its closest North Dakota relatives, the olive-backed and plains pocket mice, by its larger size. Its total and hind foot lengths average >180 and 22 mm, respectively; the smaller species have total and hind foot lengths of <170 mm and 22 mm, respectively. Its pelage is "harsh," orangish mixed with black hairs on the back, with white underparts. The tail has a black stripe on the dorsal surface. Like other members of the family, it has well-developed fur-lined cheek pouches. The average standard body measurements are: Total Length - 220 mm, Tail Length - 107 mm, Hind Foot - 26 mm, Ear - 13 mm. Average weight is about 48 g.

Distribution

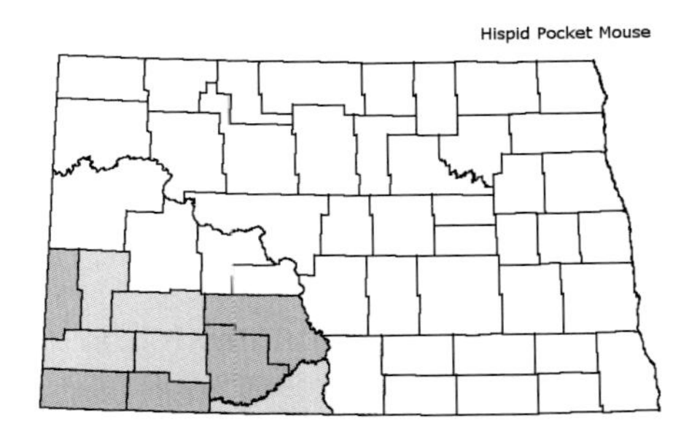

The hispid pocket mouse occurs from North Dakota south through the Great Plains into Texas and Mexico. In states neighboring North Dakota, it occurs in the western two-thirds of South Dakota, and there has been one record from extreme southeastern Montana. It was first reported in North Dakota in 1913 at a single site along the Cannonball River in Grant County. Later surveys reaffirmed its occurrence in its original reported locale, and in Bowman, Golden Valley, Morton, and Adams counties.

Habitat

Hispid pocket mice occur in a wide variety of dry grasslands and croplands. They are apparently not as closely associated with sandy soils as other pocket mice, but also occur in gravely areas with heavy soils. The few records from North Dakota and Montana have been in mixed-grass prairie, sparsely vegetated short-grass prairie associated with a prairie dog town, hayfields, and grassland associated with ponderosa pine.

Ecology and Behavior

These pocket mice are aggressive, solitary animals. They burrow in loose soil, constructing branched tunnel systems with several entrances. Sometimes a conspicuous mound is left at the burrow entrance, similar to but smaller than that left by a pocket gopher.

Hispid pocket mice are active throughout the year, and do not hibernate or store fat. During winter, they are believed to subsist on stored seeds. Like other pocket mice, they may become torpid during periods of food shortage. Seeds make up the bulk of the diet, and are cached in large quantities. Sunflower, cactus, sage, and bluestem seeds are especially important. A number of insects are also taken.

Little is known of the population dynamics of this species, especially in the northern Great Plains. Maximum longevity of other members of the genus has been reported to be four years. Hispid pocket mice are prey for a variety of predatory mammals, owls, and snakes.

Reproduction

Reproductive data on this species are very limited. Females may produce two or more litters annually during spring and summer. Litter sizes have been reported to range from two to nine, averaging 5.5. Gestation time for other members of the genus has been estimated at 25 to 26 days, and weaning at 23 days post-partum.

Status and Conservation

Hispid pocket mice are rare to uncommon in southwestern North Dakota, and are listed by the State Game and Fish Department as a Level III Species of Conservation Priority.

This pocket mouse is at the northern limit of its distribution in North Dakota. Hence, conditions in this region must be considered to be marginal for the species. Recent records of hispid pocket mice on two wildlife management areas in southwestern North Dakota demonstrate the importance of such protected tracts in maintaining species diversity.

Selected References

Bailey (1926), Dyke et al. (2004), Foresman (2001), Higgins et al. (2000), Jones et al. (1983), Jones (1993), Paulson (1988), Seabloom (2002), Seabloom et al. (1980), Shaughnessy (2016)

Ord's Kangaroo Rat

Dipodomys ordii

C. D. Grondahl, North Dakota Game and Fish Department

Description

Kangaroo rats are characteristic of the arid regions of North America, especially desert areas. The large head and eyes, and elongated hind feet and tail (hence "kangaroo rat") make it easily distinguishable from other North Dakota rodents. Dorsal pelage coloration is cinnamon-buff, while the underparts, including feet, are white. There are also white patches above the eyes and below the ears. Body weights in the northern portion of the species range average 66 grams for males and 64 grams for females. Standard body measurements (18 North Dakota specimens) averaged: Total Length - 258 mm, Tail Length - 140 mm, Hind Foot - 38 mm, Ear - 12 mm.

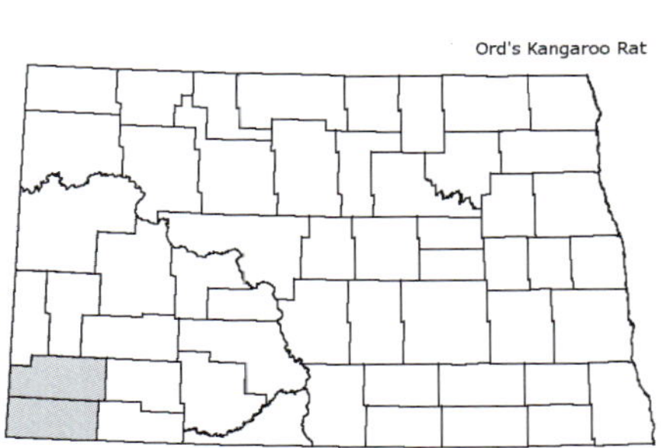

Distribution

The range of the Ord's kangaroo rat extends from southern Alberta and Saskatchewan south to southern Hidalgo, Mexico, and from central Oregon and eastern California to eastern Nebraska, Kansas, and Oklahoma. The North Dakota distribution includes limited areas of southern Slope and Bowman counties near the Little Missouri River, where it may be locally abundant.

Habitat

Ord's kangaroo rats prefer loose, sandy soils suitable for excavation of burrows and for dusting. Vegetation on such sites is usually sparse. Their burrows can be found along cutbanks and under shrubs and grassy clumps. North Dakota specimens have been taken on seeded croplands, roadsides, and upland grasslands.

Ecology and Behavior

Kangaroo rats are truly animals of the desert. They are nocturnal, spending the hours of daylight in shallow burrows which have had their entrances plugged. This behavioral adaptation allows avoidance of high surface temperatures, water conservation, and predator defense. Kangaroo rats do not hibernate, and are active throughout the year. They can live indefinitely on a diet of dry seeds, depending on metabolic water and water recovery by their extremely efficient kidneys. When available, they will take free water by scooping with the forefeet, and licking their paws.

Ord's kangaroo rat is primarily a granivore, feeding on seeds of grasses and forbs. Green and other live plant parts and animal material such as insects may be occasionally taken, especially in the spring. During foraging bouts, seeds are temporarily stored in well-developed cheek pouches, and then transferred to large food caches in the burrow system.

Population densities have varied from one to over 50 per hectare, and recorded home ranges from 0.5 to 1.3 ha. Because of their habit of plugging burrows, their presence is not always apparent, unless other sign, e.g., tracks and dusting sites, are carefully examined as well.

Ord's kangaroo rats are relatively long-lived, with a maximum longevity of 35 months in the wild. They are prey for a wide variety of predators, including foxes, coyotes, owls, and snakes. When pursued, they use their highly developed hind limbs and tails to make spectacular leaps, sometimes turning in mid air, to escape.

Reproduction

A relatively low reproductive rate typical of heteromyid rodents offsets their high longevity. Kangaroo rats in the northern parts of the species range do not breed until the second year. Onset of reproduction is generally associated with rainfall and the appearance of green vegetation. Hence, the breeding season in North Dakota commences in early spring. One or two litters are typically produced each year. North Dakota specimens with embryos have been taken as late as mid-August, indicating a second littering period in the state. Two to four young are born following a gestation period of about one month. Their eyes open at 2 weeks, and young achieve adult size at about 5 to 6 weeks.

Status and Conservation

Ord's kangaroo rat occupies an extremely limited distribution in southwestern North Dakota. Early reports discussed the species, but as of the early 1900s, no specimens had been taken. Surveys made during the 1960s and 1970s found kangaroo rats to be locally abundant.

The Ord's kangaroo rat does not appear to be a species of economic importance in North Dakota. However, it is a fascinating component of the state's faunal diversity. Although the species appears to be stable, significant land use changes within its range could affect its existence in North Dakota. Because of its extremely limited distribution in North Dakota, this is a species of special interest, deserving future monitoring.

Selected References

Bailey (1926), Brown and Harney (1993), Foresman (2001), Garrison and Best (1990), Genoways and Jones (1972), Higgins et al. (2000), Jones (1993), Seabloom et al. (1978), Svihovec (1967)

Olive-backed Pocket Mouse

Perognathus fasciatus

Apapsá–Hidatsa; Zhizhina–Dakota

Bob Gress

Description

This species is one of two "silky haired" pocket mice in North Dakota, both of which are significantly smaller than the hispid pocket mouse of the southwestern corner. Its dorsal pelage is olivaceous, mixed with black hairs. The sides have a buffy stripe, and the underparts are nearly white. There is usually a buffy spot behind the ear. As with other members of the family, it has fur-lined cheek pouches. Some North Dakota specimens have been reported to be unusually blackish. Average standard body measurements are: Total Length – 135 mm, Tail Length - 62 mm, Hind Foot - 17 mm, Ear - 8 mm. Average weight is only about 12 g.

Distribution

The olive-backed pocket mouse occurs from the southern Prairie Provinces of Canada south through eastern Montana and Wyoming, much of North and South Dakota, northwestern Nebraska, and east-

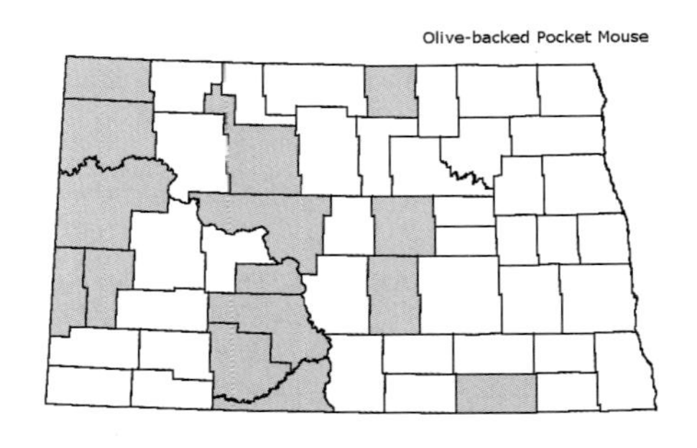

central Colorado. North Dakota records include much of the western half of the state, conforming to its South Dakota distribution.

Habitat

This species is most common in arid to semiarid grasslands with loose, sandy soils. In southeastern Montana, they occur in both grazed and ungrazed short-grass prairie, sage, and ponderosa pine-western wheatgrass. Many North Dakota records have been from haylands and moderately grazed rolling grasslands.

Ecology and Behavior

Olive-backed pocket mice are active burrowers. Tunnel systems may be as deep as 2 m, and cover an area of 6 m^2. Food may be cached off of the main tunnel, and there are separate summer and winter chambers. These animals are primarily seed-eaters; common North Dakota foods include honeysuckle, needlegrass, smartweed, butterfly weed, knotweed, and croton. They are not known to hibernate, but, like other pocket mice, are capable of entering torpor in response to food shortages.

Estimated population densities run from 0.6 to 4/ha in North Dakota and Montana. Like other pocket mice, they are solitary. Home ranges are poorly known, but individual movements of up to 66 m have been recorded.

These pocket mice occur in sparse populations, and are probably not a staple prey item for predators. Yet, they are undoubtedly taken by a variety of carnivores, especially owls. Longevity data are lacking for the species, but maximum recorded life span for other members of the genus have run from 2.5 to five years.

Reproduction

The breeding season in North Dakota is from May to July. Pregnant females have been reported from mid-June until the first of August, and there is evidence of multiple litters in a given breeding season. The period of gestation is 28 days, and average litter size is 5.4 (range two to nine). There are no known developmental data for this species, but reports of other members of the genus indicate opening of eyes at 14 to 15 days and weaning at 23 days.

Status and Conservation

The olive-backed pocket mouse occurs in sparse populations, but is fairly common throughout much of western North Dakota. These pocket mice are not considered to be economically significant. They inhabit a variety of arid grasslands, and probably do not require a specific management strategy.

Selected References

Bailey (1926), Foresman (2001), Genoways and Jones (1972), Jones (1993), Jones et al. (1983), Lampe et al. 1974), MacCracken et al. (1983), Manning and Jones (1988), Pefaur and Hoffmann (1974), Seabloom et al. (1978), Svihovec (1967), Turner and Bowles (1967), Williams et al. (1993)

Plains Pocket Mouse

Perognathus flavescens

Description

This "silky-haired" pocket mouse, like the olive-backed pocket mouse, has a grayish-buff back washed with black hairs. It has an indistinct buffy patch behind the ear, and a buffy lateral line, which is less distinct than in the olive-backed pocket mouse. The underparts are white. This is a small pocket mouse, weighing about 10 g. Average standard body measurements are: Total Length - 122 mm, Tail Length - 60 mm, Hind Foot - 17 mm, Ear - 6 mm.

Distribution

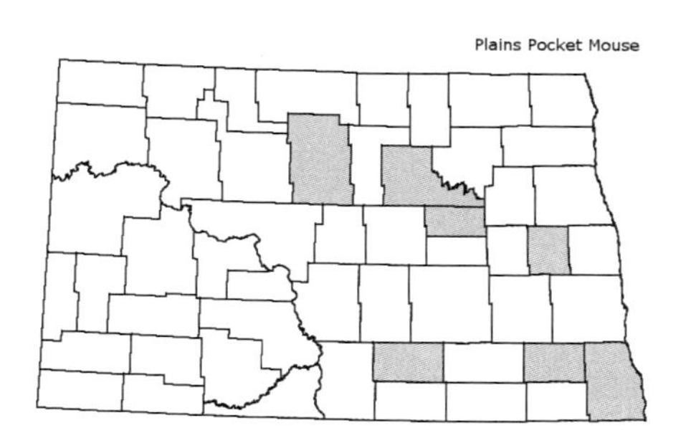

The plains pocket mouse reaches its northern distributional limit in North Dakota, extending south through eastern South Dakota and southern Minnesota through the Plains states to central Texas. Its western limits include Northern Chihuahua, Mexico, New Mexico, Arizona, and Utah. In neighboring states, it occurs in the southwestern half of Minnesota and the southeastern three-fourths of South Dakota. North Dakota reports seem to have run in a rough band from J. Clark Salyer National Wildlife Refuge (McHenry County) in the north-central part of the state to the Sheyenne National Grasslands (Ransom and Richland counties) in the southeast. There are other reports between these extremes from Benson, Eddy, Steele, and Logan counties.

Habitat

This pocket mouse seems mainly associated with sandy or sandy-loam soils in grasslands or grasslands mixed with sage or yucca. It tends to prefer looser soils than the olive-backed pocket mouse. It also occurs on disturbed sites and cultivated land, including corn and grain fields.

Ecology and Behavior

Like their close relatives, plains pocket mice construct burrows with multiple entrances. These may be up to 240 cm in length and reach of depth of 50 cm. Burrows may be plugged during the day, and have small mounds associated with the entrances.

Plains pocket mice are primarily nocturnal, but may engage in surface activity on overcast days. Like other small pocket mice, they have been described as "permissive hibernators." They do not accumulate fat prior to the cold season, but store food, become torpid, and arouse periodically to feed. Surface activity occurs between March and November, when average soil temperatures are above 9°C.

These pocket mice are primarily granivores. In Minnesota, for example, foods removed from cheek pouches and caches included seeds of wild buckwheat, sedge, foxtail, spiderwort, puccoon, sweetclover, switchgrass, ragweed, knotweed, vetch, and sandbur. Study of stomach contents has also revealed ingestion of ants.

Home ranges are very small, from 0.02 to 0.04 ha. The longest reported 24-hour movement was 47 m. These pocket mice have been observed to climb as high as 25 cm to forage on the fruiting heads of sedges.

Data on specific population density and structure are unavailable for this species. However, recent data from the Sheyenne National Grasslands in southeastern North Dakota indicate that in some situations, they may be relatively common. Longevity is unknown, but maximum recorded life span of related species has been as high as five years.

Owls are known predators, and they are undoubtedly taken by snakes and other small carnivores.

Reproduction

The breeding season in the central Great Plains is thought to run from late March through June, but is probably later at the latitude of North Dakota. Two peaks of reproductive activity have been suggested, with a lull in early July.

Recorded litter sizes have ranged from two to seven, with a mean of 4.1. There are no known developmental data for the species, but other similar size members of the genus have their eyes open at about 2 weeks, are weaned at 23 days, and attain adult weight at about 55 days.

Status and Conservation

The plains pocket mouse is listed as a species of special concern in Minnesota, and is listed as endangered in Iowa. It has been listed by the U. S. Forest Service as a species of concern on the national grasslands of the Great Plains, including North Dakota. It is rare to locally common in the northeastern half of North Dakota. The North Dakota Game and Fish Department has listed this pocket mouse as a Level III Species of Conservation Priority.

The plains pocket mouse has no known economic importance, but contributes significantly to the mammalian species diversity of North Dakota. Significant numbers of this species on J. Clark Salyer National Wildlife Refuge and the Sheyenne National Grasslands indicate the importance of such tracts in its conservation. Furthermore, considering the paucity of data on the life history and ecology of the species, its special concern status, and spotty distributional data in North Dakota, additional studies are warranted of its life history, ecology, and distribution.

Selected References

Bailey (1926), Dyke et al. (2004), Hall (1981), Hazard (1982), Hibbard and Beer (1960), Higgins et al. (2000), Jones et al. (1983), Jones (1993), Monk and Jones (1996)

Family Dipodidae
Jumping Mice

Meadow Jumping Mouse
Zapus hudsonius

Description

The jumping mice (Family Dipodidae) are easily distinguished from other mice by their very long hind legs and tails. The hind legs are much longer than the forelegs, and the tail is significantly longer than the length of the body. The meadow jumping mouse has a narrow head, small eyes, and naked ears. The fur is coarse, the back is yellowish brown with a dark central stripe, and the sides are yellowish orange. The underparts are white. Average weight is about 19 g. Average standard body measurements are: Total Length - 207 mm, Tail Length 120 mm, Hind Foot - 29 mm, Ear - 13 mm.

In northern and eastern North Dakota, the range of the meadow jumping mouse overlaps with that of the western jumping mouse. Distinguishing between the two species is difficult, and often requires comparison with known specimens. The meadow jumping mouse is smaller and darker colored, with more orange on the sides. Its tail is sharply bicolored. The western jumping mouse is larger; the fur is paler with more white or grayish hairs in the dorsal stripe. The ear is usually bordered by white hairs, and the tail is not sharply bicolored.

Distribution

The meadow jumping mouse occurs from Alaska south of tree line across Canada to Labrador. In the lower United States, its range includes the Great Plains as far south as northern Oklahoma, the deciduous forested region to the East Coast, and as far south as Georgia and Alabama. In North Dakota, its distribution is statewide.

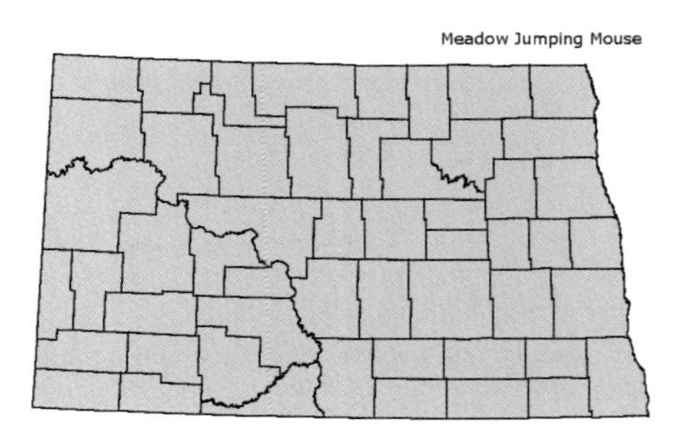

Habitat

This species does not appear to be limited by vegetation type, cover, temperature, or competition, and in this area occurs from prairie areas to deciduous forest. The one factor which seems to regulate its occurrence is moisture, and it is most abundant in non-forested, moist areas. They are found throughout the prairie-forest transition of North Dakota and Minnesota from prairie to the continuous forest, but are most abundant in prairie areas and aspen groves. In southwestern North Dakota, meadow jumping mice occur in moderately grazed rolling grasslands, deciduous forest, and hardwood draws.

Ecology and Behavior

Jumping mice are among the more common small mammals on the northern Great Plains. Their enlarged hind limbs and long tails enable them to engage in saltatorial locomotion, which, except for the kangaroo rats, is unique among regional small mammals. When disturbed, these mice may engage in a series of leaps frequently approaching 2 m in length, often changing direction, and then sit motionless until pursued again.

Meadow jumping mice are solitary animals, but do not appear to be territorial or exhibit antagonistic behavior to each other. Home ranges can be highly variable, ranging from about 0.1 to 1 ha depending on the quality of the habitat. Routine summer movements vary from 15 to 65 m from the nest. Home ranges may shift significantly, up to 1 km, possibly related to late summer drying conditions. These mice are mostly nocturnal, but activity may occur during the daylight hours as well.

Summer nests are usually above ground in dense cover, but may also be in shallow burrows, or under logs or other protective structures. The nest is usually constructed of grass or other materials such as tree leaves or wood pulp.

Meadow jumping mice are omnivorous. During spring, animal material makes up about 50% of their food, and includes primarily caterpillars and beetles. As the season progresses, seeds (especially grass seeds) and the fungus *Endogone* become more important. Meadow jumping mice are not known to store food.

The jumping mice are true hibernators, the only native mice in the region exhibiting this behavior. Seed consumption increases between mid-August and September, with body weight increasing an average of

3.2% per day; some mice eventually double their original body weight. Hibernacula (hibernation nests) are constructed in well-drained sites significantly deeper than summer nests, 0.5 m or more underground. Jumping mice enter hibernation during September and October, as soon as maximum weight has been attained. Hibernation is not a continuous state, but rather mice arouse periodically to rid themselves of waste products. Spring emergence from hibernation is in late April or early May, and appears to be stimulated by a rise in soil temperature at the level of the hibernaculum to about 9°C.

Population densities can fluctuate widely by season and by year. Average summer densities on a Minnesota study area ranged from 8.9/ha in July to 4.5/ha in September, but elsewhere in the state were as high as 29.4/ha. Meadow jumping mice have been known to live as long as two years in the wild, but these longevities are undoubtedly exceptional. Overwintering mortality (as high as 67%) during hibernation may make up the most significant losses in a population. During the warm months, survival is lower during May-June, following emergence, than during the rest of the active season. In addition, they are preyed upon by a variety of snakes, hawks, owls, and predatory mammals.

Reproduction

The breeding season begins in May and extends to late September. Jumping mice are polyestrous, producing two or three litters in a breeding season. The gestation period is 17 to 20 days, and average litter size is six (range four to seven).

Newborn young weigh less than 1 g and, like other small rodents, are naked with closed eyes and ears. By week four, they are fully furred, and their eyes and ears are open. They reach minimum adult weight of 15 g at 2 months. Young from early litters may attain sexual maturity and breed during their first summer.

Status and Conservation

The meadow jumping mouse is a common species in most moist habitats throughout North Dakota. This common species is of no economic significance. One subspecies, *Z. h. luteus*, is threatened in New Mexico, and another, *Z. h. preblei*, is of special concern in Colorado and Oklahoma. There are no known threats to the species in North Dakota.

Selected References

Bailey (1926), Foresman (2001), French and Forand (2000), Genoways and Jones (1972), Hart et al. (2004), Iverson et al. (1967), Jones et al. (1983), Quimby (1951), Seabloom et al. (1978), Tester et al. (1993), Whitaker (1972)

Western Jumping Mouse

Zapus princeps

J. Harris, American Society of Mammalogists Mammal Images Library

Description

The western jumping mouse is very similar to the meadow jumping mouse, and care must be used in distinguishing the two species. The overall coloration is very similar to the meadow jumping mouse, with a yellowish-brown back, a dark dorsal stripe, and yellowish-orange sides. However, it is slightly larger (20 g), paler, has whitish hairs in the dorsal stripe, the ears are bordered with whitish hairs, and the tail is not sharply bicolored. Average standard body measurements are: Total Length - 215 mm, Tail Length - 126 mm, Hind Foot - 28 mm, Ear - 12 mm.

Distribution

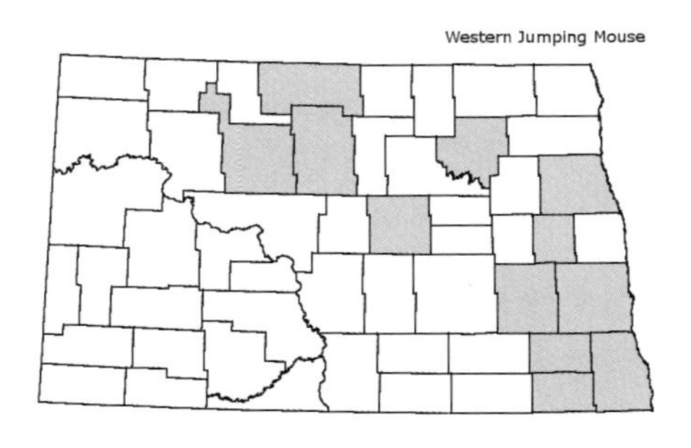

Western jumping mice occur in the Rocky Mountains from southern Yukon Territory south to Arizona. To the west, they are in eastern Oregon through the Cascades, south to the Sierra Nevada of California. They are also in the Great Plains of southern Alberta, Saskatchewan, Manitoba, northern Montana, northern and eastern North Dakota, and extreme northeastern South Dakota. In North Dakota, there are reports from the northern tier of counties, the Red River valley, and the Sheyenne National Grasslands.

Habitat

Western jumping mice frequently share habitat with meadow jumping mice, and are most common in dense grasses and herbs adjacent to streams. However, they appear to be more tolerant of warmer, drier conditions than those preferred by meadow jumping mice. In the sandhills of southwestern Manitoba, they have been found in xeric shrub and marsh-willow-meadow zones adjacent to the Souris River, aspen groves adjacent to xeric prairie, and in dune habitat. At J. Clark Salyer National Wildlife Refuge (McHenry County), they have been collected in mixed woodland/sandhill prairie habitat. In eastern North Dakota, they have been reported from tall-grass prairie habitats in Grand Forks County and the Sheyenne National Grasslands (Richland County).

Ecology and Behavior

The western jumping mouse is a relatively poorly known species, and most of the ecological research has been conducted in mountainous regions. This species, like the meadow jumping mouse, is thought to be solitary and primarily nocturnal. Home ranges in mountainous areas have been estimated at 0.2 to 0.6 ha. Nests during the active season consist of a ball of grass with one or two entrances, and are placed on the ground under the cover of heavy vegetation or in surface litter. Hibernation nests are constructed of shredded vegetation in well-drained sites about 0.5 m underground.

Western jumping mice are omnivorous, but more dependent on seeds than meadow jumping mice. Insects and other invertebrates are taken upon emergence from hibernation, but overall, more than 80% by volume of stomach contents may be seed fragments.

During late summer, mice begin laying down deposits of hibernation fat, a process taking about a month. These fat deposits may make up two-thirds of the body weight. Hibernation probably begins in late September, while arousal in the spring occurs when soil temperature reaches about 9°C. The hibernating state is not continuous; mice arouse periodically to rid themselves of accumulated metabolic wastes.

Population densities vary greatly, but have been recorded as high as 32.4/ha. Like the meadow jumping mouse, winter mortality during hibernation can reach 70% and is probably the greatest single mortality factor. About 30% of the population may survive into the third summer. Because of the time spent in hibernation, jumping mice are less vulnerable to predation than other small mammals. During the active season, they are known to be taken by snakes, hawks, owls, and weasels.

Reproduction

The breeding season probably begins in early to mid-May upon emergence from hibernation. Pregnant females have been collected between late May and mid-July and lactating females until late August. The gestation period is 18 days, and litter size ranges from four to eight (mean 5.4). There is probably only one litter per year. Juvenile females do not breed, and in one study only 41% of the yearlings bred, compared to 70% of the older females.

Status and Conservation

The western jumping mouse appears to be fairly common in northern and eastern North Dakota. The western jumping mouse is of no economic or public health significance in North Dakota. There are no known threats to the species in the state.

Selected References

Foresman (2001), Hart et al. (2004), Hazard (1983), Higgins et al. (2000), Jones et al. (1983), Whitaker (1972), Wrigley (1974)

Family Cricetidae
New World Rats and Mice
Itunkala, Itungtanka—Lakota

Sagebrush Vole
Lemmiscus (Lagurus) curtatus

T. R. Mullican, American Society of Mammalogists Mammal Images Library

Description

This is a medium-size vole, weighing about 25 g. Like other voles, it has a blunt snout, small ears, and short legs and tail. It is fairly easy to distinguish from other grassland voles by its rather long, lax fur, colored gray to brownish-gray dorsally, and very short tail, which is not much longer than the hind foot. Average standard body measurements of seven North Dakota specimens are: Total Length - 115 mm, Tail Length - 19 mm, Hind Foot - 17 mm, Ear - 10 mm.

Distribution

As the name implies, sagebrush voles occur locally over much of the sagebrush regions of North America. Their northern limits are in south-

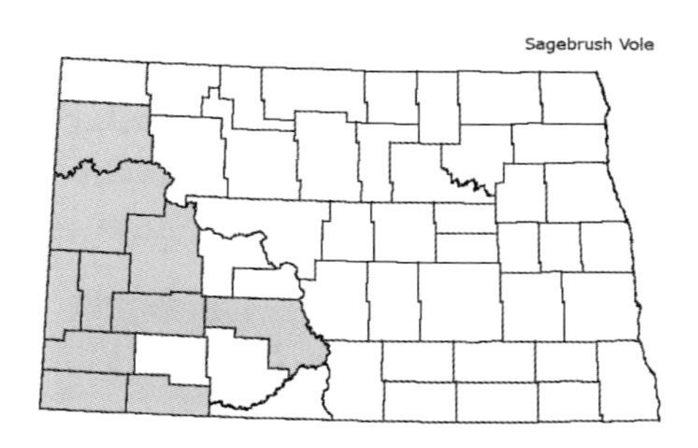

ern Alberta and southwestern Saskatchewan, and they extend south over much of the Great Plains as far as northern Colorado. To the west, they are in southern Idaho, eastern Washington and Oregon, eastern California, Nevada, and western Utah. The eastern limit of the range is in southwestern North Dakota and extreme northwestern South Dakota.

Early work recorded this species from only two localities in North Dakota, Fort Buford (Williams County) and Glen Ullin (Morton County). Subsequently, small numbers of specimens have been taken from Adams, Slope, Bowman, Billings, Golden Valley, Dunn, McKenzie, and Stark counties. Sagebrush voles probably occur in sparse numbers throughout much of southwestern North Dakota.

Habitat

Typical habitat of sagebrush voles is semi-arid prairie with loose, well-drained soil, usually dominated by sagebrush, rabbitbrush and bunch grasses. They have also been associated with yucca, brushy riparian zones, cattails, and alfalfa. In North Dakota, they have been taken in sagebrush bottoms, grazed shortgrass prairie, hardwood draws, and alfalfa hayland.

Ecology and Behavior

Sagebrush voles are somewhat colonial, and more social than many of their close relatives. Their colonies can be identified by systems of poorly defined runways in grass adjacent to rocks or the bases of shrubs. These voles also hollow out dried cattle droppings, which are used as sites of temporary refuge. Burrows, also under brush or rock piles, may be up to 460 mm below the surface, and have eight to 30 entrances. Nest chambers have multiple entrances, and are mostly lined with grass. Burrows are generally free of waste products; voles urinate and defecate outside, well away from burrow entrances.

Burrows are shared by voles of different sex and age classes. Females rarely fight, and some even nurse young of other females' litters. Fighting may occur between males and non-receptive females, among breeding males, and among females with the appearance of a strange male.

The diet of these voles is strictly herbivorous, and they will eat almost anything green. Primary foods include leaves, stems, flowers, and immature seed heads of grasses and forbs. During winter, they forage under the snow, stripping the bark and cambium layers of sagebrush, and may feed on this source almost exclusively. They do not store food, but bring freshly cut material into their burrows, where it will not desiccate. Sagebrush voles may be active throughout the 24-hour period, but activity peaks occur during evening and morning twilight hours.

Little is known of the population dynamics of this sparsely dispersed species. Peak populations have been associated with warm, dry winters, summers with above-normal precipitation, and early fall rains. A population which has peaked during early summer may have died off before fall. Sagebrush voles succumb to predation by snakes, hawks, owls, shrikes, and a number of predatory mammals. In the far West, they are known to be an important reservoir of sylvatic plague. Longevity in nature is unknown but, like their close relatives, most sagebrush voles probably die within their first year.

Reproduction

Sagebrush voles have a high reproductive rate and, at least elsewhere in their range, have been known to breed throughout the year. It is believed, however, that severe drought or cold may curtail reproduction.

Females are polyestrous, and may breed again within 24 hours after giving birth. The gestation period is about 25 days. Average litter size is five, but may vary greatly from one to 11. Newborn young, like other small rodents, weigh about 1.5 g, are naked, and eyes and ears are closed. They are fully furred at 5 days, and their eyes open between 9 and 13 days. Weaning occurs at 21 days. They are nearly adult size by 60 days, but females have been known to breed as early as 47 days of age. Males do not breed until at least 2 months.

Status and Conservation

Sagebrush voles are sparsely dispersed throughout southwestern North Dakota, and should be regarded as rare to uncommon at this edge of the species' range. Because of its peripheral status in the state, it has been designated as a Level III Species of Conservation Priority.

Because of their food habits and sparse populations, sagebrush voles are economically insignificant. Their populations act as reservoirs of sylvatic plague elsewhere in the range, but in North Dakota, there is little cause for public health concern. This could be regarded as a fire-sensitive species due to its dependence on dense grasses and shrubs such as sage and rabbitbrush. Further studies should better elucidate the population ecology and status of this poorly known species.

Selected References

Bailey (1926), Birney and Lampe (1972), Carroll and Genoways (1980), Dyke et al. (2004), Genoways and Jones (1972), Jones et al. (1983), Seabloom et al. (1980), Seabloom et al. (1978), Shaughnessy (2016), Svihovec (1967)

Prairie Vole
Microtus ochrogaster

Bob Gress

Description

This is a relatively small vole, with short legs and tail. Like most other voles, the eyes are small, and the ears are buried in the fur. The upper parts are grayish brown with a mixture of black and yellowish tips on the guard hairs, giving a grizzled appearance. The sides are paler, and the belly has a distinct yellowish cast. Average standard body measurements are: Total Length - 141 mm, Tail Length - 34 mm, Hind Foot - 18 mm, Ear - 12 mm. Body weight averages about 28 g.

The prairie vole can be distinguished from the similar meadow vole by its more grizzled appearance, yellowish underparts, and shorter tail (<38 mm).

Distribution

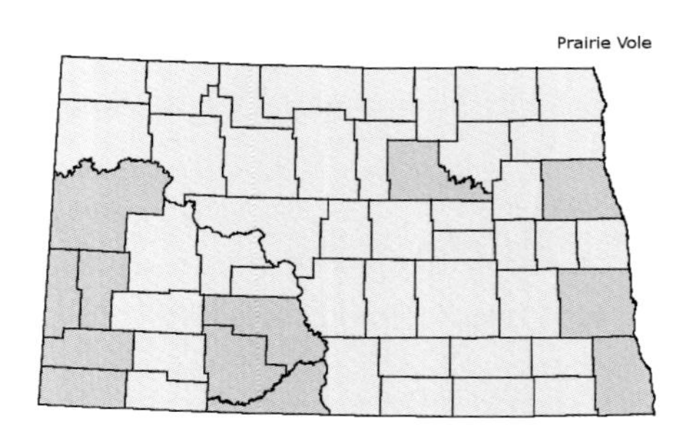

The prairie vole occurs in the central grasslands of North America. Its northern limits are in the southern Prairie Provinces of Canada, while to the west it is in eastern Montana, Wyoming, and Colorado, and northeastern New Mexico. Its southern limits are in Oklahoma, Arkansas, and West Virginia. To the east, it is found in the grasslands of Minnesota, southern Wisconsin, Illinois, Indiana, and Ohio. Clearing of forests and development of agriculture and highways have allowed further eastward expansion of its range since white settlement.

Two subspecies have been recorded in North Dakota, *M. o. minor* (relatively small and dark) east and north of the Missouri River, and *M. o. haydenii* (larger and paler) in the southwest. *M. o. minor* has been reported from numerous locations throughout eastern North Dakota, but there have been few recent records, from Benson, Grand Forks, Cass, and Richland counties. In southwestern North Dakota, *M. o. haydenii* has been recorded in Morton, Grant, Sioux, McKenzie, Billings, Golden Valley, Slope, and Bowman counties.

Habitat

Typical habitat for prairie voles is upland prairie, but they may be also found in swales, riparian areas, and croplands. In eastern North Dakota and northwestern Minnesota, they seem to prefer sandy soils and well-drained ridges, including beach ridges of Glacial Lake Agassiz. Southwestern North Dakota habitats include dry upland grasslands, moderately grazed rolling grasslands and haylands. In Theodore Roosevelt National Park, they occur in nearly all habitats, except for green ash-choke cherry. Reduction of cover by mowing, burning, and grazing practices tends to result in reduced prairie vole populations.

Ecology and Behavior

Prairie voles develop and maintain extensive runway systems. Grasses and other plants are kept clipped to the extent that occasionally bare soil is exposed. Runways are extensive, consisting of a central trunk with several branches. Burrows connect to the runway system, and contain chambers for nests and food. Burrows are typically about 50 mm deep, and may be several meters in length. Nests are typically lined with shredded grasses.

Unlike closely related species, prairie voles are highly social, forming long-term monogamous pair bonds. Both members of the pair engage in territorial defense, burrow and runway maintenance, and care of young. Aggression is high against other prairie voles of either sex and, where they are sympatric with meadow voles, the latter species is forced into moister habitats such as marshes.

Average home ranges are small, varying from 0.1 to 0.2 ha. Home ranges are reduced in size with high population densities and high herb biomass. Prairie voles can be active at any time of day or night, but their patterns may be influenced by temperature. During summer, with high daytime temperatures, they tend to be more nocturnal, while in winter, with low nighttime temperatures, there is an increase in daytime activity.

Prairie voles feed almost exclusively on grasses, sedges and forbs during spring and summer, seeking mainly young, growing leaves. Plants are first cut into sections of about 40 mm, then consumed or left in little "hay piles" in the runway. With the disappearance of green vegetation in late summer and fall, they switch to bark, tubers, roots, fruits, and seeds. These materials, along with grasses, may also be found in food caches. During fall and winter, voles may inflict considerable damage to fruit trees and ornamentals by their girdling activity. Some animal material may also be taken, and analysis of fecal pellets indicates that a high proportion of insects may be consumed in late summer and fall.

Prairie vole populations, like other members of the genus *Microtus*, undergo wide fluctuations in density, varying seasonally and over multi-year "cycles." Seasonally, populations are highest in spring (April toJune) and fall (October-November). Cycling populations fluctuate on a two- to four-year basis, with population highs averaging 130 voles/ha, and lows averaging 4/ha. Occasionally, they may be even extirpated from an area. Despite many proposed hypotheses attempting to explain vole cycles, clear understanding of these phenomena remains elusive. Certainly, the growth phase is easily explained by their high reproductive potential. Current thinking suggests that there is no single causal factor, but cyclic fluctuations may result from a combination of intrinsic factors such as spacing behavior and dispersal, and extrinsic processes such as food and predation.

Prairie voles are taken by nearly every predator with which they share range. In addition, they carry a wide range of ecto- and endoparasites. Mortality is high among juveniles, only a little over a third surviving through their first month. Survival among adults may run from 30 to 80% per month during the breeding season. During the winter non-breeding season, survival of both young of the year and adults may increase to 50 to 90% per month. However, these figures are broad averages, and survival rates will vary greatly with season, social structure, and phase of the cycle. Voles have been known to live more than one year in nature and to nearly three years in captivity.

Reproduction

The breeding season may extend throughout the year in some parts of the range, but winter breeding is unlikely in the northern states such as Minnesota and North Dakota. When winter breeding does occur, it is usually associated with mild winter conditions or prior to cyclic peaks in population density.

Monogamous breeding behavior is characteristic of prairie voles. Females exhibit induced estrus, i.e., they require the presence of a mate as stimulus for the completion of the estrous cycle. This effect is so powerful that even contact between a pregnant female and a strange male may terminate the pregnancy and cause her to come into estrus again. Female prairie voles are also induced ovulators, requiring the stimulus of copulation to bring about ovulation.

Gestation in the prairie vole is 21.5 days, and the average litter size is 3.8 (range one to eight). Females may come into a post-partum estrus. Therefore, they may simultaneously nurse one litter while pregnant with the next one.

Although monogamous, females are aggressive towards their mates, and exclude them from the nest during the period of parturition. They are allowed back in the nest about one day later, and contribute equally to parental care with the exception of lactation.

As with other members of the genus, newborn young weigh about 3 g, are blind, naked, and the ears are closed. Brown fur appears at day 2, eyes open at day 5 to10, and they begin taking solid food after day 10. Weaning is complete at 3 weeks. Females attain sexual maturity at 35 days, and males at 42 to 45 days.

Status and Conservation

The prairie vole is a common, but not abundant, species in southwestern North Dakota; it is rare to uncommon in the east. It seems to have declined since early surveys in the state. In neighboring Minnesota, it is listed as a species of special concern.

In some parts of their range, prairie voles can inflict significant damage to agricultural crops, ornamental trees, orchards, and range land, especially during their peaks of abundance. In North Dakota, however, they do not seem to have become abundant enough to be of economic significance. Rather, in the eastern half of the state, the concern may be more for their conservation.

Selected References

Bailey (1926), Genoways and Jones (1972), Hazard (1982), Hopkins (1983), Jones et al. (1983), Keller (1985), Lysne (1991), Madison (1985), Naylor (1987), Nellermoe (1983), Pugh et al. (2003), Seabloom et al. (1978), Stalling (1990), Svihovec (1967), Tait and Krebs (1985), Wolff (1985)

Meadow Vole

Microtus pennsylvanicus

Hi^{n}tu^{n}ka–Dakota; Gipápuli, Bidábaho itáhu–Hidatsa; Sakch–Arikara

C. D. Grondahl, North Dakota Game and Fish Department

Description

With an average adult weight of about 40 g, the meadow vole is the largest vole in the region. Its summer pelage is dark brown on the back, with lighter sides, and grayish underparts. Like other voles, the eyes are small and the ears are buried in the fur. Average standard body measurements are: Total Length - 155 mm, Tail Length - 39 mm, Hind Foot - 19 mm, Ear - 12 mm. These voles can be fairly easily distinguished from prairie voles by the meadow vole's dark brown (vs. grizzled) fur with grayish (vs. yellowish) underparts, and longer tails (>38 mm).

Distribution

The meadow vole has the widest distribution of any North American member of the genus. It occurs above tree line in Alaska and Canada south to the northern and eastern regions of the United States. It is also in the Rocky Mountain states as far south as New Mexico and north-

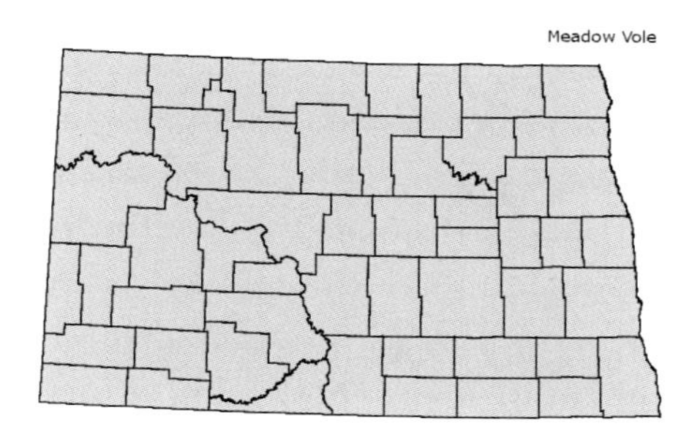

eastern Arizona. Isolated populations occur in central Washington and northern Chihuahua.

Two subspecies occur in North Dakota. The larger, darker *M. p. drummondi,* is found north and east of the Missouri River, while the smaller, lighter *M. p. insperatus,* is southwest of the Missouri River. Meadow voles are common and abundant in grasslands throughout North Dakota.

Habitat

Meadow voles occur in a wide variety of grassland types, but are most abundant in mesic habitats with dense grass cover. Preferred habitat in the prairie-forest transition of North Dakota and Minnesota is tall-grass prairie, but they can also reside in young aspen groves with grassy undercover. Meadow voles are frequently abundant in moist edges around prairie marshes, and may even be observed swimming in open water among clumps of emergent vegetation. When sympatric with prairie voles, they may be restricted to these moist habitats. They may be abundant in agricultural settings, where they are especially associated with forage crops. In southwestern North Dakota, meadow voles are common in haylands, light and moderately grazed grasslands, and green ash draws.

Any reduction of litter and standing crop biomass is detrimental to meadow vole populations. For example, in southwestern North Dakota, meadow voles were absent from both winter- and summer-grazed green ash draws, but were present in ungrazed draws. Fire results in greatly reduced populations, but fire-caused mortality is low, and voles tend to move into adjacent unburned habitat. Reestablishment of meadow voles following burns is relatively rapid, and populations may recover to pre-burn levels within 12 to 16 months.

Ecology and Behavior

Like the prairie vole, the meadow vole lives mainly in well-developed surface runways, primarily in dense grasses and sedges. In wet situations, spherical nests, about 10 to12 cm in diameter, are constructed of grasses and plant fibers on the surface, but in well-drained soils, burrows may be utilized. They are often located under boards, logs, or other surface debris. Frequently used runways exhibit piles of grass cuttings and accumulations of droppings.

Meadow voles, unlike prairie voles, do not have a complex social organization. Their mating system has been described as polygynous or promiscuous. Females defend territories, but males do not. Home ranges of males tend to overlap with female territories and with home ranges of other males. Although not territorial, individual encounters between males result in seasonally varying levels of aggression, highest during the onset of breeding and lowest with fall cessation of reproduction. During fall, female territories may break down and large communal groups of both males and females may develop. Home ranges vary from 405 to 3,480 m^2 for males and to 160 to 3,115 m^2 for females. Factors, other than sex, influencing home range size include season (larger in summer), population density (larger at lower density), and habitat (larger in marshes).

Activity may occur during all hours of the day and night. Activity periods and levels can be very labile, influenced by light levels, vegetative cover, changes in temperature, population density, and during dispersal. Dispersal tendencies vary with sex (males during winter and females during summer), and increase with reproductive development and population density.

Meadow voles feed heavily on grasses and sedges, especially young growing parts during spring, hearts at the base of the plants during summer, and seeds during fall and winter. During winter, they also utilize bulbs, roots, tubers, bark, and buds. Forage crops, including white clover, alfalfa, and red clover are also highly preferred. Fungi (endogone), insects, and animal remains may also be eaten. Especially during high population densities, meadow voles may inflict serious damage to orchards and ornamental trees and shrubs by girdling.

Like many other voles, meadow vole populations frequently exhibit "cyclic" fluctuations over periods of two to five years. Populations can irrupt from less than two to nearly 200 voles per ha, often within the same year. During the increase phase of the cycle, reproductive success is high, body size increases, and many individuals disperse to colonize new habitats. Demographic features of peak years include poor juvenile survival and inhibition of sexual maturity. Poor juvenile survival continues through the decline phase, which may be as precipitous as the increase. Causal factors regulating these population crashes are poorly understood, but current thinking suggests a combination of intrinsic factors such as spacing behavior, dispersal, and stresses related to high density, and extrinsic processes such as food and predation.

Mortality is high in meadow vole populations, and varies by age/sex groups, seasonally, and from year to year. Juvenile mortality may be as high as 80 to 90% per month, but may stabilize to as low as 30% per month with adulthood. Among adults, females tend to survive longer than males. It has been suggested that more species of predators take meadow voles than any other American mammal. These include 12 snakes, 27 birds, over 20 predatory mammals, and several large fish. Predation, disease, and inclement weather all contribute to high vole mortality but, as cited above, none of these factors have been documented to account for their periodic population crashes.

Reproduction

Breeding may begin around the last week of March and continue into late November. During cyclic peaks, a few voles may breed throughout the winter. Unlike the prairie vole, mating is promiscuous or polygamous, with males attracted to receptive females by their sense of smell. Females require the presence of a breeding male to come into estrus (induced estrus), and the mechanical stimulus of copulation to ovulate (induced ovulation). Like the prairie vole, the phenomenon of induced estrus is so powerful that the presence of a strange male is sufficient to cause a pregnant female to abort and come into estrus again.

Litters of two to eight young are born following a gestation of 21 days. Adult females may go into a post-partum estrus, mate again, and therefore can be simultaneously nursing one litter while pregnant with the next. Newborn young, like most small rodents, are hairless, their eyes and ears are closed, and weigh 1.6 to 3 g. They are fully furred by day 7, and eyes and ears are open by day 8. Young are fully weaned by 2 weeks. Young females become sexually mature and mate at 1 month of age, while males are mature at 45 days. Consequently, the reproductive potential of the meadow vole is truly remarkable. The record for reproduction was a captive female which produced 17 consecutive litters in a single year, for a total of 83 young.

Status and Conservation

Although populations fluctuate greatly, meadow voles are regarded as common to abundant throughout grassland and wetland habitats in North Dakota. They are especially abundant in areas which have not been recently burned or grazed.

Because of their abundance and food habits, meadow voles are often regarded as significant agricultural pests. Forest clearing, irrigation, and introduction of new forage species have all contributed to vole population expansion, and fence rows and road rights of way have facilitated population movement into vacant areas. Crops damaged by meadow voles include vegetables, grains, alfalfa, and other hayland. Fruit crops, tree plantations and ornamentals may be killed by girdling activity.

Approaches to control of meadow vole populations have included reduction of cover by "clean" farming and use of herbicides, protection of valuable plants by screening, protection of their natural predators, and direct control by trapping and use of rodenticides. Use of repellants has not been shown to be very effective. Direct control methods may be effective in the short term but, without accompanying cultural approaches affecting vole habitat, depopulated areas may be rapidly recolonized. Furthermore, carcass residues of rodenticides may have an impact on vole predators and the surrounding ecosystem.

Other land use and management goals should also be considered before meadow vole control is attempted. Meadow voles comprise a major food base for many mammalian and avian predators, and significantly reduced populations may result in increased predation pressure on other species such as ground-nesting birds.

Selected References

Bailey (1926), Genoways and Jones (1972), Hazard (1982), Iverson et al. (1967), Jones et al. (1983), Naylor (1987), Pugh et al. (2003), Reich (1981), Seabloom et al. (1978), Seabloom et al. (1999), Svihovec (1967), Tester (1965), Tester and Marshall (1961), Walley (1970), Wolff (1985)

Southern Red-backed Vole
Myodes (Clethrionomys) gapperi

Paul Smiths, American Society of Mammalogists Mammal Images Library

Description

Like other voles, the southern red-backed vole has small eyes and ears, a short tail, and short legs. It has a distinct reddish-brown back, brown sides, and gray underparts. Winter fur is longer and lighter colored than during the summer. The ears are more conspicuous than in other voles, and the tail is about a third of the length of the body. Average standard body measurements are: Total Length - 134 mm, Tail Length 37 mm, Hind Foot - 18 mm, Ear - 14 mm. These voles weigh about 24 g.

Distribution

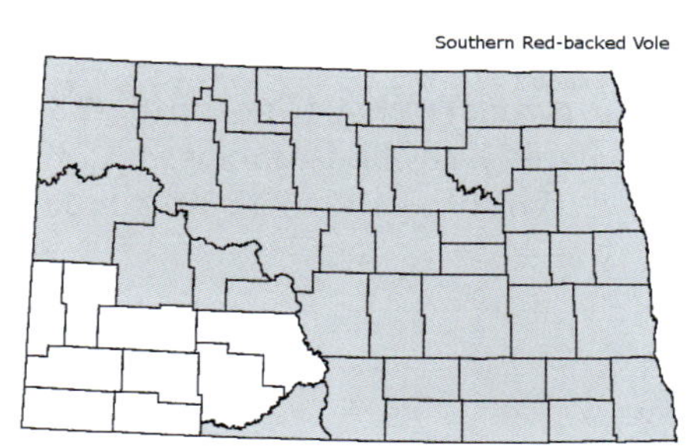

In the north, southern red-backed voles occur below the tree line from northern British Columbia across Canada to Labrador. In the western United States, the range includes the Olympic and

Cascade ranges and the Rocky Mountains as far south as New Mexico and Arizona. To the east, they are in northeastern Montana, the Black Hills of South Dakota and Wyoming, North Dakota, the Great Lakes states, New England, and the Appalachian chain as far south as northern Georgia.

Red-backed voles occur in appropriate habitat throughout North Dakota, except most of the southwestern counties. They are common in the Turtle Mountains, Pembina Hills, Devils Lake region, and the riparian woodlands of the Red and Missouri rivers and their tributaries. In the southwest, they have been recorded in small numbers in Oliver,Mercer, Dunn and McKenzie counties.

Habitat

These are primarily forest animals, and are most common in coniferous, deciduous, and mixed forests with abundant letter, brush piles, rotten logs, and stumps. In central Minnesota, they prefer white cedar forest with sparse to medium herbaceous cover. In the prairie-forest transition of North Dakota and Minnesota, they are in coniferous forest, continuous deciduous forest, aspen groves, and riparian woodland, but are least abundant in riparian woodland, possibly due to periodic flooding.

In eastern North Dakota, red-backed voles are among the most common small mammals in the forested regions of the Turtle Mountains, Pembina Hills, and Devils Lake area. They are also found in wooded river valleys, tree plantations around farmsteads and in shelterbelts, in dense cover along coulees, all areas where there is abundant litter and deadfall. Western North Dakota habitats include brushy undergrowth along the Missouri River, deciduous forests of the Killdeer Mountains (Dunn County), hardwood draws, and riparian woodland.

Ecology and Behavior

In moist habitats, red-backed voles build shallow tunnels in leaf litter and around rotten stumps and downed trees, but in drier situations, there is little sign of their presence. They build simple, globular nests of grass and other plant materials, located in natural cavities or abandoned burrows of other species.

Red-backed voles are somewhat omnivorous in their diet, switching foods with seasonal availability. During spring, they feed heavily on buds and other growing plant parts, then shifting to fruits, seeds, and fungi as the season progresses. Endogone and other fungi may be a staple during summer and fall. Some insects and other invertebrates are taken as they become available. During fall and winter, the diet consists mainly of seeds, roots, and tree bark. Seed caches may be established in the fall, or voles sometimes raid the hoards accumulated by red squirrels.

Red-backed voles are active at any time of day or night, but alter their patterns seasonally. Spring activity in Minnesota is mostly during the early morning hours, shifting to around midnight by July. In late summer, there are early morning and evening peaks, while in late fall, activity shifts to the pre-midnight hours. During winter, activity may be equal at all hours of light and dark.

These voles are not gregarious and, except for females with young, do not usually form persistent social groups. During winter, however, small colonies may be formed. Prior to the spring breeding season, levels of aggression are minimal and red-backed voles may even expand their range into grassland habitat and coexist with meadow voles where competitive exclusion occurs during other seasons of the year. During the warm months, meadow voles prevent colonization of grasslands, and red-backed voles exclude meadow voles from forest habitat. For example, in North Dakota, meadow voles and red-backed voles have been observed within a few meters of each other in their respective habitats along a sharply demarcated forest-grassland edge during summer.

Home ranges of red-backed voles in Minnesota average about 0.12 ha. These may expand during winter under snow cover, in part contributing to their occasional seasonal coexistence with meadow voles in adjacent grassland habitats.

Population densities may be as high as 65 per ha, but typically average about 10 per ha. Peak densities occur during late summer and fall, followed by a gradual decline through the winter. Red-backed vole populations do not seem to undergo cyclic fluctuations characteristic of other boreal small mammals. The period from fall freeze-up to spring thaw appears to be most critical to survival, and this severe climate coupled with predation by a wide variety of birds and mammals can result in very high mortality. The maximum recorded longevity is 20 months, but few voles survive longer than a year.

Reproduction

Red-backed voles are promiscuous breeders, and do not form lasting pair bonds. They are polyestrous, have a post-partum heat period, and thus have a high reproductive potential offsetting their high mortality. The breeding season typically begins in March, and may last about seven months.

The gestation period is 18.5 days, and mean litter size in the laboratory is 4.6. Newborn young weigh less than 2 g, are blind, hairless, and toothless. They can crawl by day 5, and are furred at about a week. Their eyes open at 2 weeks, at which time they begin taking solid food. Weaning is completed around day 17. They are sexually mature at 2 to 4 months.

Status and Conservation

Red-backed voles are common to abundant in wooded areas throughout eastern North Dakota, and seem to occur in small, scattered populations south and west of the Missouri River.

These voles usually are found in unpopulated areas, and seldom come into direct contact with people. Although they may occasionally damage young deciduous trees, their economic impact is negligible. They may destroy significant numbers of insects, and provide an interesting contribution to the faunal diversity of North Dakota.

Selected References

Bailey (1926), Brown (1971), Butsch (1954), Genoways and Jones (1972), Gunderson (1962), Gunderson (1950), Hazard (1982), Hibbard (1972), Iverson et al. (1967), Iverson and Turner, (1972b), Jones et al. (1983), Merritt (1981), Nellermoe (1983), Seabloom et al. (1980), Skaley (1967)

Bushy-tailed Woodrat
Neotoma cinerea

C. D. Grondahl, North Dakota Game and Fish Department

Description

Woodrats (packrats) are about the size of the Norway rat, but have soft, fine fur, and relatively large ears and eyes. This species is the largest member of the genus, and one of North Dakota's larger rodents. Body weights of bushy-tailed woodrats average 419 (379 to 456) grams for males, and 302 (234 to 359) grams for females. Coat color is usually grayish, with whitish feet and underparts. The species is named for its heavily furred tail, the hairs of which may average 30 mm in length. Standard body measurements (11 North Dakota specimens) averaged: Total Length - 316 mm, Tail Length - 133 mm, Hind Foot - 42 mm, Ear - 34 mm.

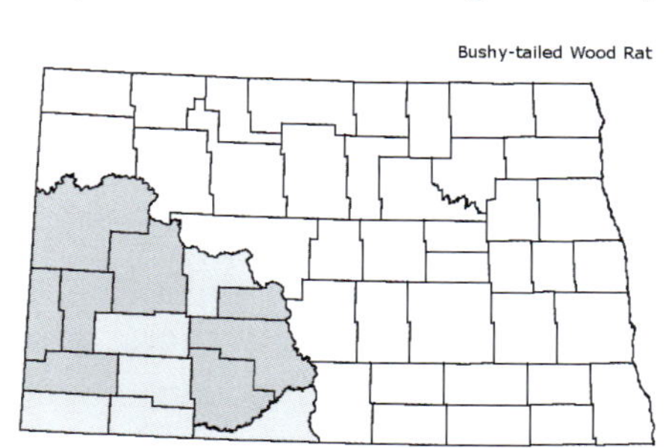

Distribution

Bushy-tailed woodrats occur primarily in mountainous terrain of the western United States and

Canada. They also extend into broken terrain of the northern Great Plains, including southwestern North Dakota south and west of the Missouri River. Early records indicated the species to be scattered throughout the Badlands region, but nowhere abundant. More recently, specimens have been reported from Slope, Billings, Golden Valley, and Dunn counties.

Habitat

In southwestern North Dakota, preferred habitat includes the ponderosa pine area of Slope County, cliffs and rocky outcrops, hilly scoria, and toe slopes. They are also known to occupy abandoned buildings.

Ecology and Behavior

Dens are typically constructed in rough terrain, under cover of outcrops, crevices, caves, or abandoned buildings. Rough topography largely determines their occurrence. Hence, population density may be limited by suitable den sites, typically resulting in a patchy distribution. Dens may consist of large masses of dried vegetation, sticks, bones, rocks, and other debris. The den contains one or more cup-shaped nests. Frequently, man-made products, such as bottle caps and broken glass, are incorporated into the den, a habit for which they have earned the nickname "packrat."

Bushy-tailed woodrats eat a wide variety of vegetation, primarily the leaves of forbs and woody plants and conifer needles. They also eat fruits, seeds, fungi, and some insects.

Population density is highly variable and somewhat dependent on the availability of den sites. Frequently, naturalists find that woodrat sign is far more abundant than woodrats.

A wide variety of ectoparasites infest bushy-tailed woodrats, largely because of the amount of time spent in the den. They are susceptible to plague, trypanosomes, and spotted fever. They are also preyed upon by a variety of nocturnal mammalian and avian predators.

Bushy-tailed woodrats maintain nocturnal activity patterns throughout the year. Although they are regarded as relatively unsocial, they frequently occur in loose assemblages. Reported home ranges average about 6 ha for males and 3 ha for females. Their breeding habits have been described as ranging from promiscuous to polygynous (more than one female per male). Scent marking is important in delineating territories, asserting male dominance, and in courtship. Scent posts may be used traditionally for many years, accumulating thick incrustations of dried urine.

Reproduction

Bushy-tailed woodrats do not breed until yearlings. Females are polyestrous, and in captivity have produced as many as seven litters per year, but reports in the wild have indicated less than three. The breeding season begins in early April, with litters produced from May to August. The estrus cycle lasts five to seven days, and a gestation period of 29 days is followed by a post-partum estrus. Litter size ranges from three to five. Longevity in captivity can be as high as three to five years, but is almost certainly much shorter in nature. Annual survivorship in the northern portion of the species' range has been reported as 31 to 34%.

Status and Conservation

The bushy-tailed woodrat is relatively common, probably maintaining sparse but stable populations in southwestern North Dakota.

The species has no significant economic importance. There are no apparent current threats to its existence in North Dakota.

Selected References

Bailey (1926), Foresman (2001), Genoways and Jones (1972), Monty and Emerson (2003), Seabloom et al. (1978), Smith (1997)

Muskrat

Ondatra zibethicus

Zih-zirukka – Hidatsa; Si^{n}kpé – Lakota;
Shantshuke – Mandan; Citakh – Arikara

C. D. Grondahl, North Dakota Game and Fish Department

Description

The muskrat is the largest "microtine" rodent, closely related to the voles and lemmings. It is a large, heavy bodied, semiaquatic rodent, named for its musky odor emitted especially during the breeding season. The Great Plains subspecies *(O. z. cinnamominus)* occurs in North Dakota, and is smaller and lighter-colored than others occurring to the east and north. Weights of North Dakota adults average 860 grams for males and 780 grams for females. Coat color is dark brown, overlain with dark guard hairs; sides are more reddish, and underparts have silvery-gray. The soft underfur is nearly waterproof, and its trapped air may constitute more than one-fifth of the animal's dry volume. Characteristic aquatic adaptations include a valvular mouth and ears, a laterally flattened scaly tail, and partially webbed, hair-fringed hind feet. Eyes and ears are relatively small. Standard body measurements (12 North Dakota specimens) average: Total Length - 500 mm, Tail Length - 217 mm, Hind Foot - 71 mm, Ear - 17 mm.

Distribution

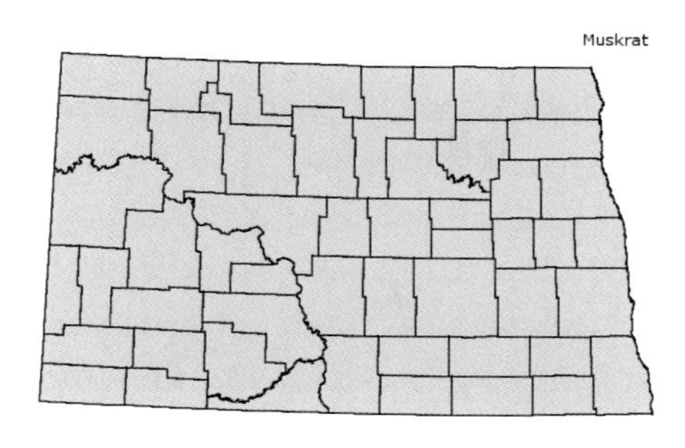

Muskrats occur throughout North America from Alaska, Yukon, and the Northwest Territories south to the Gulf of Mexico, wherever suitable water conditions exist. Introduced populations also occur in Europe and South America. Their North Dakota distribution is statewide, in lakes, rivers, wetlands, and man-made impoundments. They are most common in wetlands of the eastern half of the state having abundant emergent vegetation.

Habitat

Any aquatic habitat may contain muskrats. However, they favor and are most numerous in semi permanent to permanent wetlands containing abundant emergent vegetation (bulrushes and cattails). These deeper wetlands, in addition to providing adequate food and cover, are less susceptible to drought and winter freezeouts, and hence more conducive to large populations.

Ecology and Behavior

Muskrats are regarded as a dominant species in the wetland ecosystem. Their house-building, burrowing, and feeding activities alter the species composition, abundance, distribution, and decomposition of aquatic vegetation, resulting in significant impacts on wetland ecology and associated wildlife.

Any of three types of structures may be inhabited by muskrats, depending on conditions. Dwelling houses, the most apparent, are conical structures constructed from emergent vegetation, other aquatic plants, and mud. They may rise 4 feet or more above the water, have one or more nest chambers, and multiple submerged entrances. Smaller "feeder houses" or "pushups" may be constructed in the vicinity of dwelling houses. These typically provide shelter for animals feeding on the ice surface during winter. Occasionally, muskrats are observed having their litters in nests on rafts of floating emergent vegetation. Muskrats will often excavate burrows with submerged entrances when water depths and bank conditions are suitable. In North Dakota, entire spring/summer populations may live in bank dens. As water levels recede, muskrats attempt to maintain their burrows by extending the entrances out into standing water. Eventually, the population may shift from bank dens to house construction.

Muskrat houses provide germination sites for semiaquatic plants. They are also used by other species of marsh wildlife. For example, waterfowl and other wetland birds use them for resting, basking, and nesting, while mink commonly take up residence in houses.

Muskrats feed on a wide variety of aquatic vegetation, but are heavily dependent on roots and basal portions of emergents, such as bulrushes and cattails. During summer, feeding typically occurs on constructed rafts of stalks of emergents. Occasionally, feeding activity becomes intense enough to result in severe "eat-outs," the removal of all or most emergent vegetation. When stressed by shortages of their preferred foods, muskrats will also feed on animal material, including mollusks, crayfish, fish, and other muskrats.

As with other microtines, muskrats may exhibit wide population fluctuations. Such fluctuations have been suggested to operate on about a 10-year cycle, but many environmental factors can enter in as well. Thus, precise periodicity of these fluctuations is obscure. In the semiarid Great Plains, lack of snow cover may result in freeze-out conditions, which preclude under-ice foraging, and increase vulnerability to predation and intraspecific strife. Such conditions in North Dakota have resulted in almost total decimation of local populations.

Principal causes of mortality include predation (especially by mink, foxes, and raccoons), trapping, intraspecific strife, starvation, and various diseases. Tyzzer's (Errington's) disease is a hemorrhagic disease common in muskrat populations. Healthy animals appear to be able to carry the bacterial spores with no ill effect, but under stress, whole populations may be decimated. A variety of other diseases can also infect muskrats, including tularemia, pseudotuberculosis, and leptospirosis.

While muskrats can be active at any time, peaks of activity occur during late afternoon and late evening hours. Home ranges are small, with most activity occurring within 15 m of a lodge. Territorial behavior during the breeding season may be vigorous, especially among females, while during other seasons, muskrats may be very social. Extensive spring dispersal occurs, resulting in repopulation of vacant habitats. A second dispersal peak may occur during the fall, associated with animals being forced out due to declining water levels. These spring and fall dispersals result in extensive mortality.

Reproduction

The muskrat has been described as "loosely monogamous," but polygamous and promiscuous breeding behaviors have also been reported. The breeding season is largely restricted to the spring and early summer months, beginning in the northern states in early April. Average gestation lasts 25 to 30 days. In North Dakota, most litters are born between May 1 and June 10, with a sharp decline after the end of July. Average litter size in North Dakota is seven, with females producing two (occasionally three) litters in a season. Young are weaned at 4 weeks of age and, depending on date of birth, may reach adult size by the following winter.

Status and Conservation

The muskrat is one of the state's most common furbearers, and may be found wherever suitable water conditions exist. Stable populations require stable water conditions and suitable emergent vegetation. Yet prairie marshes are dynamic ecosystems and muskrat populations fluctuate accordingly.

Past muskrat harvests have provided a significant supplementary income for rural North Dakotans. The highest recorded harvest was 325,000 pelts taken in 1945. More recently, with declining fur markets and diminished interest in trapping, North Dakota harvests have declined, currently averaging 30,000 to 40,000 pelts per year. Nationwide, however, the species continues to be a valuable furbearer, generating $10 million to $20 million annually.

High muskrat populations can have a significant effect on wetlands with "eat outs," resulting in loss of emergent vegetation. Farm crops are occasionally damaged by their foraging. Their burrowing activities may cause damage to roads, dikes, and river banks. Control measures include trapping, shooting, poisoning, water level manipulation, and "rip-rapping" of banks to reduce burrowing. Nevertheless, when in moderate numbers, the species remains an integral component of healthy wetland ecosystems.

Selected References

Adams (1961), Erb and Perry (2003), Higgins et al. (2000), Errington (1963), Sather (1958), Seabloom (1958), Seabloom and Beer (1963), Willner et al. (1980), Wobeser (1985)

Northern Grasshopper Mouse

Onychomys leucogaster

Michtika, Michtik-tak–Mandan

R. B. Forbes, American Society of Mammalogists Mammal Images Library

Description

Grasshopper mice are heavy-bodied animals with short tails. Weighing an average of 42 g, they are about twice as heavy as deer mice. The fur is distinctly bicolored, gray to brownish gray above, and the underparts are white. Average standard body measurements are: Total Length - 157 mm, Tail Length - 43 mm, Hind Foot - 22 mm, Ear - 14 mm. Northern grasshopper mice might be confused with deer mice and white-footed mice *(Peromyscus)*, but are easily distinguished by their larger size, stout bodies, and short tails. Two subspecies have been recorded in North Dakota, a larger, darker *O. l. leucogaster* in the eastern half of the state, and a smaller, paler *O. l. missouriensis* in the west.

Distribution

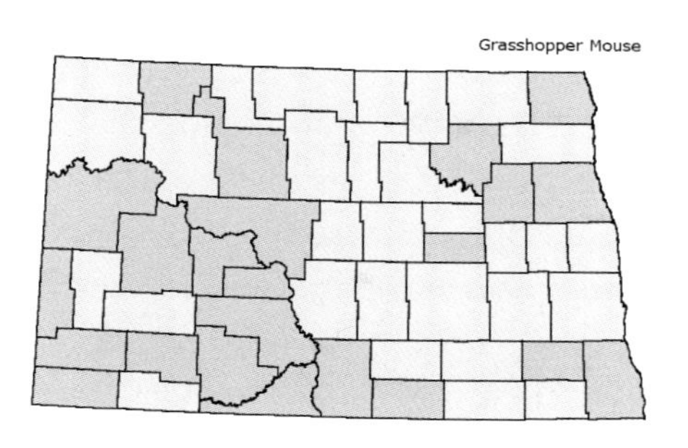

Northern grasshopper mice occur from the southern Prairie Provinces of Canada south to northern Mexico. They are found in grasslands and deserts of the United States from western Minnesota and northwestern Iowa on the east to northeastern California on the west. In North Dakota, there are scattered records from grassland habitats over much of the state.

Habitat

Although occurring as far east as the tall-grass prairies of western Minnesota, northern grasshopper mice are primarily animals of semi-arid grasslands and shrublands. They seem to prefer areas with sandy soils, which allow frequent dust bathing to maintain their pelage. In the Great Plains, they tend to occur more in areas of mixed grasses and forbs, compared to open areas, sage, or solid grass, but are also commonly associated with croplands. Preferred microhabitats for grasshopper mice include areas with high densities of pocket gopher mounds, which attract their arthropod prey. Common North Dakota habitats include haylands and moderately grazed rolling grasslands in the southwest, and rolling sandhill prairies of the Sheyenne National Grasslands in the southeast.

Ecology and Behavior

Grasshopper mice are unusual among regional small rodents for their carnivorous habits. Nearly 90% of the diet may consist of insects and other animal prey, predominant groups including grasshoppers and crickets, beetles, and caterpillars. Other small mammals may be taken, making up about 3% of the diet. They are capable of killing adult deer mice, meadow voles, pocket mice, and others. A small proportion of the diet during the warm months is plant material, including leaves of forbs, grasses and sedges, and seeds. During winter, however, they may switch to a diet of mostly seeds, which are cached in and around the nest.

Northern grasshopper mice construct four types of burrows for nests, retreats, food caches, and miscellaneous purposes such as defecation and territorial marking. Nest burrows are shallow, U-shaped with three entrances, and have a centrally located nest chamber. These are excavated by a mating pair. The other burrow types are relatively simple structures. Retreat burrows are used when animals are frightened, while cache burrows are filled with seeds and covered with sand. Defecation and territorial burrows are the smallest, under 5 cm long. Sand bathing sites are usually associated with territorial burrows.

The species is nocturnal and rarely active during daylight. Activity is greatest during dark nights with cloud cover, and least under full moonlight. During their active periods, grasshopper mice can be quite vocal, including sharp squeaks when threatened and long shrill whistles when hunting or mating. Their calls are easily heard by humans, leading some naturalists to refer to the "mouse that howls."

Northern grasshopper mice are highly territorial, although specific territory size is unknown. Territories are defended by scent marking, vocalization, and direct aggressive behavior. Intruders of the same or different species are actively pursued and, if persisting in their intrusion, may be attacked and killed by a bite to the base of the skull. Grasshopper mice have been known to kill intruders up to three times their own size.

There are few detailed studies of the population dynamics of northern grasshopper mice. Home ranges are large for their size, estimated at about 2.3 ha, probably characteristic of their predatory habits. Data on population densities are also lacking, but many surveys indicate that when present, the species may be common, but not abundant. Populations are not thought to be regulated by predation, but grasshopper mice are known to be taken by some hawks and owls. Foxes and coyotes are thought to kill, but usually not eat them because of their musky odor. Northern grasshopper mice have been known to live as long as four years in captivity, but under natural conditions probably do not survive more than one or two years.

Reproduction

Grasshopper mice form male-female pair bonds, but it is unknown if this is a form of monogamy or chance encounters of individuals living in sparse populations. Courtship behavior is complex, involving mutual chases, naso-naso and naso-anal sniffing, grooming, and somersaults prior to copulation. Both sexes actively defend the nest site, and males participate in retrieval of young to the nest.

The earliest recorded breeding in the Great Plains is March 15, but in North Dakota, the breeding season undoubtedly begins somewhat later. Pregnant females have been collected in North Dakota between the end of May and the first of August. Northern grasshopper mice are polyestrous, producing three to six litters per year.

The gestation period for nonlactating females is about four weeks, and five weeks for lactating females. Average litter size is 3.6, varying from one to six. Like other small rodents, newborn young are naked, and the eyes and ears are closed. They begin to move out of the nest by day 4, and have acquired fur by about a week. They begin to take solid food at 10 days, and their eyes and ears are open at about 2 weeks. Sexual activity may begin at 3 months in females and 4 months in males, but it is doubtful that they breed until the following spring.

Status and Conservation

Northern grasshopper mice are fairly common in southwestern North Dakota but, except for the Sheyenne National Grasslands, seem to be rare in the eastern half of the state.

This sparsely distributed species does no economic damage and, in areas where it does occur, probably destroys large numbers of insect pests. It is rare in much of eastern North Dakota, and in neighboring Minnesota has been suggested as a species of special concern.

Selected References

Bailey (1926), Genoways and Jones (1972), Hazard (1982), Jones et al. (1983), Lysne (1991), McCarty (1978), Seabloom et al. (1978), Svihovec (1967)

White-footed Mouse

Peromyscus leucopus

Wiyashpena – Dakota

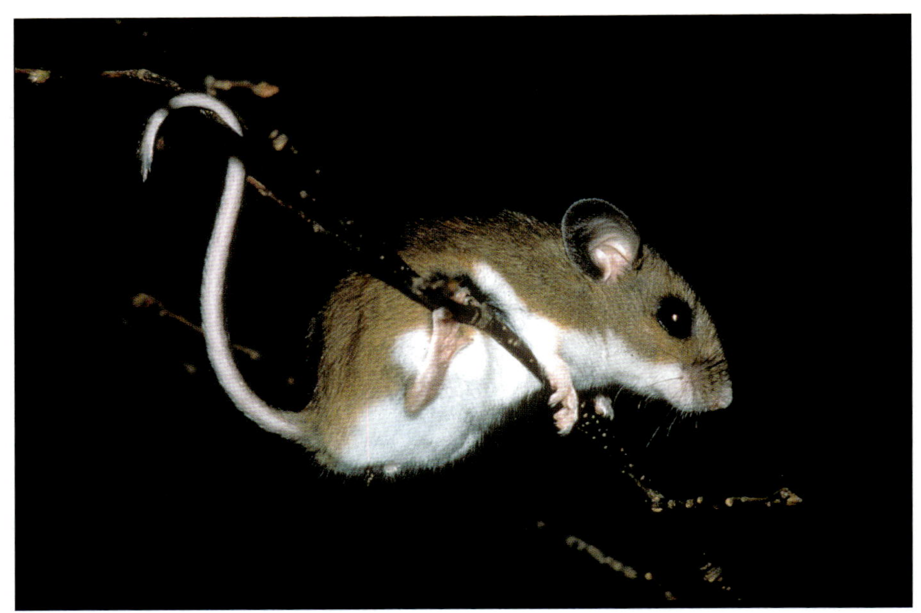

L. L. Master, American Society of Mammalogists Mammal Images Library

Description

White-footed mice are very similar in body form and coloration to deer mice which, in spite of differing habitat preferences, frequently live in close proximity. In contrast to the voles and lemmings, these mice have large eyes and ears, longer legs, and long tails. Adults have brown backs and white underparts, while the backs of juveniles are grayish. The tail is "indistinctly" bicolored, in contrast to the "distinctly" bicolored tail of the deer mouse. Average standard body measurements are: Total Length - 173 mm, Tail Length - 75 mm, Hind Foot - 22 mm, Ear - 16 mm. Body weights average about 30 g. On the Great Plains, white-footed mice can be distinguished from prairie deer mice by their longer, indistinctly bicolored tails, but comparison with known specimens is highly recommended. Identification of specimens from the prairie-forest transition of North Dakota and Minnesota is very difficult because of intergrades with longer-tailed deer mice of the eastern boreal forests.

Distribution

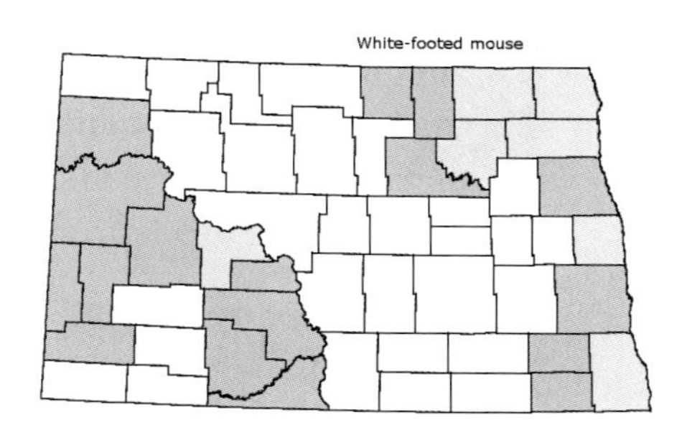

The northern range limit of the white-footed mouse extends from southeastern Alberta, southern Saskatchewan, North Dakota, Minnesota and the other Great Lakes states to the eastern seaboard. To the west, it is in eastern Montana, northeastern Wyoming, southeastern Colorado, and portions of New Mexico and Arizona. It extends into eastern Mexico as far south as the Isthmus of Tehuantepec and Yucatan peninsula. White-footed mice are in much of the eastern United States, except for the Gulf Coast plain.

White-footed mice occur throughout the forested areas of North Dakota, but not in large numbers. In the east, there are reports from wooded areas adjacent to the Red River and its major tributaries, the Turtle Mountains, and the Devils Lake region. In the west, there are scattered records from wooded areas along the Missouri River and its tributaries.

Two subspecies of the white-footed mouse have been reported in North Dakota. A darker form, *P. l. noveboracensis*, is in the far eastern portion of the state, continuing with populations in Minnesota and eastern South Dakota, while the paler *P. l. aridulus* occurs in the remainder of the state.

Habitat

Preferred habitat of white-footed mice is generally deciduous forest, woodlots, and brushy areas. Occasionally, small numbers may be found in grassy fields, but these may be dispersing young. Recent studies have indicated that highest population densities occur in smaller forest patches, possibly related to a relatively greater amount of edge or greater complexity of the understory. The availability of major food sources, such as acorns, also influences the abundance of white-footed mice.

In eastern North Dakota and northwestern Minnesota, they occur in riparian forest along the Red River and in well-developed aspen groves having an herbaceous or brushy understory. Specimens from the Turtle Mountains (Rolette County) and Devils Lake area (Ramsey County) have been taken in mature deciduous forest. Southwestern

North Dakota habitats include riparian forests along the Missouri River, deciduous forests of the Killdeer Mountains (Dunn County), and upland breaks and green ash-dominated woody draws in the Badlands. Grazing regimes apparently influence their occurrence in southwestern North Dakota.

Ecology and Behavior

White-footed mice are nocturnal animals, their activity periods closely correlated with sunrise and sunset. They are semiarboreal, readily climbing in trees and shrubs for foraging and escape. When traveling on the ground, they establish well-defined trails and travel areas, frequently over a half meter wide. Travel lanes usually follow logs, brush piles, stumps, or other significant cover. Nests are constructed of shredded plant material, and may be nearly 20 cm in diameter, and located under logs, stumps, or rocks, in hollow trees, or even in abandoned squirrel or bird nests.

Home range size is highly variable, a function of season, sex, age, food supply, and population density. The largest home ranges are during the breeding season, and the smallest during winter. Males typically have larger home ranges than females. Typical home ranges vary from <0.1 to 0.5 ha. There is some evidence that female white-footed mice defend territories, at least during the breeding season.

White-footed mice consume a wide variety of seeds, nuts, insects, and other small animals as they become available. Seeds and insects sometimes make up over 70% of the diet. Acorns are an important food source and abundant crops can result in high populations of white-footed mice. White-footed mice, along with other members of the genus, undergo spontaneous daily torpor, an adaptive mechanism resulting in energy savings, especially during periods of low temperatures or food shortages. These bouts last several hours, occur mostly in the morning, and result in significantly depressed body temperatures and heart rates.

Although populations may vary greatly annually and seasonally, they are not usually subject to the extreme fluctuations seen in other species, such as the voles. Reported densities range from 0.4 to 26.9 mice per ha, with the largest populations generally occurring during fall. Based on existing records, North Dakota populations are probably at the low end of the scale. While white-footed mice have been known to live several years in captivity, few survive as much as a year in nature. Mortality in the nest can be as high as 31%, while 58% of the juveniles may be lost within two weeks of weaning. Losses of subadults and adults can be nearly 50% within two weeks of capture, and annual overall mortality has been estimated at about 96%. Factors regulating population size have included food supply, territoriality (primarily by females), and reduced recruitment of breeding adults resulting from declines in reproduction or increased juvenile mortality. Predation is high on white-footed mice, and they provide a staple food source for snakes, owls, hawks, weasels, skunks, foxes, and bobcats.

Reproduction

In the northern portions of the range, breeding is seasonal, with peaks during spring and fall. Diminished breeding during mid-summer has been attributed to a high proportion of juvenile females in the population, or to a slump in breeding by all females. During these breeding peaks, females come into estrus every six days, and also exhibit a postpartum estrus. Litters of three to five young are born following a gestation of 23 days. Gestation may be delayed by as much as two weeks in lactating females. Newborn young weigh less than 2 g, are naked, and the eyes and ears are closed. Their eyes are open by 12 days, and their juvenile pelage is complete by 3 weeks. Weaning may begin at 19 days, but not be completed until about a month. By 6 weeks, young have achieved 90% of adult body mass. Females go into their first estrus at around 45 days. Hence, considering their high rates of mortality, most of the fall breeding peak probably involves young of the year animals. .

Status and Conservation

Based on existing reports, white-footed mice are considered to be present, but uncommon, in forested areas throughout North Dakota. Because of their small numbers in North Dakota, white-footed mice are of little economic significance. They may get into buildings or interfere with forest reseeding, but also consume large numbers of insects. The species is known to carry Sin Nombre Virus, a strain of hantavirus known to cause serious human illness. However, the incidence of infection of this virus in North Dakota white-footed mouse populations has been so low that its significance to human health is not as serious as in the closely related deer mouse *(P. maniculatus)*, the major carrier.

Selected References

Bailey (1926), Falls (1968), Genoways and Jones (1972), Iverson et al. (1967), Jones et al. (1983), Lackey et al. (1985), Layne (1968), Naylor (1987), Seabloom et al. (1978), Seabloom et al. (1999), Shaughnessy (2016), Stickel (1968), Terman (1968)

Deer Mouse

Peromyscus maniculatus

Tepa-uti – Omaha

L. L. Master, American Society of Mammalogists Mammal Images Library

Description

Deer mice are among the most beautiful of North Dakota small mammals, and very similar in overall appearance to the forest-dwelling white-footed mice. The upper parts are brown to grayish-brown in adults, and gray in juveniles. The under parts, including feet, are white. The tail is "distinctly" bicolored, dark brown above and white below. Body weight averages about 22 g. Average standard body measurements are: Total Length - 152 mm, Tail Length - 62 mm, Hind Foot - 18 mm, Ear - 15 mm. In the prairie-forest transition of eastern North Dakota and Minnesota, these deer mice intergrade with a boreal forest subspecies that has larger ears, feet, and tail. These deer mice are difficult to distinguish from the deciduous forest-dwelling white-footed mice, except for their distinct tail bicoloration. Therefore, comparison with known specimens is recommended.

Distribution

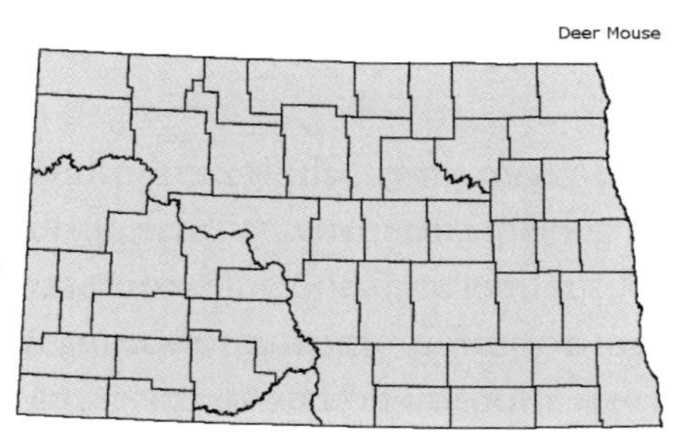

The deer mouse is one of the most widely distributed mammals in North America. It occurs in southeastern Alaska and all of Canada south of the tree line. Its southerly limits are in Baja California and the Isthmus of Tehuantepec of Mexico. It occurs throughout the continental United States, except for the eastern seaboard south of New England, and the southeastern states.

Deer mice occur throughout the grasslands of North Dakota, and may well comprise the state's most common and abundant species. They have been collected from nearly every locality of the state, except for the heavily forested areas and wetlands. They may also be found in woodland edges, tree plantings, and aspen groves.

Two subspecies of deer mice occur in North Dakota, a smaller, darker *P. m. bairdii* in the east and a larger, paler *P. m. nebrascensis* in the west. There is no sharp demarcation between these two, rather a gradual gradation in body size and color between the more moist east and semiarid western portions of the state.

Habitat

As their continent-wide distribution implies, deer mice inhabit a wide variety of habitats, from the boreal forests of the far north to the desert grasslands of the Southwest. However, the occurrence of a given subspecies may be limited to a specific set of habitat conditions. The two subspecies in North Dakota prefer early successional stage grasslands, including areas disturbed by fire or grazing. In the southwest, they are often associated with prairie dog towns. In eastern North Dakota, deer mice are abundant in roadside ditches and areas disturbed by grazing, cultivation, or fire. However, this habitat specificity seems to break down in the prairie-forest transition because they can also be found in river-bottom forests and aspen groves. Populations increase following grazing and fire due to invasion from surrounding areas. Numbers decrease one or two growing seasons following fire with the accumulation of ground litter. Preferred habitats in southwestern North Dakota include haylands, grazed hardwood draws, moderately grazed rolling grasslands, and upland grasslands.

Ecology and Behavior

Like the white-footed mouse, deer mice are nocturnal, with most of their activity occurring between the hours of sunset and sunrise at all seasons of the year. Adult activity appears to be greatest during the early hours of darkness, while that of subadults is more variable. It has been suggested that foraging may dominate the early hours of activity, but examination of stomach contents has revealed that most actual feeding is during the later stages of the activity cycle. Much of their foraging activity may involve the caching of foods in protected sites. A single mouse may cache up to 3 liters of seeds for winter use. Like other species of *Peromyscus*, deer mice undergo spontaneous daily torpor, enhancing their survival during periods of low temperatures and food shortages.

As stated above, prairie deer mice prefer open habitat. They do not tend to use developed runways like many other small mammals. Nests may be constructed under rocks, logs, or other debris, or in the abandoned burrows of other species. The ball-shaped nest consists of grass, other shredded plant material, and fur.

Deer mice are omnivorous, their primary diet consisting of insects and seeds. Up to 60% of the summer diet may be in the form of grasshoppers, beetles, caterpillars, and other insects. Seeds of weedy species, grasses, and grains make up the bulk of the winter diet. Nuts, small fruits, and fungi may also be included in the diet, but green plant material is only a minor component.

Home range size can vary from 0.03 to 1.2 ha, annually, by season, habitat, and population density. Home ranges can have considerable overlap. Individuals may even share nests during winter. Deer mice exhibit a strong fidelity to their home ranges, and homing ability to return to them when displaced. Reports of territorial behavior are highly variable, ranging from complete home range overlap, to no overlap (implying territoriality), territorial pairs, and nest sharing by lactating females and their litters.

Population density estimates of deer mice living in northern prairie areas range from 0.7 to 22.2 mice/ha. One study in western North Dakota estimated 6.8 mice/ha. Highest densities are typically in the fall, resulting from spring recruitment, and spring-born females engaging in their first reproduction. Mortality is very high in deer mouse populations, and studies indicate that only about 1% may survive into their second year. Potential life span in captivity can approach three years or more. As with other small mammals, mortality can be attributed to many factors. Many species of predators, including snakes, hawks, owls, and small- to medium-sized predatory mammals, account for many losses. Food shortages, combined with the severe winters on the northern Plains, are additional significant factors contributing to mortality.

Reproduction

Deer mice typically have breeding peaks in spring and early fall, with breeding beginning in March and ending in October. Females are polyestrous, and may have up to three or four litters per year. Males are often accepted into the nest, and may assist in caring for the young. The gestation period is about 23 days or longer if the female is nursing a litter. Litter size averages about four, but may range from one to nine. Newborn young weigh less than 2 g, are naked, and their eyes and ears are closed. Their eyes are open and they are fully furred by 2 weeks. Most young are weaned by 3 weeks, and achieve adult size by 6 weeks. By this time, most have dispersed, except during fall and winter, when they may remain with the mother for up to two months. First estrus occurs at about 9 weeks. Hence, females born during the first littering peak are fully capable of reproduction by late summer or fall.

Status and Conservation

Deer mice are common and abundant throughout North Dakota. Their seed-eating habits may interfere with plantings, and they readily get into stored grains. However, their insectivorous behavior may well offset any significant crop damage. They readily enter and establish nests in inhabited buildings, where they are regarded as pests.

Deer mice are primary reservoirs of Sin Nombre Virus, a hantavirus highly pathogenic to humans. In North Dakota, infection rates in deer mice averaged 16 to 19%, but in some localities were as high as 40%. Human infection is primarily via inhalation of dried urine and fecal material carrying the virus. Although this respiratory disease is not easily contracted, it is very serious and mortality is high, especially if it is untreated. As of 2004, there had been seven cases in North Dakota and 363 in the United States. Preventive measures include the wearing of protective masks and rubber gloves when cleaning mouse-infested buildings, and disinfecting nests, fecal material, and dead animals before removal. Mice can be controlled in buildings through the use of traps and rodenticides. Further information can be obtained from the National Center for Infectious Diseases.

Selected References

Bailey (1926), Dice (1940), Eisenberg (1968), Falls (1968), Genoways and Jones (1972), Hall (1981), Hazard (1982), Hibbard (1972), Hooper (1968), Iverson et al. (1967), Jones et al. (1983), Layne (1968), Naylor (1987), Nellermoe (1983), Seabloom et al. (1999), Seabloom et al. (1978), Stickel (1968), Svihovec (1967), Terman (1968), Tester and Marshall (1961), Walley (1970)

Western Harvest Mouse
Reithrodontomys megalotis

Bob Gress

Description

Weighing only 12 to 15 g, the western harvest mouse is one of the smallest rodents in North Dakota. Its dorsal fur is grayish brown, and the underparts, including the feet, are grayish white. The tail is furred and bicolored. Average standard body measurements are: Total Length - 136 mm, Tail Length - 62 mm, Hind Foot - 16 mm, Ear - 12 mm. Superficially, harvest mice might be confused with pocket mice, which are about the same size but are lighter brown and have external cheek pouches, or house mice which are larger, totally gray, and have naked, scaly tails. The species is very similar to the plains harvest mouse, but is larger, has a tail longer than the head and body, and the tail is less distinctly bicolored.

Distribution

The western harvest mouse occurs from southern Saskatchewan, Alberta, and British Columbia south over much of midwestern and western United States into Mexico as far as the Isthmus of Tehuantepec. The eastern limit of its distribution is western Indiana. In North Dakota, it has been recorded over much of the southern half of the state, as far north as Dunn and Oliver counties in the west, and Cass County in the east.

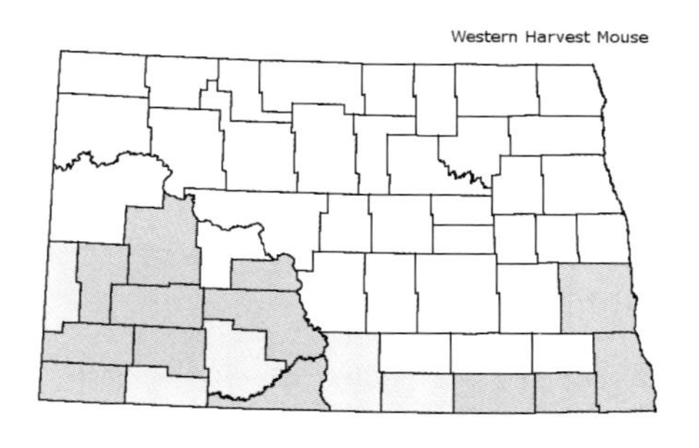

Habitat

Western harvest mice inhabit a wide variety of habitats, characterized mainly by dense vegetation, and only seem to avoid dense forests and dry uplands. They appear to be most abundant in ungrazed or lightly grazed grasslands which have not been subjected to burning. In North Dakota, they have been found in hayland, retired cropland, lightly grazed rolling grassland, cottonwood bottoms, lowland meadows, dense brush, and cattails.

Ecology and Behavior

Nests of western harvest mice have been described as "neat little nest balls on the ground or on low bushes or weeds — rarely over 8 to 10 inches above the ground." The nest, rarely over 12.5 cm in diameter, consists of a compact ball of dry grass lined with soft fibers, having a small opening at one side.

Western harvest mice eat mainly seeds, but some leaves and insects, including beetles, weevils, spiders, and moth larvae are also taken. Small grains, such as oats, barley, and wheat are taken, but probably not in significant quantities.

Like many rodents, harvest mice are strictly nocturnal, and seem to be most active on moonless, rainy nights. These mice engage in significant movements for their size. Live trapping studies indicate movements of 70 to 102 m between capture points. They do not appear to be territorial. Females with litters tolerate males, and group huddling for heat conservation is common.

Population densities have been estimated to average four to five per ha, but have also been estimated to increase to 60 per ha in dense vegetation. Little is known about longevity in this species, but females appear to reach senility (based on reduced litter sizes) at about 45 weeks of age, and few mice are thought to live past 1 year. Western harvest mice are preyed upon by numerous snakes and predatory mammals and birds.

Reproduction

Western harvest mice breed year-round in the warmer parts of the range, but in more northern areas they breed from spring to fall. In North Dakota, pregnant females have been collected from mid-June to late July, but there are no known studies of harvest mice in the state during other seasons. Females are polyestrous, and in captivity have produced as many as 14 litters in a single year.

Young are born following a gestation of 23 to 24 days. The average litter size is four, ranging from two to six. As with other small rodents, newborn young are pink, hairless, eyes are closed, and weigh about 1 to 1.5 g. They are furred within a week, eyes and ears are open at 11 days, and they are weaned at 24 days. Females are sexually mature at 4 months, a significantly longer maturation period than other small mammals such as deer mice.

Status and Conservation

The western harvest mouse is relatively common to abundant in southwestern North Dakota. However, in the east it is absent to rare. In one study, only 1 specimen was collected in a trapping effort of 14,236 trap nights in agricultural lands of Cass and Richland Counties, and Clay County, Minnesota. The western harvest mouse is extremely rare in western Minnesota, and it has been suggested that it be considered threatened in that state.

Selected References

Bailey (1926), Genoways and Jones (1972), Hazard (1982), Hoofer et al. (1999), Jones et al. (1983), Nellermoe (1983), Seabloom et al. (1978), Svihovec (1967), Webster and Jones (1982)

Plains Harvest Mouse

Reithrodontomys montanus

Description

Like the western harvest mouse, the plains harvest mouse is one of our smallest rodents, weighing only 10 to 13 g. The upper parts are pale brown, while the under parts are yellowish gray and white. There is typically a broad stripe on the back, darker than the surrounding pelage, and buffy patches behind the ears. This species is distinguished from the very similar western harvest mouse by its smaller size, the tail shorter than the head and body, the dorsal stripe, a very narrow dorsal stripe on the tail, and buffy post-auricular patches. Average standard body measurements are: Total Length - 124 mm, Tail Length - 55 mm, Hind Foot - 16 mm, Ear - 12 mm.

Distribution

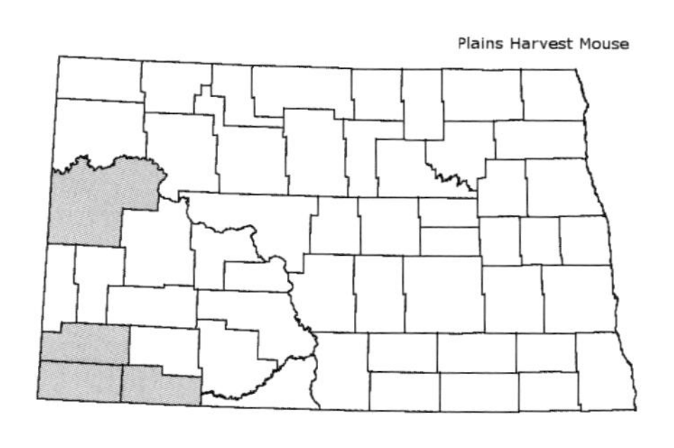

The plains harvest mouse is primarily a species of the central and southern Plains of North America, extending from southwestern North Dakota south into the Mexican states of Chihuahua, Sonora, and Durango. Its western limits are in southeastern Arizona, while to the east it occurs throughout Nebraska, and most of Oklahoma and Texas. Until recently, plains harvest mice were not known to occur in North Dakota. However, it is now known from scattered locations in Slope, Bowman, Adams, and McKenzie counties.

Habitat

Plains harvest mice are said to inhabit open grassy areas, preferring short grasses and herbs. In rocky, sparsely vegetated sites, they use surface rock for cover. They also occur in small-grain fields and sagebrush-yucca associations. Most North Dakota specimens have been taken from grassland habitats.

Ecology and Behavior

Although plains harvest mice and western harvest mice are sympatric throughout much of the Great Plains, they appear to be ecologically isolated from each other. Plains harvest mice tend to inhabit more sparsely vegetated areas, while western harvest mice occur in a wider variety of densely vegetated habitats, avoiding only dry uplands and forests.

Insects (especially grasshoppers) and a variety of plants are consumed by plains harvest mice. Seeds of buffalo grass and switchgrass are primary foods, but other species, including seeds of blue grama, Indian grass, cactus fruits, and seeds and flower heads of sunflowers, broomweed, ironweed, and snow-on-the-mountain, also contribute to the diet.

Nests are small balls of grasses, about 10 by 6 cm in size, suspended above ground by taller grasses. In short-grass or sparsely vegetated situations, they may be on the ground, in tin cans or other objects, or under logs. Fine material, such as milk weed or thistle down, may line the interior of the nest.

Like other harvest mice, the plains harvest mouse is probably strictly nocturnal. It does not hibernate, but remains active throughout the year. Home ranges of under 1 ha are typical.

Because of the difficulty in identification of skeletal remains to the species level, there are few data on specific predators of plains harvest mice. They are undoubtedly included in the diets of most small predators, including raptors, foxes, coyotes, weasels, and snakes. Longevity data are also sparse but, like most small mammals, the majority of individuals cannot be expected to survive past their first year. Capture-recapture data have indicated a maximum life span of about 14 months.

If information on the closely related western harvest mouse is comparable, females could be expected to reach senility at 14 weeks of age.

Reproduction

The onset of breeding in the northern part of the species range is unknown, but in the central Great Plains, gravid females have been reported in every month except December and January. Further south, they are thought to breed throughout the year. In North Dakota, breeding probably occurs during spring, summer, and fall. Females are polyestrous, littering every 21 to 27 days.

Litter sizes average about four, ranging from one to nine. Like most small mammals, newborn young are altricial; they are blind and naked, weighing about 1 g. They are furred by day 6, eyes are open by day 8, and weaning occurs at 2 weeks. Adult size is achieved at 5 weeks, and sexual maturity at 2 months.

Status and Conservation

Little is known of the status of the plains harvest mouse in North Dakota, except for scattered records from the 1960s in Slope and Bowman counties. Subsequent surveys during the 1970s did not encounter the species. Therefore, it should be regarded as rare to uncommon in extreme southwestern North Dakota.

Like the western harvest mouse, the plains harvest mouse is an innocuous species, contributing to the faunal diversity of North Dakota. Its status needs clarification in southwestern North Dakota through further surveys of potential habitat.

Selected References

Genoways and Jones (1972), Hoofer et al. (1999), Jones et al. (1983), Seabloom and Shaffer (2005), Seabloom et al. (1978), Seabloom et al. (1999), Shaunessy (2016), Wilkins (1986)

Family Muridae
Old World Rats and Mice

House Mouse
Mus musculus

L. L. Master, American Society of Mammalogists Mammal Images Library

Description

House mice are about the same size as deer mice and white-footed mice, weighing from 13 to 35 g. However, they are totally gray to gray-brown, their underparts being only slightly lighter than their backs. Their eyes are relatively small, ears are naked and prominent, and their tails are very sparsely haired, appearing scaly. Average standard body measurements are: Total Length - 160 mm, Tail Length - 75 mm, Hind Foot - 18 mm, Ear - 12 mm.

Distribution

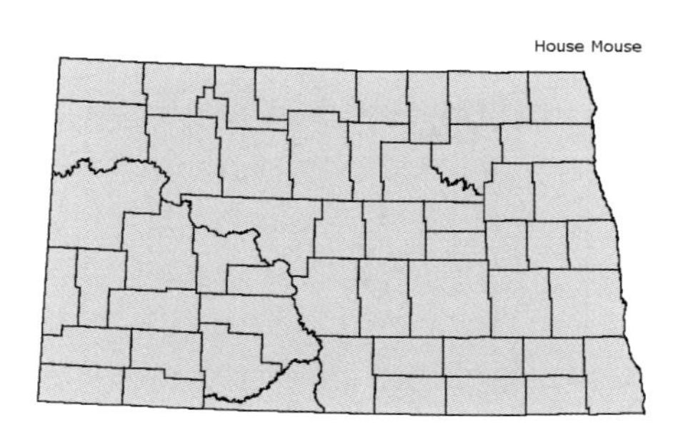

Having originated somewhere between the Mediterranean region and China, house mice are now world-wide in distribution, following human settlement and trade routes. They arrived in North America during white settlement, and now occur throughout Canada, United States, and Mexico.

House mice quickly became established in North Dakota, following white settlement, and are now common and abundant throughout the state, though usually associated with habitation.

Habitat

Wild populations of house mice can live in a variety of habitats, including forests, savannahs, rocky areas, and grasslands. In North Dakota, during the warm months, they have been collected in grasslands, hay fields, and grain fields. However, survival during the severe winters of the northern Plains is contingent on their proximity to habitation. Hence, year-round populations are found associated with dwellings, barns, grain bins, and similar structures.

Ecology and Behavior

House mice living in the wild have a number of behavioral and demographic characteristics distinct from those living associated with humans (commensals). Both may live in the same area, but in northern climates, only commensal populations are likely to survive year-round. During the warm months, wild house mice build nests in rocky crevices or underground burrows. Burrow complexes have several chambers for nests and food storage, and several entrances. When in buildings, commensal mice build their nests behind rafters, in hollow walls, behind wood piles, under stored hay, or in any protected sites near a food supply. Nests are constructed of paper, rags, or other soft material, and lined with finer shredded material.

Most activity occurs during dawn and dusk, but there are additional minor peaks during the hours of both light and dark. In the laboratory, with strict light/dark regimes, they are primarily nocturnal, but also show significant activity during the light cycle.

House mice are omnivorous, feeding on many kinds of plant and animal materials as available. Seeds, roots, stems, and leaves are all taken, including agricultural products such as corn, soybeans, and grains. They readily feed on insects, including beetle larvae, caterpillars, and cockroaches. Any accessible human foods and other products such as paste, glue, and even soap, may be taken as well.

Home ranges differ greatly between commensal and wild house mice. Commensal mice move only a few meters on a daily basis, and may have a home range as small as 10 m^2. In wild populations, home ranges may be hundreds or thousands of m^2, and dispersing individuals have been known to move up to 2 km. Territorial behavior may occur in commensal house mice, but is less pronounced in wild populations. When territories occur, they are established by a dominant male and incorporate one or more females. If there are multiple females within a territory, they may establish a loose dominance hierarchy, but are less aggressive than the males. Territories are defended by both males and females. Occasionally, one or more subdominant males are allowed in a territory, but others are forced into nonterritorial areas. As young mice mature, they are forced to disperse.

Commensal populations are generally more stable and may have densities as high as 10/m^2, while wild populations may be relatively unstable and have much lower densities of about one/100 m^2. Although wild populations typically have much lower densities than commensal populations, they have a high tendency for irruptions to plague proportions approaching 100 mice/ha. Mice in wild populations also have a greater tendency for dispersal. Mortality and population turnover is higher in wild compared to commensal populations. Mortality of young prior to independence is 60 to 70%; overall, mortality may run 30% per month, and overwintering losses may be as high as 90%. Longevity of captive individuals averages about two years, but may run as high as six years. While many wild and domestic predatory birds and mammals take house mice, low temperature has been cited as the greatest factor limiting survival. Epizootic diseases, such as pneumonia and streptococcal infections, are the only significant mortality agents in high-density populations.

Reproduction

Wild house mice usually breed seasonally, but most commensal mice (the usual situation in the northern states) breed throughout the year. Females are polyestrous, and typically produce five to 10 litters per year. Females also undergo a post-partum estrus, and are capable of breeding again 12 to 18 hours after giving birth. The gestation period is 19 to 21 days or longer if the female is nursing a litter. Litter size averages five or six, but may range from three to 12. Newborn young are naked and blind at birth, and weigh about 1 g. They are furred and their eyes are open at 10 days to 2 weeks. Weaning occurs at 3 weeks, and young attain sexual maturity at 5 to 7 weeks.

Status and Conservation

House mice are common and abundant in North Dakota wherever there is human habitation and adequate food resources. Wild populations may occur seasonally during the warmer months.

House mice have rapidly spread throughout the world, following human settlement. They are significant agricultural pests, consuming stored grains and further contaminating more with their excreta. In houses, they damage woodwork, upholstery, and clothing. Like many wild mammals, they are carriers of a number of diseases infectious to humans, such as Salmonella, rickettsial pox, and tularemia, but overall are considered relatively unimportant as public health risks.

Large-scale eradication of house mouse populations is nearly impossible because of their reproductive potential and adaptability. However, they can be locally kept to acceptable levels by vermin-proofing stored foods, sanitation, and removal of debris and materials favored for nesting. Direct control is accomplished through the use of mouse traps and continuing judicious use of rodenticides, such as anticoagulant baits.

Selected References

Bailey (1926), Berry (1981), Blackwell (1981), Bronson (1984), Genoways and Jones (1972), Hall (1981), Hazard (1982), Mackintosh (1981), Nowak (1999), Seabloom et al. (1978), Seabloom et al. (1999), Svihovec (1967)

Norway Rat
Rattus norvegicus

Dr. Antonio Ferreira, © California Academy of Sciences

Description

The Norway rat is easily distinguished from all other North Dakota rodents. Weighing up to 485 g, it approaches the size of the bushy-tailed wood rat *(Neotoma cinerea)*, but is easily identified by its more slender appearance, coarse brown fur, and long scaly tail. Its appearance is similar to that of the house mouse, but 10+ times larger. Average standard body measurements are: Total Length - 390 mm, Tail Length - 175 mm, Hind Foot - 41 mm, Ear - 20 mm.

Distribution

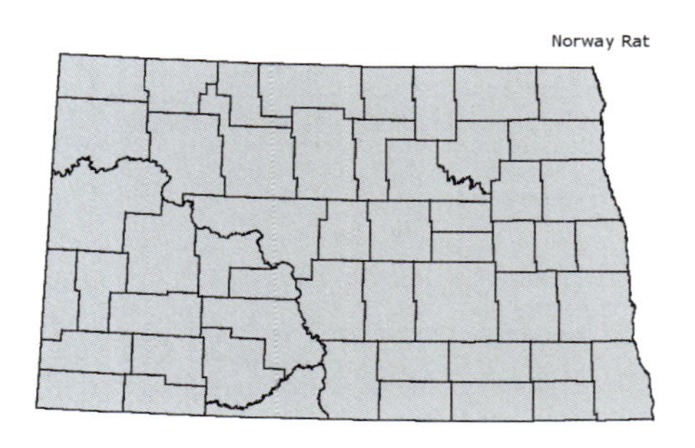

Like the house mouse, the Norway rat has spread throughout the world, following human trade and settlement. It arrived in Europe during the mid-16th century, and was first reported in North America around 1750. It took another 150 years, however, for it to spread throughout the United States. Like the house mouse, it is now continent-wide, primarily associated with human habitation.

The Norway rat is believed to have first entered North Dakota via steamboats on the Missouri River during the 1830s, and quickly became established as a pest wherever foods were stored. The arrival of the railroads in the 1880s provided other arrival routes throughout the state. Norway rats are now found throughout North Dakota.

Habitat

Norway rats are almost invariably associated with humans. In urban areas, they are found in older buildings, trash-filled areas, land fills, and sewers, while in the rural environment they are associated with barns and outbuildings, grain storage bins, and feed lots. These rats are ground-dwellers, constructing long, branching burrows under buildings or debris. When in buildings, they tend to occupy basements and the lower floors. They have been occasionally captured in heavy cover and shelterbelts near farmsteads, and have been reported taken by predators. In contrast to the house mouse, however, Norway rats have not been taken in any of the systematic surveys in the state in recent years, an indication of their close association with human development and activities.

Ecology and Behavior

In mild climates, Norway rats are capable of living year-round in feral situations, establishing extensive burrow systems along stream banks and agricultural fields. However, in more temperate regions, they are almost invariably associated with humans, seeking shelter in dwellings, farm structures, dumps, and trash-filled alleys. They are truly omnivorous, taking all types of human foods, stored grains, birds and their eggs, small mammals, and even soap and hides. They are known to attack larger animals, and even people.

The social structure of Norway rat populations consists of two parallel systems with dominant and subordinate components. Dominant males defend territories containing several females. Other males are excluded, and a male mates only with the females within his territory. The females within a territory collectively maintain nests, nurture their young and participate in territorial defense. When mature, young are forced out of the territory and into the subordinate component of the population. This component is non-territorial. When a female comes into estrus, she is followed and mated by many males, resulting in high stress, and poor reproduction and nest maintenance. Such density-dependent processes may be important factors limiting population size.

Norway rats are nocturnal and active throughout the year. Home ranges vary from about 0.5 to 1.8 ha. In both urban and rural settings, rats may live for long periods within an area of only about 30 to 40 m. However, individuals have been known to move up to 3 km from their home ranges while foraging.

Population densities vary greatly, but under favorable conditions can be as high as 1,000 to 2,000/ha. In a study of a landfill site, the maximum population was estimated at 3,000 to 4,000 rats in a 300 by 600 m area. Mortality within the adult component of the population has been estimated at 17% per month, and 91% per year. Average life span is about two years in the wild, and rats have lived up to four years in captivity.

Habitat is fundamental in controlling population size, and predation and competition are density-dependent factors operating within the constraints of habitat. Certainly, in the northern Plains with its severe winters, rat populations are limited to habitat provided by humans. Predation pressure, primarily by domestic cats and dogs, will become more intense with increasing population size. Competition will tend to increase with increases in density as territorial animals defend more favorable nest sites and available food sources, resulting in higher incidence of failed reproduction and mortality in the subordinate component of the population.

Reproduction

Norway rats are capable of breeding year-round, but reproductive peaks in many areas are in spring and fall. However, in the Great Plains, fertility can be highest in the spring, and drop in the summer with the emergence of juveniles. By the onset of cold weather in the fall, reproduction likely ceases. In buildings and milder climates, reproduction may continue throughout the year.

Female Norway rats are polyestrous, and may have up to 12 litters per year. They go through a post-partum estrus shortly after giving birth, and hence may be nursing one litter while pregnant with the next.

Average litter size is about eight, ranging from two to 22. Newborn young are naked and blind, and weigh about 5 g. By 2 weeks, they are furred and their eyes are open. Weaning occurs at about 3 weeks, and young are sexually mature at 2 to 3 months.

Status and Conservation

Norway rats are probably not as abundant in North Dakota as in many other states because of its climate and sparse human population. Nonetheless, rats are common on farms and in urban areas, where populations can build to high levels.

Norway rats inflict tremendous economic damage through their activities, amounting to billions of dollars annually in the United States alone. They consume and contaminate vast amounts of stored human and animal foods. They gnaw into electrical wiring, causing fires, and have even been known to gnaw into lead pipe and concrete. They are known to carry about 40 diseases, including plague, typhus, salmonella, and tularemia. Rat-borne diseases are thought to have taken more human lives than all of the wars and revolutions ever fought. They are known to kill poultry, livestock, and game birds. There are about 14,000 direct attacks by rats on humans annually in the United States alone.

The obvious and most successful strategy for controlling Norway rat populations is through habitat management. Sanitation, rehabilitation of urban neighborhoods, renovation of old structures and rat-proofing of stored foods have been demonstrated to result in precipitous declines of populations. Trapping, use of rodenticides, and predation by domestic cats and dogs provide some additional control, but these approaches are not generally effective substitutes for habitat management.

Selected References

Bailey (1926), Calhoun (1963), Davis (1953), Foresman (2001), Hazard (1982), Jones et al. (1983), Nowak (1999)

Family Erethizontidae
New World Porcupines

North American Porcupine
Erethizon dorsata

Pahi–Mandan, Dakota; Pahin–Lakota; Apadi^{n}–Hidatsa; Suunu–Arikara

C. D. Grondahl, North Dakota Game and Fish Departmernt

Description

The American porcupine is a large, heavy-bodied rodent, the only representative of its genus in North America. Weights are highly variable, depending on habitat, ranging from 3.5 to 18 kg. Porcupines are stout-bodied animals with relatively short legs and tails. Climbing adaptations include long claws, textured soles of the feet, and a muscular tail with stiff bristles. The body is covered with three kinds of fur, a thick, dark underfur, long white- or yellow-tipped guard hairs, and quills, which range in length from 2 to 10 cm. The hollow, barbed quills are on the back, sides, limbs, and tail, and are used defensively, driven into would-be predators. Average standard body measurements of Minnesota specimens are: Total Length - 686 mm, Tail Length - 180 mm, Hind Foot - 93 mm, Ear - 32 mm.

Distribution

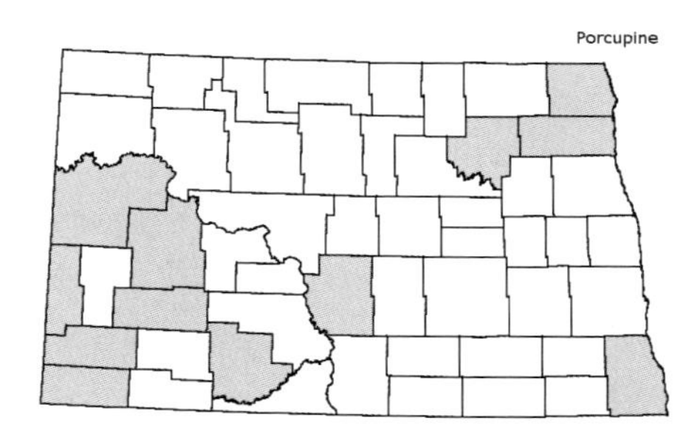

Porcupines occur throughout most of Alaska, Canada south of the tree line, the lake states and New England, and much of the western United States and extreme northern Mexico.

Two subspecies of porcupine occur in North Dakota. *E. d. dorsata*, weighing an average of 6.8 kg and having white-tipped guard hairs, enters eastern North Dakota from Minnesota and Manitoba. It occurs in relatively sparse numbers in woody and brushy areas east of the Missouri River. There are records from Pembina and Hamilton (Pembina County), Grafton (Walsh County), Devils Lake (Ramsey County), and Wahpeton (Richland County). They have also been observed along the Missouri River at Bismarck and Riverdale.

The "yellow-haired" porcupine, *E. d. bruneri*, occurs in the Great Plains, extending into southwestern North Dakota, and is larger (average weight about 11 kg) than *E. d. dorsata*, and has yellowish guard hairs. There are reports of this porcupine from McKenzie, Dunn, Stark, Slope, Bowman, Golden Valley, and Grant counties. It is fairly common throughout much of the southwestern quarter of the state.

Habitat

Throughout much of its range, the porcupine is associated with coniferous forest habitats, but when conifers are unavailable, they can utilize deciduous trees as well. The few eastern North Dakota reports are from riparian woodlands along the Red River and its tributaries and the forested Devils Lake area. They are also known from open areas, riparian forests, and spruce-aspen-oak savannah of southwestern Manitoba. In southwestern North Dakota, yellow-haired porcupines are widespread, especially in areas with woody vegetation, including ponderosa pine, deciduous forest, hardwood draw, riparian woodland, and river and upland breaks habitats.

Ecology and Behavior

Although they have poor vision and are relatively slow and awkward on the ground, porcupines are remarkably well-adapted for an arboreal lifestyle. Through use of their long claws, textured feet, and stiff bristled tails, they can climb rapidly and are difficult to dislodge.

Except during the mating season, porcupines exhibit a solitary lifestyle, and both sexes are territorial. Recorded home ranges of 8 to 73 ha have varied by sex, (males larger), habitat (larger in more open country), and season. During winter, movements are more restricted, and their solitary habits may break down to some degree. Sharing of dens by several porcupines has been observed during winter, especially in situations with few available den sites.

Dens may be in rock crevices, in hollow trees, abandoned buildings, or under logs or deadfalls, and may be identified by large accumulations of fecal pellets. Dens are not lined with insulating materials, as are those of other rodents. In the absence of a den site, porcupines may use the dense cover of a single conifer tree, such as a spruce, for refuge.

During the warm months of the year, porcupines feed on a wide variety of deciduous buds, leaves, forbs, and herbs. Much of the summer food is at ground level, and includes roots, stems, leaves, berries, catkins, seeds, flowers, nuts, aquatic plants, and grass. Cultivated crops may be taken as well. With the onset of winter, the diet shifts to the inner bark of trees and conifer needles. In northern Minnesota, the most common tree species used include white pine, bigtooth aspen, northern red oak, American linden, and white spruce. In eastern Montana, common winter foods of prairie porcupines are American elm, plains cottonwood, skunkbush sumac, chokecherry, and Wood's rose. In addition to their normal diet, porcupines (especially pregnant and lactating females) exhibit a strong craving for salt during the summer months. Plywood, rubber items, and tools such as axe handles are chewed and ingested. Wooden campground and picnic area outhouses may be nearly destroyed for their salt content.

Porcupines seem to exhibit varying patterns of activity. Some reports indicate strongly diurnal patterns, some nocturnal, and some irregular. Total hours of activity are generally less in winter than in summer.

Population densities exhibit great variation, depending on geographic location, habitat type, season, and mortality factors. There is also some indication of population "cycles," with peaks 12 to 20 years apart. Reported densities have ranged from 0.4 per km^2 in Michigan to 42 per km^2 in Massachusetts. Population density in prairie habitat in Montana was reported at six per km^2.

Porcupines are long-lived animals, a feature which counterbalances a low reproductive rate. Average longevity is about five years, with a recorded upper limit of 21 years. Causes of mortality include starvation, injuries, human interactions, disease, and predation. Although porcupines carry heavy loads of parasites, and viral and bacterial diseases, these are not considered to be major limiting factors. A number of larger predators, including lynx, bobcat, coyotes, wolves, wolverines, and great horned owls, occasionally take porcupines but, because of their arboreal habits and quill defense, their effect is minimal. The only predator shown to be effective in controlling populations is the fisher. Where fishers were reintroduced, porcupine populations were estimated at 0.4 per km^2, compared to 3.5 per km^2 in areas where they were absent. Other significant mortality factors include hunting and poisoning in areas where they are considered pests and highway mortality where porcupines venture onto roads as a source of salt.

Reproduction

Porcupines have a low reproductive potential, a function of slow rate of development, long gestation, and production of only a single young per reproductive cycle. The breeding season occurs in the fall, when males expand their home ranges to guard receptive females. Porcupines exhibit a mate-defense polygynous mating system, with dominant males having larger home ranges and significantly greater reproductive success than subordinates. Young are born in April or May, following a gestation period of 209 to 217 days. The single newborn young weighs about 450 g, has its eyes open, and is fully furred and quilled. It is totally dependent on its mother's milk for about two weeks, then begins taking solid food, but is not completely weaned until about 4 months. Sexual maturity in females is attained by their second autumn.

Status and Conservation

Based on existing reports, porcupines are uncommon in eastern North Dakota. Western yellow-haired porcupines seem fairly common in suitable habitats south and west of the Missouri River.

Porcupines have been regarded as harmful to forests, tree plantations, agricultural crops, and various man-made structures. Although trees are often damaged, they are seldom killed by girdling, and economic losses may be less than those caused by fire, insects, and foraging by other animals. Occasionally, domestic animals are injured by quills. Man-made structures and items, such as buildings, signs, rubber components of machinery, and sweat-impregnated tool handles, canoe paddles, etc., may be gnawed for their salt.

There have been various attempts to control porcupines, mostly unsuccessful. At one time, bounty payments were common, but these have been long abandoned. Use of various poisons, such as strychnine and arsenic, has been discontinued for environmental reasons. Metal sheathing and electric fencing may protect individual trees or limited areas. The fisher has been reintroduced in a number of states where it had been extirpated. These reintroductions seem to have provided the most effective measure to date in holding porcupine populations to acceptable levels.

Selected References

Bailey (1926), Genoways and Jones (1972), Hazard (1982), Hendricks and Allard (1988), Hibbard (1972), Jones et al. (1983), Marshall et al. (1962), Roze and Ilse (2003), Seabloom et al. (1978), Tenneson and Oring (1975), Woods (1973), Wrigley (1974)

Order Lagomorpha—
Rabbits, Hares and Pikas

The lagomorphs include the hares, rabbits, and pikas of the mountainous West. This is a small order, having just two living families. There are only 69 species, occurring worldwide except for Australia, New Zealand, Madagascar, southern South America, and the Caribbean and Pacific Islands. A few species, e.g. the European rabbit, have been introduced on islands where they had not occurred naturally. Rabbits and hares are easily recognized, with their characteristic posture, enlarged ears, large hind feet, and small tails. The skull of a lagomorph is superficially similar to that of a rodent, with chisel-like incisors, followed by a large space (diastema) and an array of grinding cheek teeth. However, the resemblance only reflects their similar herbivorous food habits and they are not believed to be closely related. Distinguishing morphologic features of the skull include a second pair of peg-like upper incisors just behind the first, and many perforations (fenestrations) in the side of the rostrum.

One family occurs in North Dakota, the Leporidae. Five species of leporids have been reported in the state.

Family Leporidae
Hares and Rabbits

Snowshoe Hare

Lepus americanus

Waziyata mastinska–Lakota

George Rohde

Description

Weighing up to nearly 2 kg, the snowshoe hare is significantly larger than the cottontails of the region, but smaller than the white-tailed jackrabbit, the only other hare in North Dakota. It also has proportionately smaller ears and larger hind feet than the white-tailed jackrabbit. Its summer pelage is rusty brown, sometimes flecked with black hairs, white hind feet, and white edging around the ears. The dorsal surface of the tail is black. The underparts, including the tail, are white. During winter, this hare turns completely white, except for black tips on its ears. Average standard body measurements are: Total Length - 429 mm, Tail Length - 32 mm, Hind Foot - 118 mm, Ear - 66 mm.

Distribution

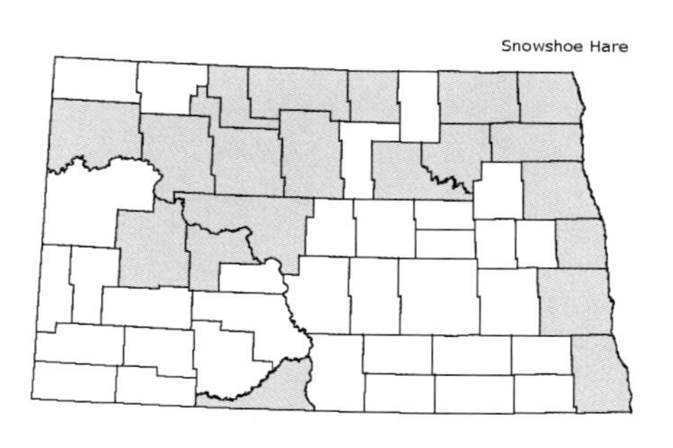

The snowshoe hare occurs through the boreal forests of much of North America from Alaska and Canada south through the Cascade/Sierra ranges, the Rocky Mountains to New Mexico, and the Appalachian Mountains to North Carolina and Tennessee. It is relatively common in much of the northern half of neighboring Minnesota. In North Dakota, snowshoe hares have been recorded in forested areas of the Turtle Mountains (Rolette County), Pembina Hills (Cavalier County), Devils Lake area (Ramsey, Benson counties), Killdeer Mountains (Dunn County), and valleys of the Red, Mouse, and Missouri rivers.

Habitat

The snowshoe hare is an animal of the boreal forest, especially areas having a dense shrub layer providing protective cover and browse. Especially favored habitats include alder swamps and burned areas which have been regrown with woody vegetation. In North Dakota, most observations seem to have been in heavy timber of birch-aspen stands, and riparian woodland with significant shrubby undercover.

Ecology and Behavior

These are relatively solitary animals, but may have overlapping home ranges and may occasionally be seen in groups when at high population densities. They are nocturnal, becoming active at dusk, and spend the daylight hours in "forms," shallow depressions under heavy cover.

Home ranges can vary from 2 to 8 ha in size, depending on the amount of cover and food supply. The smallest home ranges are in heavy cover, but under food shortages, they may wander up to 8 km.

Snowshoe hares feed on a wide variety of plants, but prefer succulent green vegetation when available during the warm months. Summer foods include grasses, sedges, horsetails, ferns, strawberry, clover, and dandelion. During winter, they switch to buds, twigs, and branches. Common winter foods in North Dakota include aspen, birch, willow, and buffaloberry. They will also feed on conifers when available.

The snowshoe hare is well-known for its cyclic fluctuations in relative abundance, with population peaks occurring ranging from seven to 17 years. During peak abundance, population densities as high as 23 per ha have been recorded, a 23-fold increase over the cyclic low. Winter food shortage may precipitate the population decline, supplemented by high predation and low reproduction. With recovery, a period of recovery may last from two to five years. Additional environmental and physiological factors may also affect the period and magnitude of the cycle. In Alberta, annual survival of adults during the cyclic peak was about 0.5, and during the low was 0.1 to 0.2. Survival of juveniles from birth to winter during the cyclic peak was about 0.4, and during the low about 0.1.

This species provides an important prey base for a large number of predatory mammals and birds. Indeed, in the far north, the abundance of some predators, e.g., lynx, is contingent on population levels of snowshoe hares.

Reproduction

The breeding season begins in March, and lasts into July or August. An average litter of three to four young is born following a 35-day gestation. The female does not excavate a burrow or construct a nest, but rather delivers her young in an open form under heavy vegetation. In contrast to the cottontails, newborn young are precocial, i.e., they are fully furred, have their eyes open, can maintain their own body temperature, and require little parental care. The young leave the form within about a day, but remain together in the vicinity for a daily nursing. Weaning occurs in about 28 days. In the Midwest, there are three to four litters per year.

Status and Conservation

Except for the heavily forested Turtle Mountains and Pembina Hills, the snowshoe hare is uncommon in North Dakota, and carries no special protected status. It can provide an important prey for furbearers such as foxes and coyotes. Aspen management to favor ruffed grouse populations likely favors snowshoe hares as well. During cyclic highs, large numbers of hares may inflict considerable damage to conifer plantations such as Christmas tree farms.

Selected References

Bailey (1926), Foresman (2001), Genoways and Jones (1972), Hazard (1982), Jones et al. (1983), Martin et al. (1951), Murray (2003), Pietz and Tester (1983)

White-tailed Jackrabbit

Lepus townsendii

Warchu–Arikara; Ma^{n}sti^{n}ska–Dakota; Mastinca–Lakota

C. D. Grondahl, North Dakota Game and Fish Department

Description

This is the largest hare in North Dakota, weighing up to 4,300 g. During summer, white-tailed jackrabbits have grayish-brown upper parts, while the under parts and tail are white, excluding the throat. Prior to winter, this jackrabbit undergoes a molt and turns completely white, except the black tips of the ears. It can be distinguished from the snowshoe hare by its larger size, and proportionately longer ears and smaller hind feet. Average standard body measurements are: Total Length - 610 mm, Tail Length - 85 mm, Hind Foot - 145 mm, Ear - 108 mm.

Distribution

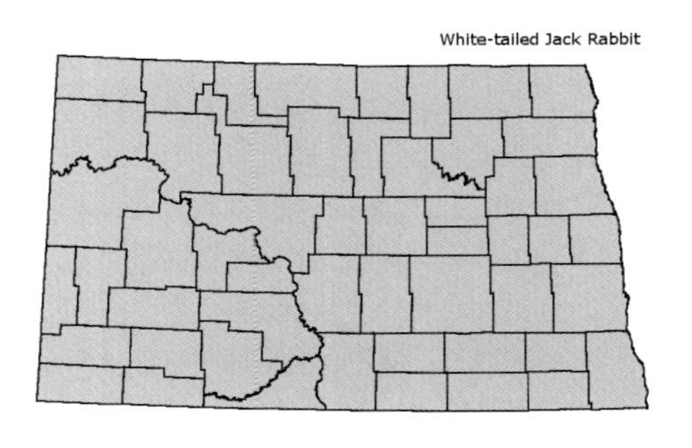

The white-tailed jackrabbit is a species of the Great Plains, occurring from the southern Prairie Provinces of Canada south to extreme northern New Mexico. To the west, it occurs in eastern Washington, Oregon, and California, while historically its eastern distributional limits were the prairie-forest ecotone of Minnesota and western Iowa. Forest clearing and agricultural practices seem to have resulted in a northern expansion in Canada, and an eastward expansion through much of Minnesota, Iowa, and Wisconsin. It occurs throughout North Dakota, except in heavily forested areas such as the Turtle Mountains and Pembina Hills.

Habitat

This is truly a species of the open plains, thriving in both tall- and short-grass prairie. It occurs in grasslands, farmland, and around urbanized areas, especially parks, golf courses and air ports. It is not associated with woodlands, except during severe winters, when jackrabbits may concentrate along woodland edges and in shelterbelts.

Ecology and Behavior

During most of the year, this jackrabbit is solitary, and is not known for its sociality. During winter, however, large groups, estimated at up to 600 animals, may assemble in protected areas presumably for purposes of heat conservation. Beginning in late winter, mating groups consisting of several males and a single female may be observed.

Like the snowshoe hare, the white-tailed jackrabbit does not normally excavate a burrow or construct a nest, instead utilizing a "form," a shallow depression under some protective cover. During severe winters, however, jackrabbits may congregate in large numbers along woodland edges and shelterbelts, and excavate extensive burrow systems in deep snow.

These animals are primarily nocturnal, and are not often seen during daylight unless disturbed. Feeding activity begins in late afternoon, reaches a peak around midnight, and usually ceases well before sunrise. Reduced levels of activity have been associated with bright moonlight and known presence of major predators. Daylight hours are primarily spent resting in forms, but if disturbed, jackrabbits will occasionally seek refuge in old burrows of other species.

The summer diet of the white-tailed jackrabbit is dominated by forbs and grasses. By contrast, the winter-spring diet shifts heavily to shrubs, primarily rabbitbrush and fringed sage, especially buds, shoots, and green bark.

Home range data on this species are limited, but home ranges of about 9 km^2 may be expected. Furthermore, movements from summer home ranges into winter concentration areas have been estimated at up to 8 km.

Estimated population densities have been highly variable, ranging from two to 43/km^2. Population structure exhibits a significantly higher proportion of females, and averages about 65% juveniles. Juvenile mortality may be as high as 75% during winter. Average longevity is about 1.5 years, and rate of population turnover is estimated at 3.4 years.

Principal causes of mortality in white-tailed jackrabbit populations are predation, weather, and human-related factors (hunting, road kill). Because of their speed, adult jackrabbits appear to be relatively safe from most predators, but a wide variety of predatory mammals, birds, and snakes may take the vulnerable young. North Dakota data indicate red fox, golden eagle, snowy owl, and coyote as significant predators.

Weather extremes may have profound effects on jackrabbit populations. Wet weather, flooding, and late-winter storms affect both mortality and reproduction. Conversely, high populations are usually associated with dry periods.

Reproduction

Onset of breeding and reproductive output can vary with environmental conditions, increasing during years of low snow cover. The breeding season in North Dakota extends from late February through mid-July, with four well-defined peaks of breeding activity in March, April, June, and July. During the first three peaks of activity, all adult females become pregnant, but less than a third during the final peak. Parturition follows a 42-day gestation, resulting in a mean litter size in North Dakota of 4.6 (range one to nine).

Like the snowshoe hare, newborn white-tailed jackrabbits are born in a relatively open form, and are precocial, i.e., have their eyes open, are well furred, and have some temperature control. They begin foraging at about 2 weeks of age, but are not completely independent until 2 months of age. Adult weight is attained at 2 to 3 months.

Status and Conservation

The white-tailed jackrabbit is common throughout the prairie areas of North Dakota, but normally more abundant in the southwest. Its overall abundance can be greatly affected by the severity of winter and early spring weather.

Although not listed as a game species in North Dakota, jackrabbits have been hunted for sport, fur, and animal feed. During the first half of the 20th century, jackrabbit drives were conducted in the state, resulting in an average annual harvest of 320,000 animals.

This species continues to have potential as a game animal and furbearer. Because of its high reproductive potential, past large-scale hunts did not appear to affect its status in the state. It is also an important prey species for larger avian and mammalian predators. Large winter concentrations can inflict significant damage to tree plantings.

Selected References

Adams (1961), Bailey (1926), James (1967), James and Seabloom (1969a, b), Jones et al. (1983), Lim (1987)

Desert (Audubon's) Cottontail
Sylvilagus audubonii

P. Myers, American Society of Mammalogists Mammal Images Library

Description

The desert cottontail is a fairly large rabbit, weighing 750 to 1,250 g. Its pelage is similar to that of the eastern and mountain cottontails, but somewhat paler, shorter, and less dense. It can be distinguished from the other regional cottontails by the ear length of >65 mm. Average standard body measurements are: Total Length - 414 mm, Tail Length - 45 mm, Hind Foot - 96 mm, Ear - 66 mm.

Distribution

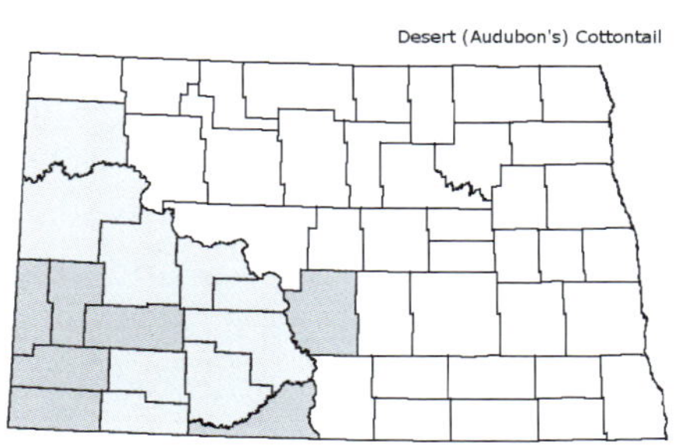

This species occurs over much of the arid portions of western North America, from northern and eastern Montana south to central Mexico, east into the western portions of the Plains states, and west to the coastal areas of southern California. It is absent from Washington, Oregon, Idaho, and northern Utah. In North Dakota, desert cottontails have been reported from Burleigh, Sioux, Billings, Golden Valley, Slope, and Bowman counties.

Habitat

These rabbits are most common in dry upland habitats, including barren upland breaks and grasslands with sagebrush, pines, and Rocky Mountain juniper. Habitat includes true badlands characterized by steep terrain, rock outcrops, and bentonite clay, all sparsely vegetated. In North Dakota, they are most common in upland breaks and upland grasslands, but may also occur in river terraces, hardwood draws, river breaks, and rolling grasslands. There is evidence that the desert cottontail can occupy a fairly wide variety of habitats, but is limited by competition from the eastern cottontail (competitive exclusion). Thus, in southwestern North Dakota, the eastern cottontail occupies the more densely vegetated hardwood draws, river bottoms, and river breaks, but cannot colonize the sparsely vegetated upland grasslands and upland breaks occupied by the desert cottontail.

Ecology and Behavior

The desert cottontail exhibits similar behavior to that of the eastern cottontail. It is primarily a crepuscular species, most active during the hours of dusk and dawn. Its activity centers on a home range varying from 1 to 6 ha, somewhat larger than that of the eastern cottontail. These larger home ranges are not surprising considering the sparse nature of its habitat. Inactive periods are spent in a burrow, while during active periods, shelter may be sought in a form, a small, cleared area adjacent to some type of cover.

Like the eastern cottontail, this rabbit is not gregarious, but females have been observed feeding together. It may be territorial, engaging in marking activity by depositing fecal pellets at prominent sites. Population densities range from two to 16 per ha.

Desert cottontails living in open country primarily feed on grasses and sedges. Depending on availability, however, they will also forage on a wide variety of woody and non-woody plants.

Mortality is high, and probably comparable to the eastern cottontail. One study indicated a maximum longevity of 19 months, and the majority of cottontails succumbing during their first year. Anecdotal accounts indicate that severe winters in southwestern North Dakota can decimate populations over large areas. As with other rabbits, desert cottontails are prey for many mammalian, avian, and reptilian predators.

Reproduction

The breeding season in North Dakota extends from mid-April through the end of July, and consists of five synchronous peaks of breeding activity. Testicular activity is maximal in June, and undergoes a sharp decline in July. Like the eastern cottontail, the gestation period is about 28 days. Potential mean litter size based on corpora lutea counts is 6.6. However, an 11% preimplantation loss results in an average litter of 5.9 at parturition. This is somewhat higher than reported litters of three to four in southern parts of the species range.

The young have their eyes open by day 10, and begin leaving the nest at about 2 weeks. However, they remain near the nest for nearly three weeks. In southern parts of the range, sexual maturity may be attained in 80 days, but in North Dakota, breeding probably does not occur until the following spring.

Status and Conservation

Desert cottontails are common in the Badlands of extreme southwestern North Dakota. However, populations may undergo drastic fluctuations contingent on severity of the winters.

This species, with its high reproductive potential and capability of living in sparsely vegetated rugged terrain, is not considered to be one of special concern. It is affected by cattle grazing, and is most abundant under moderate grazing practices. It is not listed as a game species in North Dakota.

Selected References

Bailey (1926), Bergeron (1977), Bergeron and Seabloom (1981), Chapman and Litvaitis (2003), Chapman and Willner (1978), Foresman (2001), Genoways and Jones 1972), Higgins et al. (2000), Jones et al. (1983), Seabloom et al. (1978)

Eastern Cottontail

Sylvilagus floridanus

Wahboos – Chippewa; Ma^{n}sti^{n}-sapana – Dakota; Mo^{n}stinga – Omaha

C. D. Grondahl, North Dakota Game and Fish Department

Description

The eastern cottontail is the largest (900 to 1,300 g) of the genus on the northern Great Plains. It has a dense fur, which grades from brownish to grayish, intermixed with black hairs. The underparts are white. Average standard body measurements are: Total Length – 420 mm, Tail Length - 50 mm, Hind Foot - 98 mm, Ear - 56 mm. It is similar in appearance to two other cottontails occurring on the northern Great Plains, the desert (Audubon's) cottontail and the mountain (Nuttall's) cottontail. Its shorter ears (<61 mm), and black dorsal hairs should distinguish it from both.

Distribution

This species is widely distributed throughout eastern North America from southern Canada south through Mexico, Central America, and into northwestern South America. It has also been introduced into

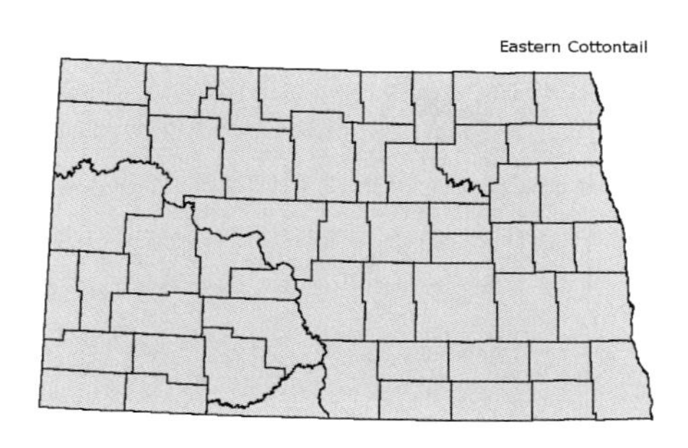

Washington and Oregon. It appears to be a relatively new arrival in North Dakota, accompanying white settlement. The first report of the species in the state was in 1890, followed by numerous records in the first two decades of the 20th century. It is now found statewide, including the southwestern corner, where it is sympatric with the desert cottontail. In its expansion, it may have displaced the mountain cottontail, which had once been recorded in areas of the Badlands now occupied by this species.

Habitat

The eastern cottontail is a habitat generalist, historically occurring in a wide variety of situations from deserts to boreal forests. More recently, it is often associated with human activity, and is common in farmland, shelterbelts, woodland edges, shrubby habitat, and urban settings. In southwestern North Dakota, it is associated with river bottoms, hardwood draws, and river breaks characterized by dense vegetation, mainly shrubs and trees. In this portion of the state, it appears to be ecologically isolated from the sympatric desert cottontail, which is predominant in upland breaks and upland grassland habitats.

Ecology and Behavior

Eastern cottontails prefer dense vegetation, including brush and brush piles for nesting and escape cover. Their established runs can be easily seen in such situations, especially during winter. Nests are excavations 10 to 20 cm deep and 15 to 20 cm long, and lined with grass, leaves, and plucked fur.

These are relatively solitary animals but do exhibit dominant-subordinate interactions associated with the breeding season. Their activity patterns are generally crepuscular, with peaks at dawn and dusk. Moonlit nights may also increase levels of activity. Home ranges of eastern cottontails vary from 1 to 3 ha, varying with age and season, and may overlap during the non-breeding season. These cottontails are not known to be territorial.

Populations have been estimated as high as 9 to 10/ha in the eastern United States, and in Kansas, a population was estimated at about

6.5/ha. Anecdotal observations in North Dakota indicate that cottontail populations can undergo extreme fluctuations, probably contingent on winter weather conditions.

Eastern cottontails feed on a wide variety of vegetation. During summer, they seem to prefer herbaceous plants, especially grasses and legumes. Around human habitation, they can cause significant damage to gardens. Winter foods include buds, twigs, and bark of woody plants. Eastern cottontails, like other rabbits and many rodents, ingest certain fecal pellets (coprophagy) produced in the caecum, which provide an important source of vitamins and minerals that would otherwise be lost.

Total mortality is high. A young cottontail has only a 10% chance of living through its first year, and only 1% survive more than two years. Causes of mortality include predation, disease, severe winters, accidents, and hunting. Many mammalian and avian predators take cottontails. Important predators in North Dakota include all medium and large hawks, owls, crows, foxes, and coyotes. In one study, predation accounted for nearly 43% of known cottontail mortality. Bacterial and viral diseases include tularemia, Tyzzer's disease, and staphylococcosis. As stated earlier, severe winters with heavy snow may result in high mortality. In the eastern United States, the eastern cottontail is an important game species, and large numbers are taken by hunters. In North Dakota, however, it is not listed as a game species.

Reproduction

The breeding season in North Dakota is believed to begin between late March and the first week of April. Two initial peaks of breeding have been estimated, around April 19 and May 16, with additional breeding continuing through July. Gestation averages 28 to 29 days. Potential North Dakota litter size, based on corpora counts, averages 7.05. This is followed by a 10% preimplantation loss, resulting in an average realized litter size of 6.3. There is a tendency for decreased litter size as the season progresses.

Cottontail newborn young are altricial, i.e., they are essentially naked, their eyes are closed, have poor temperature control, and require considerable maternal care. Their eyes open at about 1 week, and they are able to leave the nest at 2 weeks of age. A small incidence of juvenile breeding has been observed in more southern parts of the United States, but is unlikely at the latitude of North Dakota.

Status and Conservation

The eastern cottontail is common throughout much of North Dakota. In general, it appears to be more abundant in the southern portions of the state and in urban areas. However, it is subject to extreme fluctuations in relative abundance. It is an important game species elsewhere in its range, but is not considered so in North Dakota. As a habitat generalist with a high reproductive potential, it is not believed to be a species of special concern. It can be a nuisance species in agricultural areas, gardens, and landscape settings, necessitating some local control activity.

Selected References

Bailey (1926), Bergeron (1977), Bergeron and Seabloom (1981), Chapman et al. (1980), Chapman and Litvaitis (2003), Jones et al. (1983), Seabloom et al. (1978), Wobeser (1985)

Mountain (Nuttall's) Cottontail

Sylvilagus nuttallii

Nis–Arikara; Itakshipisha–Hidatsa

C. Fisher, American Society of Mammalogists Mammal Images Library

Description

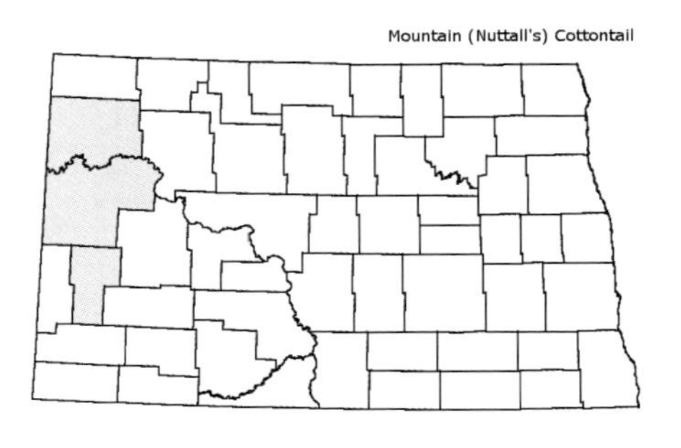

The mountain cottontail is very similar to the other two cottontails in North Dakota. It can be distinguished from the desert cottontail by its shorter (<71 mm) ears, darker coloration, and somewhat smaller hind legs. Its hind feet are covered with long, dense hair, distinguishing it from both desert and eastern cottontails. Compared to the eastern cottontail, it is smaller (total length usually <390 mm), and has more densely furred ears. Average standard body measurements are: Total Length - 373 mm, Tail Length - 48 mm, Hind Foot - 92 mm, Ear - 65 mm.

Distribution

This cottontail occurs primarily in intermountain regions from extreme southern British Columbia, Alberta, and Saskatchewan south to Arizona and New Mexico. Its western range limits are in the foothills of the Cascade and Sierra Nevada ranges, and its eastern recorded limits include extreme western North Dakota and the Black Hills of South Dakota. In North Dakota, there were early records from Williams, McKenzie, and Billings counties, but subsequent collecting efforts have failed to reveal its presence in the state. It may have been replaced in western North Dakota and southeastern Montana by the eastern cottontail, which has expanded its range, accompanying white settlement. Mountain cottontails have been collected in Wibaux County, Montana, within five miles of the North Dakota line. Hence, it may still inhabit small portions of extreme western North Dakota.

Habitat

Mountain cottontails occupy a fairly wide variety of habitats, including rocky, wooded, and brushy areas. They have also been observed around abandoned dwellings. Additionally, in South Dakota, they have been recorded in conifers and river bottoms. Habitats in Montana include willow patches, sagebrush slopes, and cultivation. Early North Dakota records were from dense thickets and brush along stream and river bottoms. Many of these habitat types are very similar to those of the eastern cottontail, adding credence to the theory that in North Dakota, the mountain cottontail has been displaced by a superior competitor.

Ecology and Behavior

This cottontail is more solitary than its close relatives, but may forage in association with others where patches of especially lush vegetation occur. It is a crepuscular species, with most of its activity in late afternoon-early evening and the early morning hours. Foraging is usually in brush, or within a few meters of brushy cover. During spring, it feeds on grasses, while later in the year, it switches to woody plants, including sagebrush and juniper. In rocky areas, mountain cottontails excavate burrows for shelter, but in sagebrush and river bottoms will live above ground and take shelter in "forms," i.e. small leveled patches with overhead cover.

As with other cottontails, they are subject to predation by a variety of mammals, birds, and snakes. Little is known of the population dynamics, including mortality and longevity, of the species.

Reproduction

Elsewhere, breeding seasons range from February to July, with up to five litters produced per year. Most reports indicate an average litter size of four to six. Young are apparently able to leave the nest when they reach a weight of about 75 g. Young females may be capable of breeding at 90 days of age, but such cases are unlikely, especially in the northern Great Plains.

Status and Conservation

The mountain cottontail is probably absent from North Dakota, having been displaced by the eastern cottontail. However, fairly recent records from extreme eastern Montana provide a possibility that it may still occur in very limited numbers. Continuing surveillance of extreme western North Dakota is warranted to better determine its status in the state. Specimens suspected of being mountain cottontails should be sent to a competent authority for identification.

Selected References

Bailey (1926), Chapman (1975), Chapman and Litvaitis (2003), Foresman (2001), Genoways and Jones (1972), Higgins et al. (2000), Jones et al. (1983), Matthews and Swenson (1982), Seabloom et al. (1978), Turner (1974)

Order Artiodactyla—
Even-toed Ungulates

The major characteristic uniting the families of artiodactyls is the unique foot structure resulting in the weight being borne by only two digits (third and fourth). The third and fourth metacarpal and metatarsal bones are fused into a single structure, the cannon bone, providing rigid support for the animal's weight. The two digits articulating with the cannon bone are capped with hoofs, or modified claws. Another major characteristic relates to their herbivorous food habits. The mouth cavity is enlarged, and the cheek teeth are equipped with crescent-shaped cusps, which provide a broad grinding surface for chewing fibrous material. Upper incisors are absent. The jaw articulation provides for lateral movement facilitating chewing. The stomach consists of four chambers providing for digestion of cellulose. The largest chamber, the rumen, begins digestion. The food mass is regurgitated and again chewed ("chewing the cud"). It is then passed into the other chambers for further digestion.

Males, and in some species females, possess horns or antlers. Horns consist of a bony core off of the frontal bone, covered by a sheath of keratinized epidermis. Horns are permanent structures, except in the pronghorns, where the sheath is shed annually. Antlers are also bony structures, but are shed annually. The growing antler is covered by a thin, haired layer of skin (velvet), which is shed when the antler is fully developed. As the animal matures, its antlers may become increasingly enlarged, branched, and in some species, such as moose, palmate.

Artiodactyls are widely distributed throughout the world, absent only from the oceanic islands, Antarctica, and the Australian region. They comprise the largest number of domestic animal species, and have long provided meat, hides, milk, and beasts of burden. There are 211 living species worldwide, with 12 species occurring in North America.

Currently, there are seven species in North Dakota, including the reintroduced bighorn, and bison, which occurs in semicaptive herds. North Dakota families include the Cervidae (deer, moose, elk), Antilocapridae (pronghorn), and Bovidae (bison, bighorn).

Family Cervidae
Deer

Moose
Alces americanus

Muswa–Cree, Ojibway; Wesucharut–Arikara; Ta–Lakota; Pachúptaptach–Mandan

Harold Umber, North Dakota Game and Fish Department

Description

As the largest member of the deer family, the moose is about the size of a horse. In North Dakota, weights of yearling and adult males average 431 (250 to 590) kg, and 398 (272 to 526) kg for yearling and adult females. Males possess large, palmate antlers weighing up to 35 kg. Antlered females are rare. Moose have a deep body, long legs, short tail, long ears, and an overhanging muzzle. Both sexes possess a "bell," a dewlap of skin and hair hanging from the throat. Coloration is blackish brown, but may fade to a lighter brown during winter. Very young moose are reddish brown in color. Ranges of body measurements are: (Males) Total Length - 2180 to 2960 mm, Tail Length - 76 to 110 mm, Hind Foot - 690 to 870 mm; (Females) Total Length - 2140 to 2860 mm, Tail Length - 90 to 122 mm, Hind Foot - 710 to 860 mm.

Distribution

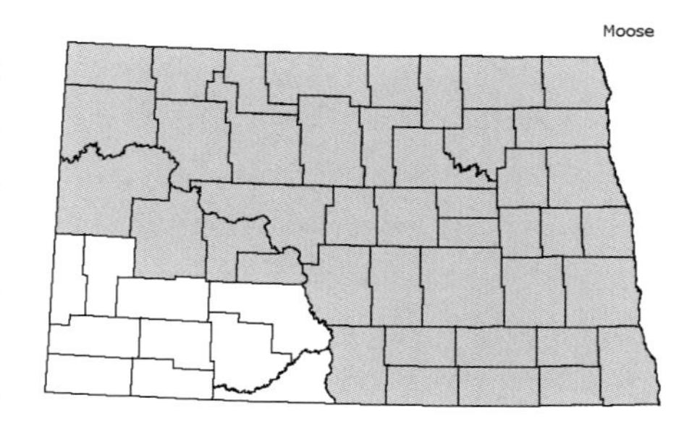

Moose occur throughout the boreal forests of Eurasia and North America. In North America, they are distributed through the forested areas of Alaska, Canada, the Rocky Mountain states as far south as Colorado and Utah, North Dakota, northern Minnesota, northern Wisconsin, the upper peninsula of Michigan, and the New England states of Maine, Vermont, and New Hampshire. Highest densities occur within a band of boreal forest that extends across the continent. Occasional wanderers have been reported in the central United States as far south as Iowa and Missouri.

Historically, in North Dakota, moose were restricted to the heavily forested areas of the Turtle Mountains, Pembina Hills, and the major tributaries of the Red River. Accounts of early traders in the area indicated that they were not as abundant as elk or other big-game species, and had apparently disappeared from the state during the early years of the 20th century. By the 1960s moose had returned to North Dakota, and small numbers were occupying much of their historic range in the state. The population has expanded, but in recent years, numbers have declined in the traditional northeastern North Dakota range. However, moose numbers on the prairie have increased markedly, and they may be seen along the forested areas of the Red River and its tributaries, the beach ridges of Glacial Lake Agassiz, far out onto the drift prairie and along the Missouri River.

Habitat

Moose tend to prefer early successional forest stages with abundant deciduous browse. Primary habitat types in North America include fire-dominated boreal forest, ecotones between coniferous and deciduous forests, large delta flood plains, tundra and subalpine areas, and riparian stream-valley shrub zones. Fire often plays an important role in maintaining moose habitat because it results in increased high-quality forage. Heavy overstory cover acting to intercept snow provides important winter habitat. Hence, the aspen-dominated forests of northeastern North Dakota, along with the nearby riparian woodlands, provide primary habitat in the state. Moose on the North Dakota prairie select shrubs and available woodland habitat. Cropland, particularly sunflowers and corn, are often favored by moose in the fall.

Extreme cold is not limiting to moose because of their large bodies and heavy insulation. In fact, during winter moose begin to suffer heat stress at temperatures above -5°C, and during summer above +14°C. At temperatures above 20°C moose may exhibit panting behavior. Because of their long legs, they are able to negotiate deeper snow conditions than other big-game species, and their movements are only hindered at depths greater than 70 cm.

Ecology and Behavior

Moose browse on stems and twigs of woody plants during winter and on more succulent plant parts during the rest of the year. Over 200 species of plants are consumed, but willows, aspen, birch, and alder predominate when available. During spring and summer, aquatic plants are also consumed, when available, satisfying a high sodium demand. In North Dakota, the diet is predominantly woody browse, consisting primarily of aspen, willow and bur oak in the Turtle Mountains, and Russian olive, willow, and cottonwood on the prairie. Agricultural crops sought out by moose include corn, sunflowers, and sugar beets. Daily food consumption can approach 20 kg.

Except during the rutting season and when cows are caring for calves, moose are solitary. Daily activity is crepuscular, with activity peaks occurring around sunrise and sunset. Moose have strong attachments to their home ranges, which may vary from <1 km^2 in the winter for cows in Ontario to $>1,300$ km^2 in the summer for bulls in Alaska.

In northwestern Minnesota, winter home ranges average 3.1 (bulls) and 3.6 (cows) km^2, while during summer-fall they enlarge to 14.8 (bulls) and 17.4 (cows) km^2. In North Dakota, seasonal home ranges varied from 1 km^2 (bull) in the Turtle Mountains to 293.8 km^2 (cow) in the central part of the state. Along the Missouri River near Williston (Williams County), moose had a median home range of 8.9 km^2, while on the prairie near Kenmare (Burke County), they had significantly larger median home range of 40.1 km^2. They may use the same home range throughout the year, or may migrate between separate winter and summer ranges. Northwestern Minnesota migrations involved movements of 14 to 34 km. None of 55 radio-collared moose in North Dakota exhibited seasonal migratory movement. Moose are not known to be territorial, except possibly cows with young calves. Young moose may disperse from their natal home ranges, but the patterns are highly variable. Some may establish home ranges overlapping with those of their mothers, while others may disperse more than 200 km.

Moose emit a variety of visual, vocal, and olfactory signals. Rutting bulls exhibit a swaying gait, swinging their antlers from side to side. Mock battles consist of hitting and shredding trees, followed by displacement feeding, and wallowing. If the rival fails to yield, a fight may ensue, consisting of antler pushing. Males vocalize during the rut, described as a "croak," while females emit a long moan. Female communication with the calf has been described as a "grunt."

Estimates of moose population densities have varied from <0.1 to as high as 9.3 per km^2 in various parts of their range in North America. In northern Minnesota, reported densities were 0.3 to 9.3 per km^2. Annual survivorship of adults can be as high as 95%, depending on hunting pressure, level of predation, and other factors, but juvenile survival is considerably lower. In areas with significant numbers of large predators, most young moose may be lost within six weeks of birth. Overall, mortality of calves averages about 42%, with 85 (81 to100)% of the losses attributed to predation. In North Dakota, winter recruitment rates range from 41 to 98 calves per 100 cows.

Significant causes of mortality include hunting, predation, accidents, winter starvation, and disease. In many parts of the range, predation, especially by wolves and bears, can have a significant impact on moose populations. In North Dakota, with no such large predators, other mortality factors must be considered. Collisions with motor vehicles and trains can be the primary cause of accidental deaths in moose range. Starvation can result in losses during severe winters with heavy snow conditions, especially among young moose. Moose are hosts of a large number of parasites, only a few of which adversely affect animals already in good condition. The meningeal worm, *Parelaphostrongylus tenuis*, causes a severe neurological disease in moose. This parasite is usually carried by white-tailed deer, but is not pathogenic to them. In northeastern North Dakota, more than 30% of adult white-tailed deer are infected with meningeal worms. Continued infection in moose populations seems dependent on deer. Deer are required for completion of the parasite's life history. Winter ticks can seriously infest moose, resulting in hair loss, anemia, and loss of fat reserves. Infestations can average 32,500 ticks per moose, and may result in significant mortality. On the Agassiz National Wildlife Refuge of northwestern Minnesota, 78% of the mortality of adult cow moose and 61% of the calf mortality was attributed to disease, parasites, and starvation.

Reproduction

The breeding season, or rut, runs from early September to late October, peaking during the first week of October in North Dakota. Cows are receptive for seven to 12 days, but true estrus lasts <24 hours. Cows are polyestrous, and if not bred during their first estrus, will come into heat again in about three weeks. Moose have a tending bond mating system, in which a dominant bull defends an estrous cow until mating occurs and then goes on in search of other receptive females. Mean parturition is about 10 May, and highly synchronized, with 80% of births occurring between 26 April and 24 May. Females in good nutritional condition may breed as yearlings, but maximum reproductive potential does not normally occur before age 4. They may continue successful reproduction until about age 18. Males are physiologically capable of breeding as yearlings, but generally are unable to compete until about age 7.

Pregnancy rates among adults are high, typically >70%, in North Dakota, 95%. Young are born in late May, following a gestation averaging 231 (216 to 240) days. Twinning may occur if cows are in good physical condition. Twinning rates of fetuses from hunter-harvested cows was 73%, and observed twinning from radio-collared cows was 36%. Newborn calves are reddish, weighing 11 to 16 kg. Calves continue to nurse until fall, and are then gradually weaned. Growth is rapid, and by fall, calves may average 196 (140 to 227) kg. Maximum body size is achieved at about 5.5 years.

Status and Conservation

The North American moose population has been regarded as stable or increasing in available habitat, and is currently estimated to be between 0.8 million and 1.2 million. Since the 1960s, moose populations in North Dakota have reoccupied their historic range, and are occasionally observed far out onto the prairie. However, significant declines in moose numbers have been seen recently in northern Minnesota, southeastern Manitoba, and northeastern North Dakota. The cause of these declines is unknown, but is believed to be disease-related.

Principal management strategies for moose involve habitat manipulation and harvest regulation. Habitat management practices include logging, mechanical disturbance, prescribed fires, and application of fertilizer and herbicides. The primary intent of these practices has been to set back forest succession and stimulate browse production. However, provision of an interspersion of overhead cover is also important for concealment, interception of snow during winter, and shade during summer.

Moose are harvested in Canada and 12 states in the United States, with an annual harvest exceeding 80,000. The first season in North Dakota since 1901 was held in 1977, when permits were issued for 10 bulls, nine of which were successful. The state began issuing "any moose" permits by lottery in 1987. Since then, 131 to 175 permits have been issued annually, with hunter success averaging over 90%. In states where populations are already held in check by predation, male-only harvests are practiced, as large numbers of males have little influence on recruitment. In states where predation is insignificant, any-sex harvests work well in population management.

Selected References

Bowyer et al. (2003), Franzmann (1978), Franzmann (1981), Franzmann and Schwartz, eds. (1998), Hazard (1982), Jensen et al. (2017), Jones et al. (1983), Knue (1991), Maskey (2008), Maskey and Smith (2017), Peek et al. (1976), Phillips et al. (1973)

Elk, Wapiti

Cervus elaphus

Wapiti–Shawnee; Wah–Arikara; Ompa, O^{n}pa–Mandan;
A^{n}pa^{n}–Omaha; Upa^{n}–Dakota; Hehaka–Lakota, Madoka–Hidatsa

C. D. Grondahl, North Dakota Game and Fish Department

Description

The North American elk, or wapiti, is one of the largest members of its family on the continent, second only to the moose. In North Dakota, adult females average 225 (199 to 252) kg, and adult males average 317 (213 to 431) kg. Elk are reddish brown during summer, while in winter, the head, neck, and legs are dark brown, and the sides are grayish brown, lighter than the rest of the body. Both sexes have a dark mane and a yellowish rump patch. Average standard body measurements of adult elk were: (Females) Total Length - 2,264 mm, Tail Length - 115 mm, Hind Foot - 674 mm, Ear - 205 mm; (Males) Total Length - 2,431 mm, Tail Length - 123 mm, Hind Foot - 700 mm, Ear - 218 mm.

Distribution

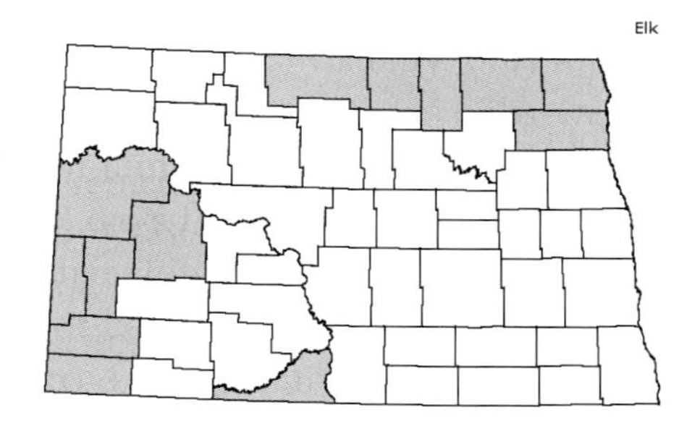

The North American elk is the same species as the red deer of Europe and Asia; hence, the distribution of the species is circumpolar. Prior to the arrival of Europeans, it was one of the most widely distributed deer on the continent. Elk occurred along the northwestern coast, in western and interior California, in the southwestern United States and Mexico, and throughout much of southern Canada and the United States east of the Cascade and Sierra Nevada mountains. They were largely extirpated from the eastern United States in the 19th century, primarily due to unregulated hunting and development of agriculture. Protection, introductions, and careful management have resulted in reoccupation of much of the historic range, especially in the West.

During the early 19th century, elk occurred throughout North Dakota, in prairies as well as forested areas. The fur trader, Alexander Henry, found them abundant along the Red River, providing a source of meat for his trappers second only to the buffalo. They remained common in the state until the 1870s. As with a number of other big-game species, they had essentially disappeared from the state before the turn of the 20th century, with the last known animal being shot in Mountrail County. Five populations now occur in North Dakota. In the 1970s, a small herd moved into the Pembina Hills (Pembina, Cavalier counties), presumably out of Manitoba. The northern Badlands herd in Dunn and McKenzie counties originated from escapees from the Fort Berthold Indian Reservation in 1977. The third population originated in 1985 with the introduction of Rocky Mountain elk to Theodore Roosevelt National Park from Wind Cave National Park, South Dakota. These elk remain in the park and its vicinity (Billings and Golden Valley counties) and currently number over 1,000 animals. In the late 1980s and early 1990s, elk began being reported in the Turtle Mountains (Bottineau and Rolette counties). These animals move back and forth across the U.S./Canadian border and number several hundred. In the late 1980s, elk appeared in the Porcupine Hills of eastern Sioux County. In 2017, an agreement between the North Dakota Game and Fish Department and the Standing Rock Indian Reservation provided for limited hunting of this population.

Habitat

As their original North American distribution implies, elk exhibit a wide habitat tolerance and adaptability. The various subspecies were able to exist in dense coastal rain forests, non-forested bottomlands of California, dry forests and chaparral of the Southwest, cool shrub forests and prairies of the Midwest, mixed conifer-hardwoods of the East, and a variety of montane habitats of the Rockies. The only habitats avoided were the western deserts and the humid Southeast. Currently, western populations inhabit coniferous forests in rugged, broken terrain and foothills, the rain forests of the Pacific Northwest, and open shrublands of the Southwest. More open areas may be used during green-up of spring and summer, while cool, shaded habitats are sought during hot weather. Cover requirements include protection against temperature extremes as well as for hiding and calving. Ideal cover includes grasslands or meadows interspersed with forests having large amounts of edge.

Recent research in the Pembina Hills (Cavalier county), Turtle Mountains (Bottineau, Rolette counties), and Porcupine Hills (Sioux County) indicates the importance of forested habitats to North Dakota elk. In the Pembina Hills, they tend to inhabit the oak-aspen forests of the Pembina Hills and adjacent farm land. Elk in the Killdeer Mountains (Dunn County) use birch-aspen forest during summer, and shift to oak-ash as well as birch-aspen in fall. In the northern badlands proper they use oak-ash in summer, shifting in fall to cottonwoods and agricultural lands along the Little Missouri River as well as oak-ash habitat. Hardwood draws and juniper draws provide important cover and forage in Theodore Roosevelt National Park during daytime in summer, while grassland habitats are used for grazing during the twilight hours. During winter, use shifts from wooded areas to grasslands, which are used throughout the day. Unlike early reports and the northern badlands herd, riparian (cottonwood) habitats along the Little Missouri River do not receive extensive use.

Ecology and Behavior

Diets of elk are highly variable, reflecting their wide range of habitats. Some populations are primarily grazers, while others depend more on browse. Generally grasses and forbs are more important in spring and early summer, and woody browse during winter. In northeastern North Dakota, fall diet consists primarily of corn (60%), grasses (21%), and shrubs (12%). Winter diet consists of about 24% shrubs,

18% grasses, and 50% corn. Chokecherry and serviceberry are the most commonly used shrubs, and sedges and wheatgrasses the most common graminoids. The high use of corn reflects local agriculture and its use in wildlife food plots. In the Killdeer Mountains, summer diets are fairly equally proportioned among forbs, shrubs, and grasses, while in the northern Badlands, shrubs are used more heavily. Wheatgrasses, buffaloberry, winterfat, and western snowberry are among the most important species. Compared to the northern Badlands, elk in Theodore Roosevelt National Park consume fewer shrubs and forbs. Grasses and browse make up >95% of the annual diet, and five species, needle and thread, threadleaf sedge, western wheatgrass, winterfat, and snowberry, comprise about 70%. Browse use peaks in mid-summer and falls off during the fall, when elk resume grazing. Use of browse and forbs increases again in the spring.

Elk are highly gregarious, but their sociality varies by sex, season, and population. Cow-calf bands range in size from <seven in densely forested areas to >400 in open habitats. Cow-calf groups tend to remain together throughout the year, except during the calving season, when cows and newborn calves isolate themselves for up to three weeks. Adult bulls often associate in small groups, with considerable interchange between groups. Bull and cow-calf groups tend to occupy separate areas, except during the breeding season, when rutting bulls actively search for and attempt to maintain harems of cows.

Home ranges are highly variable, functions of vegetation type and region. Estimated summer home ranges have varied from as little as 3 km^2 in California redwoods to 245 km^2 in the sagebrush steppe of eastern Idaho. In some areas, elk perform extensive seasonal migrations between summer and winter ranges, which may be as much as 80 to 100 km. In the northern Badlands and Killdeer Mountains of North Dakota, summer home ranges average about 26 km^2, enlarging in the fall to 33 km^2. Killdeer Mountains elk undergo seasonal movements between summer and fall ranges, but Badlands elk do not. Home ranges in Theodore Roosevelt National Park vary from 26 km^2 to 80.0 km^2, with no apparent differences by sex or season.

During spring and summer, elk exhibit a fairly distinct diurnal cycle of activity, bedded during the hours of darkness and alternating feeding, traveling, and resting during daylight. Feeding appears to be concentrated to the early morning and late afternoon-evening hours.

During the fall rut, however, the amount of time spent feeding and bedding decreases, and active and rest periods are dispersed throughout the day and night. Following the rut, more time is spent feeding and bedding, and overall activity begins to resemble that for spring and summer.

As with other big-game species, elk are subject to predation, disease, accidents, starvation, and hunting. Predation by wolves, bears, mountain lions, and coyotes can be a major mortality factors on calves. Of these, only coyotes are present in significant numbers in North Dakota, but they are not believed to be a significant limiting factor. However, mountain lions have been increasing in the northern Badlands in recent years. Brucellosis has significantly affected calf production and survival in some areas. Chronic Wasting Disease, a fatal disease of the central nervous system, has been detected in nearby states. Neither of these diseases has been found to date in North Dakota elk. Starvation has not been a factor affecting North Dakota elk populations. In most areas, including North Dakota, hunting and associated losses are the major mortality factors affecting populations. Life table data on hunted Rocky Mountain elk herds indicate an average life expectancy of about three years, and a maximum longevity of 15+ years.

Reproduction

The breeding season extends from September through October, occasionally running into November. Elk are polygamous, bulls beginning to gather harems of 15 to 30 cows in early September. Mature bulls are hostile to each other, engaging in bugling, digging and thrashing with antlers, and sparring. Fights are frequent. Yearling cows may be fertile, depending on their nutritional status. On average, pregnancy rates for yearlings are about 24%, while older cows may average about 86%. In Theodore Roosevelt National Park, the pregnancy rate in yearlings is about 54%, and 91% in adult cows. Cows are spontaneous ovulators, and have a cycle of about 20 days. A single calf is born in late May or early June, following a gestation period of about 250 (247 to 265) days. Twinning is rare in elk.

Newborn calves are spotted and weigh about 15 (13 to 18) kg. Calves nurse intensively for 2.5 months, but lactation begins to decline by September. Growth is rapid, with birth weight doubling within the first month. By fall, North Dakota young average 135 kg. Maximum weight is reached at about 6 years in males and 5 years in females.

Status and Conservation

Elk populations have generally increased in the western states and provinces over the past 50 years. As stated above, there are currently five populations in North Dakota. An aerial survey of the Pembina Hills herd in the northeastern corner of the state during 2009 recorded 156 individuals. During spring, 2011, 95 elk were observed in the northern badlands/Killdeer Mountains area. The largest population is in the south unit of Theodore Roosevelt National Park and currently stands at about 250. Elk in the park have a high pregnancy rate and 90% calf survival. The population has little natural mortality and is currently increasing by about 20% per year, doubling every 3 to 4 years. Elk numbers in the Turtle Mountains are difficult to assess because they frequently cross into Canada. Minimum numbers are estimated to be more than 200 animals. In December 2016, a total of 110 elk were observed in the Porcupine Hills. Statewide, elk numbers may exceed 1,000 animals.

Elk were originally abundant throughout North Dakota. Considering their adaptability, varied food habits, reproductive potential, and few natural predators, they have the potential for significant expansion of their current ranges and populations. With population increases, issues such as crop depredation, destruction of property, competition with livestock and other wild ungulates, and possible disease transmission will need to be addressed. Consequently, one of the first management questions is whether to allow further population growth, or to maintain small but huntable herds, which are not as likely to cause conflicts.

Disease transmission could be a problem where large concentrations of elk and cattle are in contact, especially during winter. Although diseases such as Brucellosis and Chronic Wasting Disease have not been reported in the state, continued monitoring is essential and efforts should be made to keep contact between elk and cattle to a minimum.

There is dietary overlap between elk and cattle during spring and fall, when both heavily utilize grasses. However, in the northern Badlands, there appears to be little overlap at the species level, with cattle primarily using bluegrasses and brome, while elk forage more on wheatgrass. In some states, expanding elk populations have been associated with declines in mule deer. No consistent trends are apparent, but considering their adaptability and more generalized food habits, high elk populations may reduce habitat quality for mule deer. Issues of range quality, competition with other ungulates, potential future starvation, and landowner conflicts are especially apparent in Theodore Roosevelt

National Park, where the population is currently more than triple the population objective, and there are few available options to reduce it to acceptable levels. Effective management in North Dakota will involve working closely with private landowners in the northern Badlands to assess acceptance levels for all interested constituents, determining desired population levels, and controlling the herd through appropriate harvests. This approach to participation with constituent groups will be important to decision making in Theodore Roosevelt National Park as well as the rest of the state.

Selected References

Bailey (1926), Blood and Lovaas (1966), Bryant and Maser (1982), Bubenik (1982), Jones et al. (1983), Kistner (1982), Knight (1970), Knue (1991), Osborn et al. (1997), Peek (2003), Sargeant and Oehler (2007), Skovlin (1982), Strassler (1996), Sullivan (1988), Taber et al. (1982), Toweill, and Thomas (2002), Westfall (1989)

Mule Deer

Odocoileus hemionus

Tsitashipisa – Hidatsa; Sinte-sapana – Dakota;
Sinte sapela – Lakota; Shunte-psih – Mandan; Takatit – Arikara

C. D. Grondahl, North Dakota Game and Fish Department

Description

The first scientific description of this species was provided by Meriwether Lewis, and in North Dakota. In North Dakota, yearling and adult male mule deer average 74 (57 to 116) kg, and females average 63 (50 to 77) kg. The upper parts are reddish brown in summer and grayish brown in winter. The forehead is dark, and there is a brown spot on either side of the nose. The rump patch is grayish white, and the underparts are white. Compared to the white-tailed deer, the ears are much larger (two-thirds the length of the head), and the tail is smaller, with short hair, and with a black tip. Ranges of standard body measurements are: (Males) Total Length - 1,370 to 1,800 mm, Tail Length - 150 to 230 mm, Hind Foot - 410 to 590 mm, Ear - 120 to 150 mm; (Females) Total Length - 1,160 to 1,800 mm, Tail Length - 110 to 200 mm, Hind Foot - 325 to 510 mm, Ear - 120 to 140 mm.

Mule deer can be distinguished from white-tails in the field by their larger ears, smaller black-tipped tails which do not "flag" back and forth when running, dichotomous branching of the antlers, and their unique bounding gait, termed "stotting."

Distribution

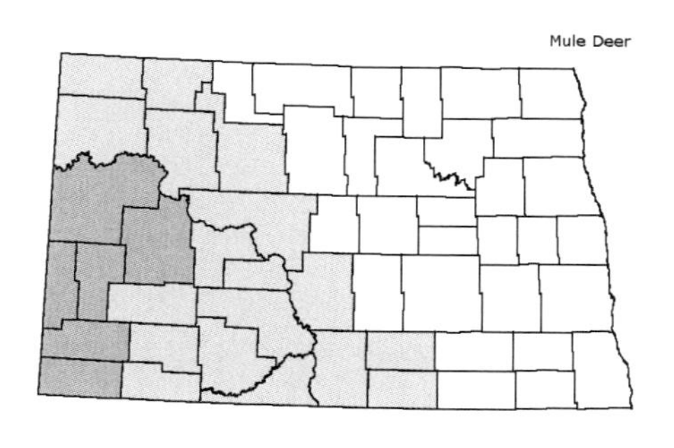

Mule deer occur throughout much of western North America, from the Pacific coast eastward to approximately the 100th meridian. The northern limit of distribution is in southern Yukon Territory, and the southern limits are in Baja California and central Mexico. The eastern limits are in western Saskatchewan, North Dakota, South Dakota, Nebraska, Kansas, and the panhandles of Oklahoma and Texas.

During the early years of white settlement, mule deer occurred in upland prairies throughout North Dakota, with reports from areas around Valley City, Devils Lake, Pembina, and the Turtle Mountains, as well as the west. By the turn of the 20th century, however, they had been nearly exterminated from the state, likely the result of development of intensive agriculture, overgrazing, and unrestricted hunting. With protection, populations recovered, and mule deer are now common in much of western North Dakota, especially in the Badlands and near the western tributaries of the Missouri River. Occasionally, individuals wander east of the Missouri River.

Habitat

In general, mule deer prefer semiarid, open forests, and shrub lands associated with rough terrain. In the Badlands of North Dakota, brush complex vegetation types, especially buck brush and brush transitions with green ash stands, are important for foraging does during summer and fall. North-facing slopes are also preferred, but grasslands are avoided. In southeastern Montana, hardwood forests, along with pine/juniper woodlands, are also heavily used during summer. During winter, mule deer select rough topography, including badlands, juniper breaks, ponderosa pine forest, and riparian habitats surrounded by sagebrush grasslands. Warmer slopes which are protected from wind appear to be selected for bedding.

Ecology and Behavior

Mule deer consume a wide variety of plant species, and require approximately equal proportions of woody and herbaceous forage. Nearly 800 species are known to be consumed by mule deer, primarily shrubs and forbs. In eastern Montana, grasses and sedges are mainly consumed during spring, while forbs are more important during late summer. Woody browse makes up over half the rumen contents throughout the year, but is especially important during fall and winter.

The social structure of mule deer is similar to that seen in whitetails. The basic social unit consists of two or more generations of related females and their offspring. The maternal group is dominated by an older female that has been reproductively successful. The group is relatively stable throughout the year until the summer fawning season, when does and young fawns isolate themselves from others. Maternal groups then reform during late summer and fall. Other groups consist of small assemblages of males of various ages and yearlings of either sex wandering together. During the cold months, when food resources are restricted, mule deer may be forced to aggregate in large groups in common feeding areas.

Activity is crepuscular, with peaks occurring at dawn and dusk. In southwestern North Dakota, during summer and fall, activity peaks occur at sunrise and just before sunset, with little activity between 10 a.m. and 4 p.m. CST. On hot days, mid-day activity consists mainly of comfort movements and moving to avoid direct sunlight. Bedding sites are often in shaded arroyos and in wooded areas. With cooler temperatures during fall, mid-day activity levels appear to increase, and a greater proportion of the active time is spent feeding.

Mule deer tend to confine their activity to discrete home ranges, which remain fairly stable from season to season and year to year. In North Dakota, average summer and fall home ranges of adult and yearling does average 275 (160 to 690) ha, with no significant differences in size between age or season. Home ranges of yearling and adult bucks are significantly larger. Mule deer may undergo dispersal from their natal home ranges. Rate of dispersal in southwestern North Dakota and southeastern Montana appears to be low.

As with most big-game species, mule deer populations were at an all-time low in the early 20th century but, in response to protection and careful management, have largely recovered over most of their primary range. Populations fluctuate in response to changes in annual recruitment and mortality. Populations in neighboring eastern Montana have ranged from one to $6/km^2$, while in North Dakota, estimates for Theodore Roosevelt National Park averaged $3/km^2$, and 10 to $12/km^2$ in the Kendley Plateau area south of Belfield (Billings County).

Maximum longevity in the wild has been recorded as 19 to 20 years, but average longevities are undoubtedly much lower, varying in time and place. In southwestern North Dakota, estimated maximum age of harvested does was 13 years, and bucks at 10 years; however, the vast majority of bucks are 4.5 years-of-age or less. In eastern Montana, annual survival rates of males have varied from 39 to 59%. In North Dakota the annual survival rate for adult does is about 86% and overwintering (December – May) juveniles survival was 68%. The leading cause of adult mortality was predation (32%), and of juveniles was malnutrition (22%). In northern areas, severe winters can be critical for mule deer, and losses of 33 to 50% of all animals are possible. North Dakota long-term data indicate that cold winters are particularly important when harsh conditions extend into April fawn recruitment. Additionally, there can be a lag effect from back-to-back severe winters. Hunting mortality can be regulated to maintain stable populations, encourage growth, or mitigate losses to other factors. In North Dakota, predators include coyotes, bobcats, feral dogs, golden eagles, and occasionally mountain lions. Coyotes are probably the principal predators, mainly taking fawns, but their impact on deer populations is uncertain. As much as 85% of fawn losses during summer and fall have been attributed to coyotes, and 95% of fawn losses during winter. However, survival to early winter has also been correlated with summer forage production, which may influence rates of predation. Hunting mortality is thought to have little effect on the natural mortality of adult females, but can be the major cause of mortality for adult males. Although mule deer are known to carry a variety of diseases and parasites, their role in population control is not well understood and there has been no documentation of major impacts on populations. Chronic Wasting Disease (CWD) is a fatal disease affecting the central nervous system of deer, moose, and elk. There have been nine confirmed cases of CWD in North Dakota mule deer (Grant and Sioux counties), and it has been found in South Dakota, Minnesota, Montana, and Saskatchewan. Collisions with vehicles occur as well, but in the sparsely populated areas of western North Dakota, their influence on populations is minor.

Reproduction

The breeding season typically occurs in November and December, peaking in mid November, but may extend later into the winter. Most does become capable of breeding at 1.5 years. Fawn pregnancies are uncommon. The estrous cycle is 22 to 28 days, with does being receptive for 24 to 36 hours. Bucks are reproductively mature at 1.5 years, but are generally unable to compete for does until they are older. Mule deer are polygynous, their breeding system regarded as a "tending bond" type in which a dominant buck tends an estrous doe until copulation or displacement by another buck.

Fawns are born between late May and mid July (peaking in mid June), following a mean gestation of 203 (183 to 218) days. Older does typically bear twin fawns, but yearlings usually carry only a single fetus. In eastern Montana, reproductive rates average from 1.5 to 1.9 fetuses per doe. Newborn fawns are spotted, and weigh about 3 kg. Weaning begins at 5 weeks, and is completed at around 16 weeks. Growth continues into adulthood, with maximum weights attained at 120 months for bucks and 96 months for does. Currently, the observed fall recruitment rate in North Dakota averages about 90 fawns per 100 does.

Status and Conservation

Mule deer populations have recovered over much of western North America, and are hunted as game animals throughout their range. By 1900, only a handful of mule deer were left in remote corners of North Dakota due to unregulated market hunting. Between 1920 and 1931, all deer hunting was outlawed in the state. Mule deer numbers remained low until the 1940s. Since then, there have been hunting seasons every year beginning in 1954. In 2008, more than 9,800 mule deer were harvested in the state.

Two management strategies are important for maintenance of healthy mule deer populations in southwestern North Dakota, population management through hunting regulation, and management of critical habitat. The North Dakota Game and Fish Department has well-established hunting units, and regulates numbers of permits per unit annually. The numbers of permits are governed by previous years' harvest data, spring and fall aerial surveys, estimated annual recruitment, and other factors.

The maintenance and protection of woody vegetation is critical to mule deer populations in North Dakota. Ash-hardwood stands and adjacent brushy areas receive heavy use by lactating does and fawns during the first few weeks following birth. This cover continues to be important as fawns develop, providing bedding sites which are relatively secure from predation. During severe winter weather, deer tend to seek bedding sites in juniper stands, where snow cover, temperature, and wind are moderated by the canopy cover. Hence, management practices enhancing mule deer habitat must consider mitigating heavy grazing and soil compaction in woody habitats, restriction of mechanical and chemical shrub control and protection of winter cover provided by stands of juniper. Truck traffic associated with energy development tends to reduce levels of mule deer activity within 100 m. Therefore, planned road construction which avoids woody habitat may result in less impact and allow its use by mule deer during critical periods.

Selected References

Anderson and Wallmo (1984), Bailey (1926), Ciuti et al. (2015), Fox (1989), Jackson (1990), Jensen (1992), Jensen (1988), Jones et al. (1983), Knue (1991), MacCracken and Uresk (1984), Mackie (1970), Mackie et al. (2003), Swenson et al. (1983), Wallmo (1981),Wood (1986), Wood (1988)

White-tailed Deer

Odocoileus virginianus

Tachtsha – Dakota; Sintehanska – Lakota;
Tstita-taki – Hidatsa; Mahmanaku – Mandan; Ta-paht – Arikara

Craig Birhle, North Dakota Game and Fish Department

Description

The white-tailed deer is highly variable in body size, reflecting a wide geographic range and variety of habitat types. North Dakota adult white-tails average 59 (41 to 94) kg for females, and 76 (45 to 110) kg for males. The upper parts are generally reddish brown during summer, shifting to grayish brown in the other seasons. There are white bands around the muzzle and eyes, and a white patch on the throat. The underparts, including the belly, insides of legs, and lower tail, are all white. Fawns are reddish brown with white spots, which are retained until between the third and fourth months. Ranges of standard body measurements are: (Males) Total Length - 1,700 to 2,150 mm, Tail Length - 260 to 360 mm, Hind Foot - 510 to 538 mm; (Females) Total Length - 1,340 to 2,062 mm, Tail Length - 152 to 325 mm, Hind Foot - 480 to 520 mm.

When viewed from a distance, several features distinguish white-tails from mule deer. These include their smaller ears, unbranched antler tines, conspicuous white tail, which "flags" while running, and their graceful loping gait while running.

Distribution

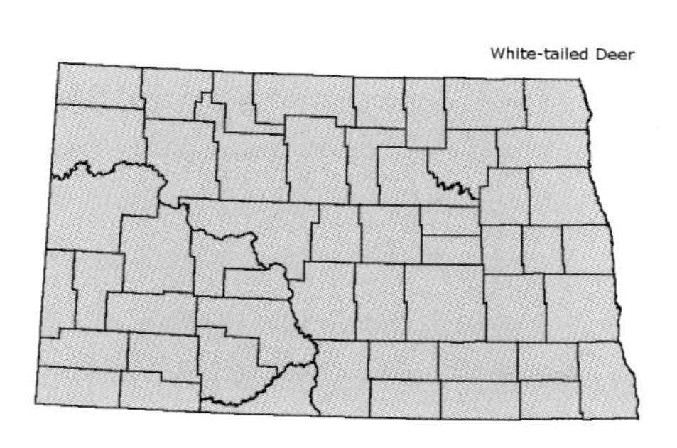

The white-tailed deer is very widely distributed in the Western Hemisphere, extending from extreme southern Yukon and Northwest Territories in Canada, throughout most of southern Canada, the United States, Mexico, Central America, and into northern South America. White-tails occur in 45 of the contiguous United States, being absent or rare only in Utah, Nevada, and California. Subsequent to white settlement, they have extended their range northward in Canada and into upland prairies of the Great Plains, in response to logging and forest fires setting back forest succession, along with the development of agriculture.

Prior to white settlement in North Dakota, white-tailed deer occurred primarily in the forested areas and riparian woodlands of the river systems. As settlement progressed, overhunting nearly eliminated them from the state. With protection in the first half of the 20th century, populations began to recover and even expand into previously unoccupied areas following the development of agriculture. At present, white-tails occur throughout North Dakota, even in urban settings.

Habitat

White-tailed deer are habitat-generalists, able to survive in a wide variety of habitats, from the boreal forests of Canada and the United States to tropical rain forests of Central and South America. Important white-tail habitat in eastern North Dakota is typical of the northern Great Plains, consisting of lowland draws and swales adjacent to river systems. Woody draws, tree plantings, marshes, cropland, and Conservation Reserve Program (CRP) tracts are also used for cover and forage in North Dakota.

Distribution in semiarid southwestern North Dakota is especially associated with the drainages of the Missouri and Little Missouri rivers, where white-tailed deer are most commonly observed in moderately grazed rolling grasslands, hardwood draws, riparian woodlands, and deciduous forests. In adjacent southeastern Montana, white-tailed deer select select riparian areas associated with agriculture, and dense stands of ponderosa pine.

Snow depth, quality, and duration can affect white-tailed deer mobility and availability of forage. When snow cover exceeds 30 to 40 cm, they may retreat to protected river valleys with significant overhead cover acting to intercept snow. These "deer yards" may attract large concentrations of deer, which use them traditionally, resulting in forage depletion and potential starvation.

Ecology and Behavior

White-tailed deer are highly are highly adaptable and accept a wide variety of foods. Generally, browse, mast (acorns), and forbs make up the majority of the diet. However seasonal shifts in forage abundance and quality, along with metabolic demand, play an important role in diet selection. In central North Dakota, woody browse makes up >25% of the diet throughout the year. In addition to browse, the diet is dominated by various crops during winter-spring, especially corn, sunflowers, and wheat. Forbs and grasses become more important in summer and fall.

Two basic social groups have been identified in white-tailed deer, maternal groups and fraternal (bachelor) groups. The typical maternal group consists of an adult doe, her yearling daughter(s), and their fawns-of-the-year. These maternal groups remain together throughout the year, except during the fawning season, when does and their newborn fawns isolate themselves. Maternal groups may fuse into larger groups during fall and winter.

Adult males form fraternal groups of varying size. Frequently two to five males travel together, especially during winter and summer. These groups are frequently more social than maternal groups, with mutual grooming being common. Fraternal groups break up during the breeding season, and bucks become solitary, except when pursuing or tending does.

Activity is generally crepuscular, with dusk and dawn peaks, but can be influenced by a variety of variables. During winter, they tend to be more diurnal, with a peak occurring during late afternoon. In many areas, white-tails have well-defined traditional home ranges varying from 59 to more than 700 ha. In North Dakota, summer and winter home ranges of adult does average about 226 and 635 ha, respectively. They are not territorial, however, with only bedding sites and limited resources being defended.

White-tails engage in three types of movement patterns, short-distance or local, migratory, and dispersal. Short distance movements occur when seasonal weather extremes are minor, in contrast to migrations, which occur in response to predictable extreme seasonal shifts in temperature and snowfall, and involve travel to and from established areas. Dispersals are long-distance movements away from established home ranges, and usually involve young deer. In central North Dakota, seasonal movements of does range from 0.6 to 32.2 km, while those of fawns range from 2.6 to 120.4 km. The longer fawn movements almost certainly involve dispersal from the natal home range. The longest recorded movement of a male and female white-tailed deer in North Dakota was 216 and 273.6 km, respectively.

Population densities can vary greatly from area to area and year to year as functions of forage supply, winter conditions, hunting pressure, and predation. In North Dakota, densities vary from <1 per km^2 in short-grass prairie and intensively farmed areas to about >10 per km^2 in mixed grass prairie and winter concentration areas. If hay and silage intended for livestock is left unprotected, hundreds of deer from the surrounding area may congregate on one farm during a winter and cause significant depredation problems.

In the northern Great Plains, annual survival in recent years has averaged about 70%. On the Lonetree Wildlife Management Area in central North Dakota, annual doe survival ranged from 77 to 83%; survival of fawns ranged from 78 to 89%. Neonate survival is best during warm and moist June weather conditions. Along the lower Yellowstone River, Montana, annual survival of bucks was 57%, and does 73%.

As with other big-game species, principal causes of mortality include hunting/poaching, winter starvation, accidents, predation, and disease. Based on the returns of 28 radio-collared does from central

North Dakota, the principal cause of mortality was hunting (61%), followed by predation (18%), deer-vehicle collisions (11%), and starvation (11%); cause of death of an additional 17 carcasses could not be determined. Along the Yellowstone River in Montana, hunting accounted for 77 to 91% of adult mortality and 9 to 22% was attributed to predation. In North Dakota, coyote predation occurs mainly during the period from late March to early June. Starvation can cause significant losses during severe winters and when population densities are high. White-tailed deer are susceptible to a number of diseases, the predominant one in this area being Epizootic Hemorrhagic Disease (EHD). This is a viral disease transmitted by midges (Culicoides) which can cause death in a few days and result in mortality of <15 to >50% of the population. Outbreaks typically occur during late summer to early fall. EHD outbreaks in North Dakota are only known from the southwest. Chronic Wasting Disease is a fatal disease affecting the central nervous system of deer, moose, and elk. There have been two confirmed cases in North Dakota white-tailed deer (Sioux and Grant counties), and it has been found in the nearby states and provinces of South Dakota, Minnesota, Montana, Alberta, and Saskatchewan. Bovine tuberculosis, a bacterial disease primarily affecting cattle, has been found in white-tailed deer in neighboring northwestern Minnesota. Although not regarded as a limiting factor in wild deer populations, they can act as a reservoir of infection for domestic livestock.

Reproduction

In northern portions of the range, breeding typically occurs from late October to December. When diet quality is high, significant numbers of doe fawns may breed, but usually later than adults. However, most does delay breeding until they are 1.5 years old. Bucks become sexually mature at 1.5 years. During the rut, bucks exert dominance by rubs, scrapes, and other visual and olfactory signals. Overhanging branches, twigs, and bark are marked with the head and antlers. Males also use postures and eye contact in asserting dominance. If dominance is unresolved by signals, brief fights may result.

Fawns are usually born in late May or June, following a 202- (187- to 222-) day gestation. Twinning is typical if the adult does are in good condition at the time of conception. Newborn fawns are spotted, and range in weight from 1.8 to 3.6 kg. Their weight may double within two weeks and quadruple in 30 to 40 days. They may begin grazing in a few weeks, and are normally fully weaned by 10 weeks. Their spots are lost by late August or September. As stated above, females in good condition may have their first estrus at about 6 months, but males are not sexually mature until 1.5 years.

Status and Conservation

North American white-tailed deer populations have recovered from their all-time low of about 350,000 in 1900 to a current level of 20 million, comparable to their estimated levels prior to white settlement, a result of adequate protection, transplants, and land use practices which have increased cover and forage production. In some areas they have increased to nuisance levels, resulting in vehicle collisions, and damage to crops and landscaping. White-tails are currently thriving in North Dakota, a result of careful regulation of seasons, along with a series of relatively mild winters.

In recent years, North Dakota has experienced high white-tail populations, with an annual harvest approaching 85,000 and a hunter success around of 70 to 75%. The major management goal continues to be provision for an adequate harvest, not only to provide for satisfactory hunting experiences for sportsmen, but also maintaining populations below carrying capacity, enhancing survival through severe winters, and reduction of depredation.

Occasional EHD outbreaks in southwestern North Dakota continue to result in die-offs of white-tails. While it is not possible to control these outbreaks, their effects can be alleviated by adjusting allowable harvests. Serious zoonotic diseases, such as Chronic Wasting Disease and Bovine Tuberculosis, require monitoring in North Dakota deer herds because they are present in adjacent states.

Selected References

Bailey (1926), Bihrle (1996), Dusek et al. (1989), Dusek et al. (1988), Gerads (2002), Halls (1984), Harmoning (1976), Jensen (2000), Jones et al. (1983), Kernohan et al. (1994, 2002), Knue (1991), Michel, E. S., et al. (2018), Miller et al. (2003), Peterson (1984), Seabloom et al. (1978), Smith (1991), Smith (2005), Sparrowe and Springer (1970), Swenson et al. (1983)

Family Antilocapridae
Pronghorns

Pronghorn
Antilocapra americana

Koka–Mandan; Tatókana–Dakota;
Tatokala, Nigesanla–Lakota; Uchi–Hidatsa; Chka–Arikara

C. D. Grondahl, North Dakota Game and Fish Department

Description

The pronghorn is the only surviving species of its North American family. It is one of the smaller wild ungulates in North America, well-adapted for running in open country. In North Dakota, yearling and adult males average 54 (45 to 61) kg, and females average 45 (41 to 56) kg. The back is rusty brown halfway down the sides, while the rump and underparts are white. The rump has erectile hairs which can serve as a warning when danger threatens. The neck has a black mane, there are two white bands across the throat alternating with black or tan markings, and males have black markings on the snout. Average measurements of mature Alberta pronghorns were: (Females) Total Length - 1,406 mm, Height at Shoulder - 860 mm, Tail Length - 97 mm, Hind Foot - 397 mm, Ear - 142 mm; (Males) Total Length - 1,416 mm, Height at Shoulder - 875 mm, Tail Length - 105 mm, Hind Foot - 405 mm, Ear - 143 mm.

Distribution

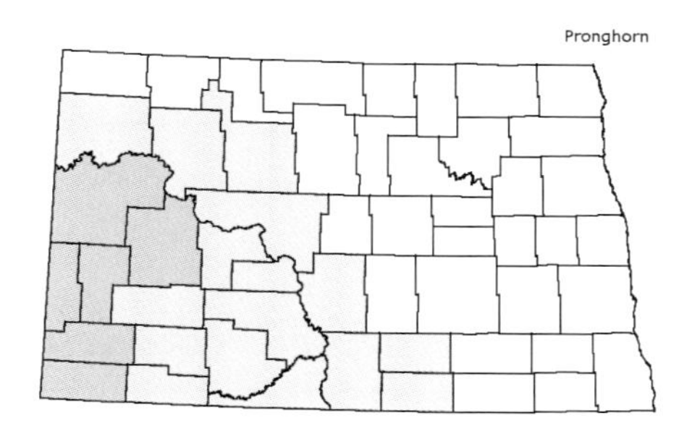

The species is restricted to the grasslands and deserts of North America. However, with the loss of grasslands resulting from settlement and farming, their range was drastically reduced. Currently, pronghorns occupy semiarid to arid grasslands and deserts where agriculture is minimal. They now occur in southern Alberta and Saskatchewan, 15 western states, and northern Mexico. The eastern limit of their distribution includes western portions of the Dakotas, Nebraska, Kansas, and the Oklahoma and Texas panhandles. There are currently about 1 million pronghorns in scattered populations, nearly half of which are in Wyoming.

Pronghorns originally occurred throughout much of the grassland areas of North Dakota, with reports as far east as the Red River valley and even scattered records from western Minnesota. Early reports indicate significant numbers from Valley City, western Cass County, Devils Lake, and the vicinity of the Pembina Hills and Turtle Mountains, as well as western North Dakota. Pronghorns were largely eliminated from eastern North Dakota by 1900, and by 1925, the state's population had been reduced to 225 animals in the southwestern corner. Subsequently, populations have recovered to the extent that pronghorns now occupy approximately the western half of the state, with the greatest numbers in the extreme southwest.

Habitat

Relying on their keen eyesight and speed to avoid predation, pronghorns require the open cover of grasslands to provide unlimited visibility. Newborn fawns are vulnerable to predation, and their favored bedding sites require concealment and long-range visibility provided by a mixture of grasses and forbs. Typical habitat contains at least 50%

plant cover including a mixture of grasses, forbs, and shrubs. During severe winters, pronghorns may select areas with increased shrub density, and increased height of certain species such as silver sage and rabbitbrush. In eastern Montana, sagebrush-grass areas provide critical habitat, but pronghorns also use Conservation Reserve Program (CRP) land, and may forage on alfalfa and growing wheat fields. In North Dakota, the eastern edge of pronghorn range, ground cover abundance comprises 1 to 19% shrubs, 1 to 42% forbs, and 26 to 88% grasses. Areas with prolonged winters and deep snow accumulations of 25 to 30 cm can be limiting to foraging.

Ecology and Behavior

Pronghorns have been described as dainty or selective feeders, foraging on a wide variety of plants. The proportions of various plant species in their diet do not reflect the proportion in the habitat. Generally, forbs are most important during summer, while grasses are taken in early spring or other periods of new growth. Browse may dominate the diet during summer and autumn. In North Dakota, the rumen content during fall may average 46% browse (e.g., sagebrush, snowberry), 33% forbs, 5% grasses, and 15% farm crops. Most studies from the northern Plains also emphasize the importance of various sage species throughout the year.

Pronghorns are social animals, normally found in groups, except when females isolate themselves during fawning and older males become solitary during summer. Group sizes vary seasonally, and are smallest during summer. Groups normally consist of females and their young, accompanied by a single mature male, and bachelor groups of 1- to 3-year-old males. Following the rut, they may coalesce into much larger mixed sex groups, which remain together throughout the winter. Both female and bachelor groups exhibit a high level of aggressive interactions in which dominant individuals displace others from prime food sources and bedding sites.

Home ranges are highly variable, changing with weather patterns, seasons, and changes in habitat. In Wyoming, ranges during summer and autumn vary from 2.6 to 5.2 km^2, while winter ranges in Montana vary from 6.5 to 22.5 km^2. Seasonal movements may be undertaken during winter in response to low temperatures and snow conditions, and during summer to moister areas with increased plant diversity and abundance. In some areas, deep snow has forced movements of up to 160 km.

Activity budgets primarily involve foraging and resting, with foraging most commonly occurring in the morning and evening and resting during mid-day. During winter, about 80% of the time is spent eating or reclining, while during the summer, the relative time spent eating declines significantly.

The primary causes of mortality in pronghorns include hunting, predation, overwintering losses, disease, and accidents. Predation by coyotes, golden eagles and bobcats can be a significant cause of mortality among pronghorn fawns. Fawn losses to predation can range from 5 to 99%, with the coyote being the most common predator. Although fawn mortality can be very high, adult mortality is often negligible in adults until age 7 in males and 10 in females. Mortality of all age classes may be very high during severe winters with heavy snow. In North Dakota, severe winters of 1964–65, 1977–78, 2009–2010, and 2011–2012 resulted in the loss of 50% to 80% of the state's pronghorn population. Such precipitous losses may also be followed by rapid population recovery under favorable conditions. The bluetongue virus, frequently carried by domestic sheep and transmitted by gnats, can also result in significant pronghorn mortality. During one year, 3,200 pronghorn deaths in Wyoming were attributed to this disease. Other mortality can result from accidental deaths, such as vehicle collisions and animals caught in fences.

Reproduction

The breeding season runs from mid-September through early October. Females usually do not come into their first estrus until 16 months of age, while males rarely mate before 3 years. If a female fails to conceive, she may have a second estrus 28 days later. Pronghorns are polygamous, but females exercise mate choice by visiting several males before mating.

Young are born following an average gestation of 252 days. As many as seven embryos have been reported, but the typical litter size is two. In North Dakota, fawns are born from mid-May to early June. Prior to birth, the doe isolates herself, and does not rejoin the female group until about three weeks after giving birth. Fawns are kept hidden, and only visited by their mothers for suckling every two to three hours.

Newborn fawns average about 3.5 (2.2 to 4.9) kg, but weights may vary as a function of range conditions, health of the female, and number of embryos. This relatively large size indicates that pronghorn show the highest known rate of maternal investment for North American ungulates. Following parturition, fawns usually remain alone except for brief periods of bonding with their mothers and suckling. They then may form sibling bonds and join fawn groups. Weaning occurs at 4 to 6 weeks, but some suckling may last until 12 weeks. Adult weight is not reached until the second winter.

Status and Conservation

Prior to white settlement, it has been estimated that at least 40 million pronghorns roamed through western North America, perhaps in larger numbers than the bison. Hunting and habitat loss resulted in their near extinction. There are now nearly 1 million in scattered populations over much of their historic range. The North Dakota population is currently about 10,000, with an annual hunter harvest approaching 1,200. Although the population has fluctuated widely from less than 1,300 to over 15,000, there have been only five years since 1951 when it was necessary to close the season.

The pronghorn is a popular big-game species, providing recreation for nearly 1,500 sportsmen in North Dakota annually. Management issues mainly center around maintenance of grassland habitat with a diverse mixture of grasses, forbs, and shrubby browse. Rangeland heavily grazed by livestock may result in depletion of forbs important to their diet. Pronghorns are disturbed by fences, which disrupt their daily and seasonal movement patterns. North Dakota pronghorn harvests are based on census and estimated recruitment within management areas, and populations appear to be sustaining current hunting pressure. Oil and gas development has caused disturbance and displacement of pronghorn. Placement of roads and wells can be problematic, particularly in areas with high-value sagebrush habitat.

Selected References

Autenreith et al. (2006), Bailey (1926), Byers (1997), Byers (2003), Cadieux (1986), Christie et al. (2017), Jacques et al. (2009), Jatnieks-Straumanis (1983), Jensen (2000), Jensen et al. (2004), Kolar (2009), Jensen and Seabloom (1993), Jones et al. (1983), Knue (1991), MacCracken and Uresk (1984), McCabe et al. (2004), Nelson (1925), O'Gara (1978), O'Gara and Yoakum (2004), Selting and Irby (1997), Yoakum (1978)

Family Bovidae
Bovids

Bison
Bison bison

Te–Omaha; Pte, Tatanka–Lakota, Mandan;
Mité–Hidatsa; Tanaha–Arikara

Robert Seabloom

Description

The bison is the largest native land mammal in North America. Adult males may weigh up to 907 kg, and females to 545 kg. The body is massive, especially the forequarters. The massiveness and height of the body is accentuated by the hump, formed by long processes extending up from the thoracic vertebrae. Coloration is generally dark brown. The pelage covering the head, neck, hump, and front quarters is long, while that covering the rest of the body is short. The tail is short, with a tufted tip. Ranges of body measurements are: (Males) Total Length - 3,040 to 3,800 mm, Tail Length - 330 to 910 mm, Hind Foot - 580 to 680 mm; (Females) Total Length - 2,130 to 3,180 mm, Tail Length - 300 to 510 mm, Hind Foot - 500 to 530 mm. Ear lengths average 150 mm in males and 120 mm in females.

Distribution

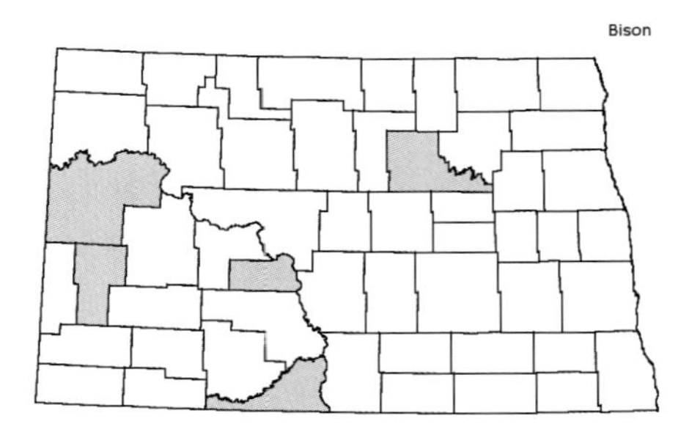

Bison originally occurred throughout much of North America from Alaska and northern Canada south to northern Mexico. The northern boundary of the Plains bison was in north-central Saskatchewan and southern Manitoba. In the United States, they extended from the prairies of eastern Oregon to northern Mexico, the Gulf of Mexico, eastward to the Appalachian Mountains, and in places to the Atlantic coast. Historic populations may have been as high as 30 million. Widespread market hunting, and possibly livestock-born diseases, resulted in their near-extinction in the late 19th century. Present-day occurrence is limited to isolated herds in parks, preserves, and private holdings.

Typical of the Plains states, vast herds of bison originally occurred throughout North Dakota. During the 19th century, they were subjected to extensive market hunting. Some hunters reported taking up to 100 per day, and one firm alone reported shipping more than 60,000 hides in one year during the 1870s. It has been suggested that even such losses could not account for the complete extermination of the bison from the northern Plains, and that the introduction of livestock-borne diseases during the 1870s and 1880s provided a more valid explanation. Except for a few individuals, bison were essentially eliminated from the North Dakota by 1884. Subsequent introductions have resulted in establishment of herds in both north and south units of Theodore Roosevelt National Park, Sully's Hill National Game Preserve near Devils Lake, several Indian reservations, and various private holdings in the state.

Habitat

Bison exhibit a strong preference for open grasslands and meadows, and their historic distribution was centered on the central grasslands and northern parklands. However, they have been known to utilize habitats from semideserts of the Southwest to boreal forests, wherever grass was available. They will also use forested areas for escape cover, thermal cover, and to escape insect pests. Because of their ability to process low-quality forage, they appear to be better adapted than domestic cattle to the warm-season grasses of the short-grass Plains. The bison has been cited as truly a "Pleistocene" mammal, well-adapted to survive harsh winters and heavy snow conditions. When foraging, they have been referred to as "animated snow plows," moving the head from side to side while feeding in several feet of snow. Bison generally show a positive response to recently burned areas, at least during the growing season for one to three years post-fire, where they are attracted to warm-season perennial grasses.

Ecology and Behavior

Bison are grazers, foraging on grasses and sedges throughout the year. Warm-season grasses are predominant in the diet on short-grass prairie, while cool-season grasses and sedges make up the majority of the diet in mixed prairie. In tall-grass prairie, graminoids may comprise 98% of the diet throughout all seasons. Sedges are most common in winter and spring, decreasing in summer and fall, reflecting their availability. Forbs are avoided, comprising <2% of the diet.

Bison are very gregarious, typically remaining in groups of 11 to 20 cows, calves, yearlings, and young bulls throughout the year. Adult bulls are solitary or remain in small groups most of the year. They begin to mix with cow-calf herds in mid- to late July at the beginning of the rut, and remain with them until late August or September. Dominance behavior occurs in all sex/age groups. Threat postures may precede fighting, and include tail elevation, broadside threats, wallowing, and lunges. Fighting behavior includes horn locking, shoving, and hooking.

Activity is primarily diurnal, with most feeding occurring during the day, while nights are spent loafing and occasionally feeding or traveling. Foraging time decreases from summer to winter, while bedding increases. During winter, there are two main daytime foraging bouts, while during summer the number of bouts increases.

Home range sizes are highly variable, depending on sex, age, and habitat quality. Estimates in large preserves vary from 177 to >1,000 km^2, and in Yellowstone National Park adult females had home ranges of 541 km^2, while in smaller preserves, reports vary from 27 to 70 km^2. Bison typically engage in daily movements of about 3 km between foraging sites. Seasonal migratory movements are common, ranging from 14 to 40 km in montane habitat to 240 km in boreal parklands. However, in the northern Great Plains, high forage quality throughout the year negated the necessity for lengthy migrations other than local movements resulting from unpredictable stimuli.

In most herds, the majority of mortality is human-caused, through commercial harvest and sport hunting. Predation by grizzly bears and wolves can be a significant factor in large parks and reserves, such as Yellowstone. Accidental drowning by animals falling through thin ice has resulted in significant mortality in the past, sometimes involving entire herds. Severe winter weather, especially early spring storms, has occasionally been a major mortality factor in Yellowstone National Park, especially affecting calves. A wide variety of diseases, such as anthrax, tuberculosis, malignant cartarrhal fever, and brucellosis, have had varying affects on mortality and reproduction. Brucellosis affects the reproductive system, causing abortions during first pregnancies.

Reproduction

The breeding season may extend from late June through September, but most copulations occur during late July and early August. During the breeding season, cows are polyestrous, having a cycle of about three weeks and sexual receptivity of one to two days. Bison exhibit a tending bond type of breeding behavior, in which the bull attempts to keep the estrus cow peripheral to the main herd before copulation. Cows may participate in mate selection by approaching high-ranking bulls. Sexual maturity in cows is usually between ages 2 and 4, but there are a few cases of conception in yearlings. Maturation of bulls is similar to that in cows, but most bulls do not breed until 6 years. A single calf is born between mid-April and late May following a gestation of 285 days. Twins are rare in bison.

Newborn calves are reddish-tan, and weigh about 20 (14 to 25) kg. They are precocious, attempting to graze by 5 days, and drinking water within a week. Nursing continues for at least seven to eight months. Calves begin to darken at 2.5 months, and weigh 135 to 180 kg at 8 to 9 months. Yearlings weigh 225 to 315 kg, and maximum weight is reached at about 6 years.

Status and Conservation

During the 1970s, there were about 65,000 Plains bison in North America, mostly in captive herds. By 1990, the population had grown to about 90,000, and is currently estimated at 580,000 to 700,000. Only about 3% of this population is in publicly owned and conservation- oriented herds. In North Dakota, the largest herds are in Theodore Roosevelt National Park, where herd size is regulated at 200 to 400 in the south unit, and 100 to 300 in the north unit. Other smaller herds occur on Sully's Hill National Game Preserve, various Indian reservations, Cross Ranch Nature Conservancy Preserve, and private ranches. North Dakota state law classifies all bison as livestock.

The bison was a keystone species on the North American prairie prior to European settlement, and its near extermination resulted in the removal of a vital component maintaining the natural prairie ecosystem. While North American herds have expanded in recent years, further recovery of natural bison populations is limited by habitat loss due to agricultural development, urbanization, and other competing land uses. Commercial bison production, diseases, and genetic issues further affect recovery policies.

Several exotic diseases have created additional problems for bison management. Bison are susceptible to cattle-borne brucellosis and bovine tuberculosis. For example, in Yellowstone National Park, bison frequently migrate out of the park during winter onto lands seasonally grazed by cattle, raising concerns by the ranching industry. Chronic Wasting Disease, carried by cattle, has been diagnosed in elk, moose, mule deer, and white-tailed deer, but not in bison, pronghorn, or bighorn sheep. Continued disease monitoring of these species is warranted.

The hunting of bison is allowed in several locations in North America, providing recreation, meat production, and revenue for wildlife programs. Hunting on public lands is permitted in three areas of the contiguous states, four areas in Alaska, and five in Canada.

Farming of bison is a rapidly growing agricultural practice, providing lean meat for private consumption. Commercial bison production occurs in all states and provinces of the United States and Canada, with herds ranging from a few head to more than 5,000.

In North Dakota, bison in Theodore Roosevelt National Park are maintained for their historic significance, tourist interest, and their role in maintaining the badlands ecosystem. North Dakota Indian reservations maintain herds for cultural and commercial purposes. The Nature Conservancy's Cross Ranch Preserve uses bison in maintenance of its native prairie.

Nonconsumptive use of bison includes their aesthetic value to tourism, their role in restoration of prairie ecosystems, and their role in cultural and spiritual enhancement on Indian lands. With continued decline of already sparse human populations on the Great Plains, there have been proposals for re-establishment of free-living bison herds on a large scale. Any such proposals would take decades for implementation.

Selected References

Bailey (1926), Forde et al. (1984), Hanson (1984), Jones et al. (1983), Koucky (1983), Meagher (1978, 1986), Reynolds et al. (2003)

Bighorn Sheep

Ovis canadensis

Ansa-chta–Mandan; Heki^{n}skagi–Dakota;
Azichtia–Hidatsa; Arikusa–Arikara

C. D. Grondahl, North Dakota Game and Fish Department

Description

Bighorn sheep are classified in the same family (Bovidae) as the bison, and domestic cattle and sheep. They are of blocky build and relatively short legged, somewhat larger than the Dall's and Stone's sheep of Alaska and northwestern Canada. Weights of North Dakota bighorns average 89 (56 to 113) kg for rams, and 57(48 to 66) kg for ewes. Coloration is grayish brown to dark brown, with a whitish muzzle, belly, and rump. The tail is short and dark brown. Ranges of standard body measurements are: (Males) Total Length - 1,321 to 1,956 mm, Tail length - 102 to 152 mm, Hind Foot - 356 to 483 mm, Ear - 100 to 130 mm; (Females) Total Length - 1,168 to 1,880 mm, Tail Length - 102 to 107 mm, Hind Foot - 279 to 432 mm, Ear - 90 to 127 mm.

Distribution

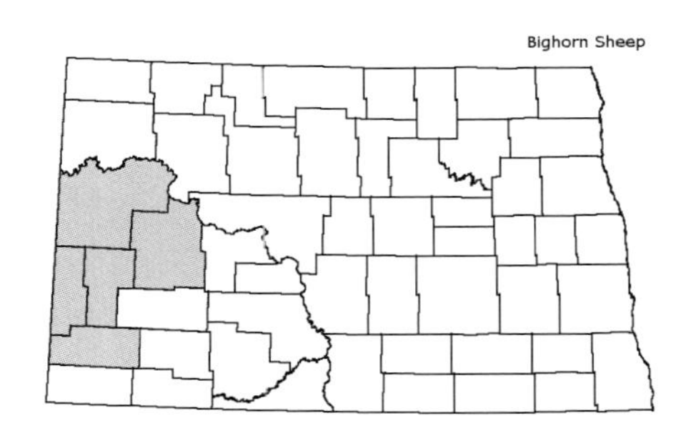

Originally, bighorn sheep occurred from the Canadian Rockies south to Mexico and Baja California. Their distribution has been correlated with low precipitation levels in winter and spring. They range from high mountain meadows to low desert mountains, foothills, and eastward to the badlands of the Great Plains. Current distribution is spotty, with many relict and reintroduced populations resulting from transplants.

Prior to white settlement, bighorn sheep occurred in badlands areas of extreme eastern Montana and Wyoming, southwestern North Dakota, and western South Dakota and Nebraska. These badlands populations were thought to comprise a separate subspecies *(O. c. auduboni)* from the Rocky Mountain bighorn *(O. c. canadensis)*, but recent research has assigned them to the Rocky Mountain subspecies. Bighorns were observed by Lewis and Clark, Audubon, and other early explorers, but went into decline with white settlement, and the last known bighorn in the state was killed around 1905. Possible causes of the extirpation include overharvest, competition with domestic livestock, and the transmission of livestock-borne pathogens to which the sheep had no immunity.

Eighteen bighorn sheep were introduced to the North Dakota badlands from British Columbia in 1956, providing the nucleus for reestablishment of the species in the state. Subsequently, additional transplants from Idaho, Montana, and Oregon augmented the original herd. Currently, there are 15 herds of 10 to 60 bighorns scattered throughout the badlands.

Habitat

Bighorn sheep occupy dry to mesic grasslands in close proximity to precipitous escape terrain used elude predators. They are found at an extreme range of elevations, from high mountains to desert grasslands.

Open grasslands and shrub-steppe communities provide bighorns with good visibility to detect predators, if they are not far from steep cliffs and canyons, which are used for escape. Most sheep activity in North Dakota occurs within 300 m of escape terrain, described as steep, rugged terrain with 80% slope and 10 m vertical rise. Grasslands with a variety of sedges, forbs, and shrubs are used for foraging. Densely forested areas are avoided, because of forage limitations and lack of visibility. In many areas, bighorns migrate between seasonal home ranges, with changes in forage availability and snow cover. However, in North Dakota only rams display seasonal movements between home and breeding ranges. Preferred habitat for bedding in North Dakota consists of non-vegetated areas near escape terrain, especially during summer. Foraging during summer is usually on shrub and disturbed habitats on toe slopes of ridges and buttes, shifting in fall to a variety of other habitats, including open juniper, xerophytic shrub, grassy slopes, and disturbed sites.

Ecology and Behavior

Bighorns are relatively opportunistic in their diet, and feed on a wide variety of plants, depending on availability. In many areas, forbs dominate the diet, followed by grasses, with shrubs making up a small percentage. However, relative amounts of these forage classes may vary among populations, sex/age classes, and individuals.

Bighorn sheep are highly social, living in separate ewe and ram groups for most of the year. Ram groups consist of males, generally over 3 years of age, while ewe groups include lambs and other offspring up to about 2 years. Ram and ewe groups occupy separate, but undefended home ranges. Group sizes vary but may reach >100 animals. In North Dakota, ram groups have ranged from 4 to 22 individuals and ewe groups from 4 to 43. Ram groups exhibit a strong dominance hierarchy, but ewe groups do not. In North Dakota, rams are known to move from one herd to other nearby herds.

Bighorns are highly social, living in separate ewe and ram groups for most of the year. Ram groups consist of males, generally over 3 years of age, while ewe groups include lambs and other offspring up to about 2 years. Ram and ewe groups occupy separate, but undefended, home ranges. Group sizes vary, but may reach >100 animals. In North Dakota, ram groups have ranged from four to 12 individuals and ewe groups from 14 to 43. Ram groups exhibit a strong dominance hierarchy, but ewe groups do not. In North Dakota, rams are known to move from one herd to other nearby herds.

Activity is diurnal, generally involving alternating peaks of foraging and loafing/ruminating. Typically, foraging peaks are at dawn and dusk, but there are seasonal shifts with the relative amount of time foraging increasing during the cold months. Home ranges of North Dakota ewes vary from about 10 to 20 km^2, but more than 50% of their time is spent on 33% of the total home range.

During their first year of life, mortality may be as high as 90%, with most lambs being lost to predation during their first few weeks of life. In North Dakota, lamb recruitment (survival to 1 year old) can be highly variable among herds, with predation and disease appearing to be major factors. Lamb recruitment is significantly higher in areas with low levels of human disturbance. Lamb recruitment has averaged about 28% (1999-2017), and the overall state population is estimated at 320 animals. Very young lambs (<3 weeks of age) are especially vulnerable to coyotes, while older lambs are highly susceptible to pneumonia. Additional influences on lamb mortality have been suggested, including weather, inbreeding depression, poor maternal nutrition, poor mothering, and human disturbance. Mortality among older sheep is relatively low, averaging 9% annually in North Dakota. Adult mortality can be strongly affected by proximity to domestic sheep. Exposure to *Mycoplasma ovipneumoniae*, normally carried by domestic sheep and goats, has caused pneumonia outbreaks resulting in significant losses of adult bighorns and many subsequent years of poor lamb survival.

Reproduction

Bighorns are monestrous, coming into heat as early as late October, and breeding can extend to early December. During the mating season, rams join the ewe bands in search of females in estrus. Rams engage in posturing, kicking, jump threats, and clashing of horns as they compete for receptive females. Rams mainly exhibit a tending bond mating tactic, in which he defends and copulates with a single female, then going on to mate with others.

Pregnancy rates are high among bighorn sheep. In North Dakota greater than 95% of ewes are successfully bred annually, indicating a high incidence of breeding and conception. Lambs are born between

April and June, following a gestation of 174 days. Usually, only single young are born, but twins have been recorded. Newborn lambs weigh between 2.7 and 4.5 kg. Ewes keep their lambs secluded from other bighorns in steep escape terrain for 1 or 2 days following birth. Growth is rapid for the first 6 months, after which it slows. Some lambs may begin grazing at 2 weeks, but weaning is not complete until 4 to 6 months. Females achieve maximum size at 3 to 4 years, but males continue to grow until about 8 years of age. If they survive their first year, median survival of bighorns in North Dakota is 11 years, with 86% mortality by 16 years. The oldest ram and ewe recorded in North Dakota were 14 and 22 years old, respectively.

Status and Conservation

The original bighorn population of North America has been variously estimated between 500,000 and 4 million sheep. Recent surveys estimate the current population at about 67,000, and bands occur in most habitats of the historical distribution. In North Dakota, since the introduction of 18 bighorns in 1956 and subsequent transplants, the population has grown slowly or remained stable. It currently stands at about 320 in 15 distinct herds. Two "metapopulations," north and south of Interstate 94, are recognized for management purposes. Management goals are to maintain each at a minimum of 125 individuals. Bighorn sheep are classified as game animals in North Dakota, with a lottery season for rams. The annual harvest has been as high as 12 rams, but in recent years harvest has been limited to four to six.

The historical decline of bighorn sheep populations has been attributed to a variety of human-caused factors, including overhunting, transmission of diseases by domestic sheep and goats, overgrazing, recreational activities, and disturbance resulting from various development activities. Management efforts have largely centered on transplants of animals into their historic range, eliminating sheep and goats from bighorn habitat, and reduction of human disturbance.

Treatment of herds infected with *Pasteurella* or lungworms by vaccination or treated baits has generally proved unsuccessful, and no vaccines for exposure to M. o exist. Consequently, prevention of contact with domestic sheep and goats has been a major goal of management.

Use of translocations continues to be an important management tool in restoring bighorn sheep populations. However, they are less successful when domestic sheep are within 16 km of bighorn use areas. It is recommended that bighorn translocations be from indigenous sources that are ectopically similar into large blocks of habitat that promote movement, and have no domestic sheep in the area.

Except in protected areas, such as parks where they have become habituated to people, bighorns are very intolerant of human disturbance, and have difficulty adapting to encroachment on their traditional use areas, especially lambing areas. Disturbances such as vehicular traffic, people on foot, and low-flying aircraft can affect foraging efficiency and activity patterns, and cause ewes with lambs to flee to less favorable habitat, rendering lambs more susceptible to predation. In North Dakota, lamb recruitment has been significantly higher in areas with low vs. moderate and high levels of human disturbance. Areas with low levels of disturbance were defined as no improved roads crossing through, and <0.05 km2 of road adjacent to primary use areas; human activity limited to local ranchers and hunters in fall; no oil wells and associated service and maintenance vehicles.

Selected References

Bailey (1926), Bleich et al. (2018), Fairaizl (1978), Feist (1997), Geist (1971), Hall (1981), Jones et al. (1983), Knue (1991), Krausman and Bowyer (2003), Sayre (1996), Shackleton (1985), Wiedmann and Hosek (2013), Wiedmann and Bleich (2014), Wiedmann and Sargeant (2014)

Species Potentially Occurring in North Dakota

The following is a brief discussion of species which may have occurred in North Dakota in the historic past, occur close by in neighboring states, or about which there are scanty or questionable data in the state. Serious naturalists should be alert to these potential additions to North Dakota's fauna.

Dwarf Shrew *(Sorex nanus)*: *S. nanus* has been recorded in nearby Carter and Dawson Coutnies, Montana. In South Dakota, there are records from the Black Hills to as far east as the Pierre area. Its external morphology is very similar to other small shrews occurring in our state. Hence, identification based on the skull is difficult and should be referred to a specialist. Dwarf shrews range from 85–103 mm in total length, and weigh 2–3 gm. These shrews seem to be most common in alpine and subalpine rocky habitats, but have also been collected in sedge marshes, sagebrush-grass, and riparian zones.

Preble's Shrew *(Sorex preblei)*: *S. preblei* is smaller, but otherwise similar to *S. cinereus/haydenii*. Like the dwarf shrew, its external morphology is very similar to other small shrews occurring in our state. Preble's shrew has a total length of 77–95 mm, and weighs 2–4 gm. In Montana, records are scattered over the state in semi-arid grassland and sage habitats. The most easterly Montana record is from Sheridan County, adjacent to Divide County, North Dakota.

Eastern Mole *(Scalopus aquaticus)*: Moles are heavy-bodied insectivores highly adapted for digging. The eastern mole weighs about 100 g, and averages 170 mm in total length. The body is covered with short, dark brown to grayish fur, but the muzzle, feet, and tail are nearly naked. The front feet are very broad and shovel-like. Like other insectivores, the eyes are tiny and there is no external ear (pinna). Bailey (1926) cited an unconfirmed report from southeastern North Dakota and a nearby record from Crookston, Minnesota. The only subsequent report for North Dakota was from a Grafton taxidermist, H. V. Williams, who prepared a specimen from the Grand Forks area in 1960 (UND collection). Moles dig shallow, temporary tunnels while foraging for earthworms and other invertebrates. They tend to prefer moist sands or loams, and avoid heavy clays and gravels. Where they occur, their sinuous burrows are often easily seen, especially in mowed areas.

Star-nosed Mole *(Condylura cristata)*: The star-nosed mole is smaller than the eastern mole, but has a longer tail. Its fur is black, and covers the tail as well as the body. The front feet are shovel-like, but smaller than those of the eastern mole. The tip of the nose has 22 pink, fringe-like papillae, which have a tactile function. Bailey (1926) believed that they occurred in the Red River valley and Turtle Mountains, and cited a report of one in captivity at Towner (McHenry County). There have been no further reports of this species in North Dakota. These moles prefer loose, moist soils, and their burrows are frequently in the vicinity of marshes and streams near rocks, tree roots, and logs.

Southern Flying Squirrel *(Glaucomys volans)*: With gray to brownish-gray silky fur on the upper parts, and a white belly, this squirrel has a very similar appearance to the northern flying squirrel. However, it is about a third smaller (62 g), and the ventral hairs are white to the base, in contrast to the lead-colored bases of northern flying squirrel ventral hairs. In Minnesota, it has been reported as far north as southeastern Polk County. In North Dakota, there is one specimen from Walsh County in the UND collection prepared by Grafton taxidermist H. V. Williams in 1957. These squirrels may be found in mature deciduous forest, where they typically nest in woodpecker holes or hollow trees. Man-made structures such as bird houses and attics may also be used, if available.

Southern Bog Lemming *(Synaptomys cooperi)*: The southern bog lemming resembles a small meadow vole with a short tail. Its total length is approximately two-thirds that of the meadow vole. Diagnostic characters include a short tail that is about the length of its hind foot. There are no records of bog lemmings in North Dakota, but they occur in adjacent Kittson, Marshall, and Polk counties of northwestern Minnesota. It has also been recorded near the Missouri River in southern South Dakota. In northern Minnesota, bog lemmings seem to prefer moist spruce bogs and the margins of springs and marshes, but are also found in moist fields with grasses and sedges, and even in upland forests. Moist forests along the Red and Pembina rivers may provide habitat for this species.

Grizzly (Brown) Bear *(Ursus arctos*; Mato, Mato-chota - Dakota; Mato, Mato unknapininde - Mandan; Waowesica - Lakota; Lachpitzi - Hidatsa; Kúnuch, Konuch-tarawis - Arikara): Prior to the arrival of Europeans, grizzly bears inhabited nearly all of western North America, and were common throughout North Dakota. Fur trader Alexander Henry reported that they were common along the Red River and its tributaries, and described them as "seen in droves" along the Sheyenne River. Lewis and Clark encountered them along the Missouri River, and Teddy Roosevelt wrote accounts of bear activities around his cabin along the Little Missouri River. Over the past century or so, in the conterminous United States, grizzly bears have been reduced to isolated areas of Washington, Idaho, Montana, and Wyoming. It is unlikely that these big bears will ever again reach North Dakota, but until recently, we said the same thing about gray wolves and mountain lions.

Caribou *(Rangifer tarandus)*: Except for the prairie, all the northern half of North America was regarded to be potential caribou range. Woodland caribou occurred in northeastern Minnesota until the 1930s, and there have been occasional individual sightings since that time. Bailey (1926) reported that although there had been no reports of caribou in North Dakota, wandering bands may have strayed into the Red River valley or Turtle Mountains in the past. Furthermore, he reported that caribou antler fragments had been found in the Turtle Mountains. Currently, the nearest population is along the east shore of Lake Winnipeg in Manitoba; hence, the probability of individuals wandering into northeastern North Dakota is undoubtedly very low.

Investigating Mammals

Unlike birds, most mammals are not readily observed in nature. Many are secretive, nocturnal, cryptically colored, and not easily seen by the casual observer. Some diurnal species, such as the squirrels, are readily seen, and identification along with other valuable data can be obtained and recorded simply through the use of binoculars and a notebook. For others, however, their presence must be determined through the use of other, sometimes indirect, approaches.

Indirect observation and interpretation of animal sign may provide the first clue to the presence of a species. Here, the naturalist's visual acuity and knowledge of animal habits are of inestimable value. Animal sign is any environmental evidence of the presence of a species. It can include tracks, scats (droppings), runways, home sites, cuttings, feeding sites, antler rubs, and wallows, to name a few. Tracks are most observable in moist sand or mud, and permanent records can be obtained by preparation of plaster of paris casts. Sometimes naturalists lay smoked paper on small mammal trails, or set up artificial scent posts surrounded by smooth sand to identify predators. One of the best references for sign identification is Murie's (1954) "Field Guide to Animal Tracks." This little guide provides precise drawings and discussion of the tracks, trails, scats, cuttings, and home sites of most North American mammals.

Prey species can be identified by analysis of the scats of carnivorous mammals and the regurgitated pellets of hawks and owls. Here, the scat or pellet is washed and sieved, and the hairs and bone fragments are carefully removed and sorted. Skull and bone fragments are examined under low-power magnification and keyed to species. Hair identification is even more specialized, and involves mounting of guard hairs on slides, and microscopic examination of the cortex and medulla. A good reference for hair identification is a book by Moore et al. (1974), "Identification of the Dorsal Guard Hairs of Some Mammals of Wyoming," which includes many North Dakota species.

The primary documentation of the occurrence of a species is based on carefully prepared museum specimens consisting of dried study skins and skulls. Most of the evidence for species description and taxonomic classification has been based on studies of skins and skulls as well, although relatively recent developments in the specialized fields of biochemistry, karyology (chromosome numbers and structure), and molecular biology (DNA and RNA) have also made major contributions.

For the naturalist working in the field, it is important to develop the skill to prepare standard study skins and skulls. Extensive instructions for study skin and skull preparation are provided in references such as Anderson (1965), Hall (1981), and Martin et al. (2001).

Human Health Considerations

Like other wildlife, wild mammals can be reservoirs for a number of diseases transmissible to humans, including rabies, tularemia, histoplasmosis, lyme disease and hantavirus. Therefore, good sanitation and common sense are essential when handling wild animals, carcasses, or animal products and excretions. Until recent years, the preparation of small mammal specimens, such as mice and shrews, was considered to be a relatively safe activity. In 1993, however, the identification of Sin Nombre Virus, a strain of hantavirus causing severe respiratory illness in humans, changed that belief. This virus is primarily carried by deer mice, and human infection is thought to be via inhalation of virus attached to airborne particles of excreta. As of 2008, there have been 10 cases of the disease in North Dakota, including six deaths. Although human infections are uncommon, deer mouse infections in North Dakota have been as high as 40%. Stringent protocols have been established for working with rodents in the field, and naturalists should refer to published guidelines (Kelt et al. 2010).

Legal Considerations

There has been increasing concern for species conservation leading to federal and state legislation protecting threatened and endangered species. Some species, though listed as game animals or furbearers, exist in such sparse populations that their seasons have been continuously closed for years. Other non-game animals have so few records, or are so poorly known, that they are listed in a state as species of special concern. The North Dakota Game and Fish Department has established a list of 100 "species of conservation priority", highlighting certain birds, mammals, amphibians, reptiles and fish meriting special consideration of their conservation needs (Dyke et al. 2004).

Scientific collecting permits are required for all protected species and may be obtained from the North Dakota Game and Fish Department. Game animals and furbearers may be taken in season by licensed hunters and trappers. Any deviation, such as out-of-season collecting, also requires a special permit.

Collecting

Many small mammals, such as mice and shrews, are easily captured with ordinary mouse traps. Rat traps may be needed for larger species, such as chipmunks and ground squirrels. Game animals and furbearers may be collected by shooting and larger traps used by fur trappers. Bats are collected with mist nets, such as those used by bird banders, and small-gauge shot shells. Special traps have been designed for moles and pocket gophers. Occasionally, road-killed animals may be in satisfactory condition for specimen preparation.

A number of baits have proven satisfactory for use with snap traps and rat traps, including peanut butter, rolled oats, bits of meat, fruits, and even bits of candy. Traps should be set near burrow entrances or in runs along logs, in grass, water edges, etc. During summer months, it is necessary to check traps early in the morning, or more than once per day, because specimens will be rapidly spoiled by insects, especially ants, or rotted by heat. If not prepared immediately after capture, specimens may be placed in small plastic bags and kept on ice or frozen.

Storage

Study skins and skulls should be stored in tight boxes or cabinets with naphthalene or paradichlorobenzene (moth repellent) to prevent insect damage. Protection from light prevents the skin from fading. Eventual placement of study skins and skulls in an established museum collection is highly recommended to make them available for serious study. ***Again, special permits are required for collecting and storage of all mammal specimens.***

Keys to the Mammals of North Dakota

A key is a written tool which aids in the identification of an animal or specimen. It consists of a series of paired, alternative statements (couplets), one of which best describes the specimen in question. Each couplet has an "a" or "b" alternative. When a decision is made, the number in the right margin indicates the next couplet to be considered. Eventually, you should reach a final couplet which includes the species or other group to be identified.

No key is perfect, and it has been said that a key works best for its author. Likewise, a key may be valid only for the geographic area where it was written. The following keys are based on published references to the mammals of North Dakota, along with research specific to the state's fauna. The most obvious characters of typical adult specimens are emphasized although, for some of the lesser known species, some rather minute features and technical measurements are necessary. Multiple characters are included as much as possible to account for variability within a species and damaged specimens. It is important that *all* characters and alternatives be carefully considered. Comparison with known specimens is highly recommended, especially when poorly known species and subtle characters are involved.

Skull structures and measurements used in the keys are illustrated in the accompanying figures, and/or defined in the glossary. Measurements are given in millimeters, and in some cases tenths of millimeters, so accurate measurement is required. With small mammals, especially skulls, use of low-power magnification may be necessary. Finally, there are separate keys for skins and skulls to account for specimens where only exterior features are available, as well as for skulls picked up in the field.

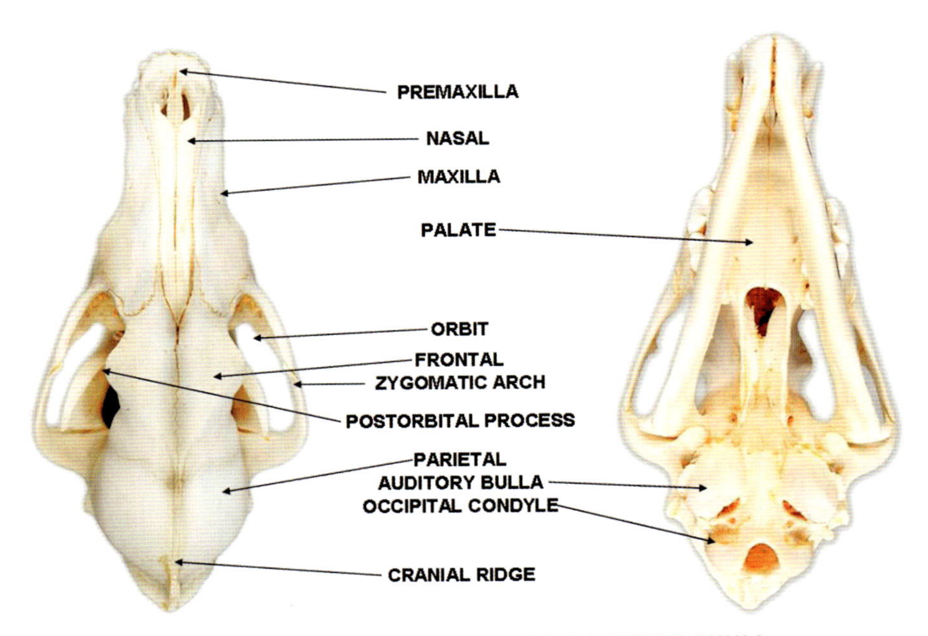

DORSAL AND VENTRAL VIEWS OF A COYOTE SKULL

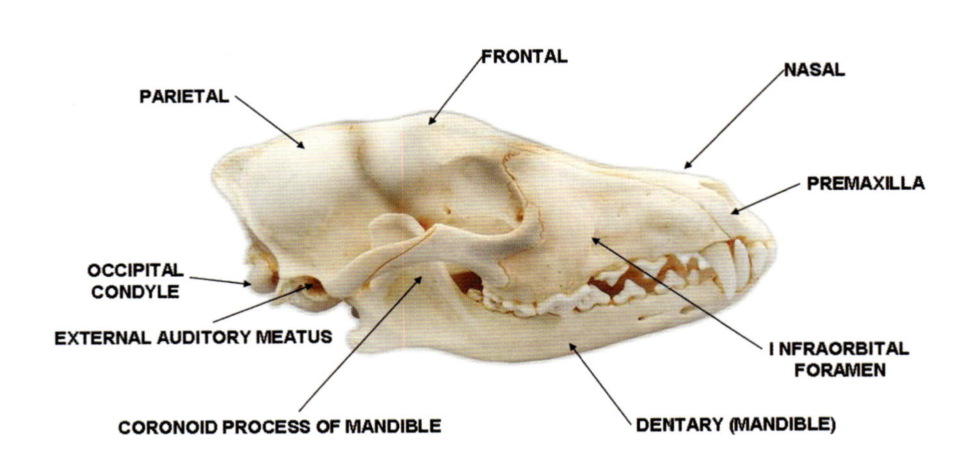

LATERAL VIEW OF A COYOTE SKULL

Photos by Craig Bihrle, North Dakota Game and Fish Department

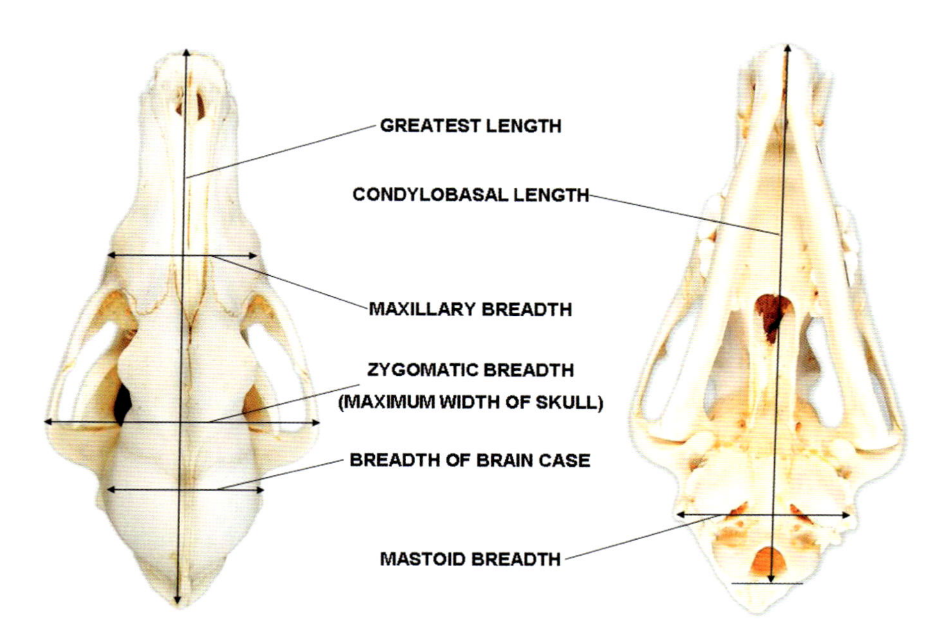

SOME MEASUREMENTS OF A COYOTE SKULL

Photos by Craig Bihrle, North Dakota Game and Fish Department

Keys to Skins and Whole Specimens

Key to Orders

1a. Feet terminating in digits modified as hooves; other digits reduced or absent..Artiodactyla (Even-toed ungulates)

1b. Feet not terminating as hooves; digits with claws.................................2

2a. Forelimbs modified as wings, with highly elongated digits connected by a flight membrane; hind limbs and tail connected by a similar membrane, the uropatagium......................................Chiroptera (bats)

2b. Forelimbs not modified as wings..3

3a. First toe of hind foot thumb-like, clawless, opposable; tail long, naked, and prehensile; marsupium (pouch) present in females .... Didelphimorphia, *Didelphis virginiana* (Opossum)

3b. All functional toes with claws, none opposable, tail not prehensile, no marsupium..4

4a. Front feet with five claws; tail not flattened or with a cottony tuft.......5

4b. Front feet with four claws or, if with five the tail is flattened or with a cottony tuft..6

5a. Pinna (external ear) present; fur with guard hairs and underfur; eyes not reduced; length of head and body more than 115 mm ....Carnivora (Carnivores)

5b. Pinna reduced and buried in the fur; fur without guard hairs; eyes tiny; total length under 115 mm ....Eulipotyphla (Soricomorpha) (Shrews)

6a. Tail a cottony tuft, much shorter than the ear; four clawed toes on hind feet; feet densely furred........ Lagomorpha (Hares and Rabbits)

6b. Tail longer than the ear; hind feet with five clawed toes; soles of feet usually naked....................................... Rodentia (Rodents)

Keys to Genera and Species

Order Eulipotyphla (Soricomorpha)

1a. Total length greater than 110 mm 2

1b. Total length less than 110 mm 3

2a. Total length averaging 129 mm; tail length averaging 25 mm; pelage slate-colored; no fringe of hairs on the hind foot .... *Blarina brevicauda* (Northern Short-tailed Shrew)

2b. Total length averaging 149 mm; tail length averaging 67 mm; pelage black with silvery underparts; conspicuous fringe of stiff hairs on hind feet .... *Sorex palustris* (Northern Water Shrew)

3a. Hind foot usually under 10 mm .... *Sorex hoyi* (Pygmy Shrew)
*This species is difficult to distinguish from *S. cinereus* and *S. haydenii*, and requires examination of the skull.

3b. Hind foot usually over 10 mm 4

4a. Total length averaging about 109 mm; fur usually tricolor, with back dark brown, sides lighter, belly much lighter .... *Sorex arcticus* (Arctic Shrew)

4b. Total length under 105 mm; fur not tricolored 5

5a. Fur gray, drab on upper parts, sides and belly whitish to buff .... *Sorex merriami* (Merriam's Shrew)
*Note there has been only one report of this species in North Dakota

5b. Fur dark brown to gray brown on upper parts, grayish white on belly 6

6a. Tail length over 36 mm .... *Sorex cinereus* (Masked Shrew)

6b. Tail length under 36 mm .... *Sorex haydenii* (Hayden's Shrew)

Order Chiroptera

1a. Ears very long, averaging 36 mm; when laid back, extend to the middle of the body ....*Corynorhinus townsendii* (Townsend's Big-eared Bat)

1b. Not as above; ears under 20 mm 2

2a. Dorsal surface of the interfemoral membrane nearly naked (except sometimes the basal fourth); color brown; hairs never tipped with white..3

2b. Interfemoral membrane thickly furred entirely or on the basal half; black hair sometimes tipped with white 9

Photo by Robert Seabloom

Dorsal views showing interfemoral membranes of *Eptesicus fuscus* **(Ef),** *Lasionycteris noctivagans* **(Ln), and** *Lasiurus cinereus* **(Lc)**

3a. Forearm more than 40 mm.................*Eptesicus fuscus* (Big Brown Bat)

3b. Forearm under 40 mm ....*Myotis* (Mouse-eared bats)

4a. Posterior border of the interfemoral membrane with a conspicuous fringe of hairs ....*Myotis thysanodes* (Fringed Myotis)

4b. Posterior border of the interfemoral membrane without a conspicuous fringe of hairs 5

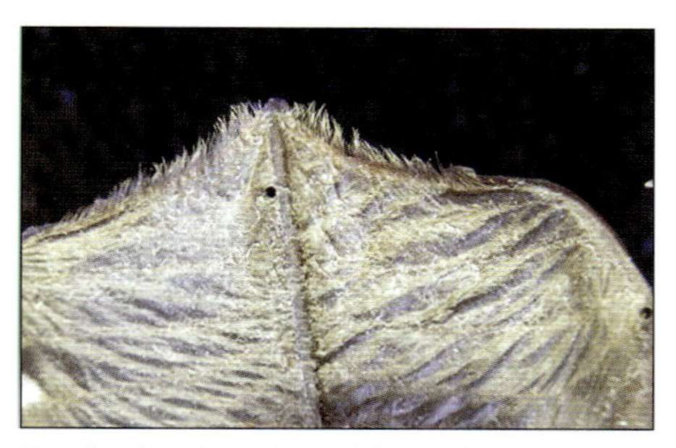
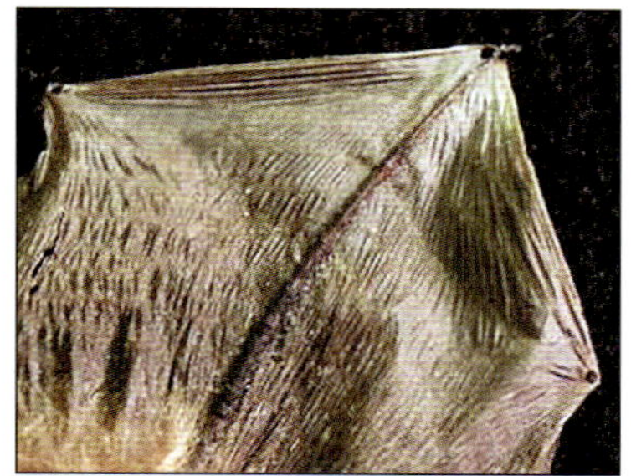

Posterior border of interfemoral membrane of *Myotis thysanodes* (left) showing fringe of hairs, compared with *M. lucifugus* (right)

5a. Dorsal pelage pale yellowish brown, contrasting with blackish ears and membranes; forearm usually under 33 mm; hind foot usually under 8 mm from heel .... *Myotis ciliolabrum* (Western Small-footed Myotis)

5b. Dorsal pelage brownish, not contrasting with color of ears and membranes; forearm more than 33 mm; hind foot more than 8 mm from heel 6

6a. Ears long, usually 18 mm or more from notch.. .... *Myotis evotis* (Long-eared Myotis)

6b. Ears shorter, usually under 17 mm from notch 7

7a. Ear when laid forward extends beyond the tip of the nose, usually 16 to 18 mm from notch; tragus usually 10 to 12 mm, more than half the length of the ear from the notch... ....*Myotis septentrionalis* (Northern Long-eared Bat)

7b. Ear when laid forward extends little if at all beyond the tip of the nose, under 15 mm; tragus shorter...8

8a. Ear usually 11 to 14 mm from notch; calcar with keel........................... ....*Myotis volans* (Long-legged Myotis)

8b. Ear usually 14 to 15 mm from notch; calcar without keel ....*Myotis lucifugus* (Little Brown Myotis)

9a. Interfemoral membrane not heavily furred to the tip of the tail (only the basal half); pelage blackish-brown and frosted ....*Lasionycteris noctivagans* (Silver-haired Bat)

9b. Interfemoral membrane heavily furred to the tip of the tail; pelage yellow-brown, or reddish, and frosted 10

10a. Body yellowish-brown, frosted with white; no white spot on shoulder... *Lasiurus cinereus* (Hoary Bat)

10b. Body red to yellow-gray; white spot on each shoulder.......................... .... *Lasiurus borealis* (Red Bat)

Order Lagomorpha

1a. Hind foot length more than 115 mm; winter pelage white....................2

1b. Hind foot length under 115 mm; winter pelage agouti..........................3

2a. Total length over 550 mm; ears long, over 85 mm from notch, relatively narrow; tail length more than 50 mm; hind foot length more than 150 mm; distinct black patch of hair on outer half of back of the ear....................*Lepus townsendii* (White-tailed Jackrabbit)

2b. Total length under 550 mm; ears relatively short and broad, less than 85 mm from notch; tail under 50 mm; hind foot under 150 mm; black hairs on the tip of the ear do not form a distinct patch................ ....*Lepus americanus* (Snowshoe Hare)

3a. Ear length more than 65 mm .. *Sylvilagus audubonii* (Desert Cottontail)

3b. Ear length under 61 mm 4

4a. Total length more than 390 mm; inside of ear relatively sparsely furred.....................................*Sylvilagus floridanus* (Eastern Cottontail)

4b. Total length under 390 mm; inside of ear relatively densely furred ....*Sylvilagus nuttallii* (Mountain Cottontail)

Order Rodentia

1a. Sharp quills on the body and tail .... Family Erethizontidae, *Erethizon dorsata* (North American Porcupine)

1b. No quills on any part of the body..2

2a. Hind feet webbed; tail scaly, flattened horizontally .... Family Castoridae, *Castor canadensis* (North American Beaver)

2b. Hind feet not webbed; tail not flattened horizontally.........................3

3a. Tail well covered with hairs, the longest 10 mm long.......................... ....Family Sciuridae 4

3b. Tail covered with short fur, or naked 14

4a. Skin between forelimbs and hind limbs loose, forming a gliding membrane.................. *Glaucomys sabrinus* (Northern Flying Squirrel)

4b. Skin between forelimbs and hind limbs not noticeably loose 5

5a. Length of head and body more than 350 mm; weight more than 1100 g; feet black; tail short and bushy .... *Marmota monax* (Woodchuck)

5b. Head and body less than 350 mm; weight under 1,100 g; feet seldom black; tail length variable...6

6a. Tail short, less than 25% of total length, black-tipped .... *Cynomys ludovicianus* (Black-tailed Prairie Dog)

6b. Tail longer, more than 25% of total length, not black-tipped 7

7a. Dorsal pelage always striped; stripes continuing on to the face 8

7b. Dorsal pelage striped, mottled, spotted, or uniform; if striped, stripes do not continue on to face 9

8a. Dorsal stripes not continuous to the base of the tail, indistinct on face; rump reddish; tail less than 40% of the total length .... *Tamias striatus* (Eastern Chipmunk)

8b. Dorsal stripes continuous to the base of the tail, distinct on face; rump not reddish; tail more than 40% of the total length.. *Neotamias (Tamias) minimus* (Least Chipmunk)

Photo by Robert Seabloom

Study skins of *Tamias striatus* (upper) and *Neotamias minimus* (Lower)

9a. Total length more than 340 mm 10

9b. Total length less than 340 mm 13

10a. Dorsal color dappled or mottled, grayish, dappled with black and white .... *Poliocitellus (Spermophilus) franklinii* (Franklin's Ground Squirrel)

10b. Dorsal color more or less uniform or grizzled 11

11a. Total length less than 400 mm; upper parts olive brown to reddish orange; distinct black stripe on side during summer .... *Tamiasciurus hudsonicus* (Red Squirrel)

11b. Total length more than 400 mm; upper parts gray or orange-brown 12

12a. Pelage predominately gray above, white below; tips of tail hairs white .... *Sciurus carolinensis* (Gray Squirrel)

12b. Pelage orange-brown above, yellowish or buffy below; tips of tail hairs orange .... *Sciurus niger* (Fox Squirrel)

13a. Dorsal pelage marked with a series of alternating dark brown, pale brown, and broken pale and dark brown longitudinal stripes .... *Ictidomys (Spermophilus) tridecemlineatus* (Thirteen-lined Ground Squirrel)

13b. Dorsal pelage uniform, color yellowish brown .... *Urocitellus (Spermophilus) richardsonii* (Richardson's Ground Squirrel)

14a. Eyes and ears small; forefeet enlarged with large claws, adapted for digging; external, fur-lined cheek pouches; tail under 35% of total length .... Family Geomyidae 15

14b. Eyes and ears not reduced, moderate to large; forelimbs not enlarged; fur-lined cheek pouches absent or, if present tail more than 44% of total length 16

15a. Dorsal pelage rich brown, glossy; any white fur on underside confined to chin; forelimbs greatly enlarged; hind foot length more than 30 mm .... *Geomys bursarius* (Plains Pocket Gopher)

15b. Dorsal pelage gray-brown, not glossy; white fur on throat and occasionally on chest; forelimbs only moderately enlarged; hind foot length under 30 mm .... *Thomomys talpoides* (Northern Pocket Gopher)

16a. External cheek pouches present; tail long ..... Family Heteromyidae 17

16b. External cheek pouches absent; tail variable 20

17a. Soles of hind feet densely haired; hind feet and legs greatly enlarged for saltatorial (jumping) locomotion; tail long, about 140 mm, with bushy tip .... *Dipodomys ordii* (Ord's Kangaroo Rat)

17b. Soles of hind feet naked; hind feet, legs, and tail not greatly enlarged 18

18a. Pelage harsh, sometimes with spiny bristles; total length greater than 180 mm .... *Chaetodipus hispidus* (Hispid Pocket Mouse)

18b. Pelage soft; pelage under 150 mm 19

19a. Dorsal color strongly olivaceous; usually a buffy spot behind the ear .... *Perognathus fasciatus* (Olive-backed Pocket Mouse)

19b. Dorsal color buffy, often overlaid with black hairs; no buffy spot........ .... *Perognathus flavescens* (Plains Pocket Mouse)

20a. Tail much longer than head and body; hind feet about one-third length of head and body ....Family Dipodidae 21

20b. Tail equal or shorter than head and body; hind feet not elongate......... ....Families Muridae, Cricetidae 22

21a. Dorsal pelage slightly grizzled; ears with border of white hairs; total length 216 to 247 mm; tail not sharply bicolored........................ .... *Zapus princeps* (Western Jumping Mouse)

21b. Dorsal pelage not grizzled; ears without border of white hairs; total length 188 to 216 mm; tail sharply bicolored .... *Zapus hudsonius* (Meadow Jumping Mouse)

22a. Total length over 440 mm; tail flattened laterally, sparsely haired; five clawed toes on front feet; hind toes fringed with stiff hairs .... *Ondatra zibethica* (Muskrat)

22b. Total length under 440 mm; tail rounded; four clawed toes on front feet...23

23a. Total length more than 250 mm 24

23b. Total length under 250 mm 25

24a. Tail nearly naked; fur short and harsh; ear under 25 mm from notch .... *Rattus norvegicus* (Norway Rat)

24b. Tail heavily furred; fur long and fine; ear more than 25 mm from notch........................*Neotoma cinerea* (Bushy-tailed Wood Rat)

25a. Length of the tail less than one-third the total length.......................26

25b. Length of the tail more than one-third the total length 30

26a. Ears prominent; dorsal coloration gray; belly pure white.................... .... *Onychomys leucogaster* (Northern Grasshopper Mouse)

26b. Ears nearly concealed by long fur; coloration brownish to reddish 27

27a. Tail under 28 mm; coloration gray-brown; sole of hind foot relatively hairy..........................*Lemmiscus curtatus* (Sagebrush Vole)

27b. Tail more than 25 mm, coloration brown or reddish; sole of hind foot not densely haired 28

28a. Back reddish, with gray sides .... *Myodes (Clethrionomys)gapperi* (Southern Red-backed Vole)

28b. Back and sides are similar, brown or grizzled...................................29

29a. Tail more than 38 mm; underparts gray.. .... *Microtus pennsylvanicus* (Meadow Vole)

29b. Tail under 38 mm: underparts washed with yellowish .... *Microtus ochrogaster* (Prairie Vole)

30a. Back grayish brown, grading to lighter brown on sides; tips of belly hairs white; tail bicolored, white below, darker above 31

30b. Back grizzled, with blackish guard hairs and underfur with a buffy band tipped with black; tips of belly hairs slaty yellowish; tail may or may not be bicolored...32

31a. Tail distinctly bicolored, edge between dark dorsal and white ventral hair distinct................. *Peromyscus maniculatus* (Deer Mouse)

31b. Tail not distinctly bicolored, dark dorsal hairs grading into white ventral hairs .... *Peromyscus leucopus* (White-footed Mouse)

32a. Tail naked, scales show plainly; pelage coarse; dorsal parts gray to gray-brown, underparts only slightly lighter ....*Mus musculus* (House Mouse)

32b. Tail clothed with short hair, scales do not show plainly; pelage silky; dorsal parts grayish brown, underparts grayish white.............33

33a. Tail longer than head and body; tail indistinctly bicolored, the dorsal stripe relatively broad, ca half the width of the tail .... *Reithrodontomys megalotis* (Western Harvest Mouse)

33b. Tail shorter than head and body; tail distinctly bicolored, the dorsal stripe relatively narrow, ca one fourth the width of the tail ....*Reithrodontomys montanus* (Plains Harvest Mouse)

Order Carnivora

1a. Five clawed toes on both front and hind feet 2

1b. First toe on front foot rudimentary and high on foot, other four functional; four clawed toes on hind feet 18

2a. Adult weight more than 90 kg; pelage black (occasionally brown or cinnamon), except for brown muzzle and occasional white spot on chest; tail rudimentary ....Family Ursidae, *Ursus americanus* (Black Bear)

2b. Adult weight under 35 kg; pelage not as above; tail apparent............3

3a. Black mask across eyes; tail with alternating buff and black rings and black tip.................Family Procyonidae, *Procyon lotor* (Raccoon)

3b. Tail without alternating rings.. 4

4a. Pelage black, with white spots or stripes .... Family Mephitidae 17

4b. Pelage variable, not as above ....Family Mustelidae 5

5a. Coloration from brown to black, with yellowish to brownish bands from shoulder to the sides, converging at the rump. May be white to yellowish throat markings. Weight 6–18 kg. ... *Gulo gulo* (Wolverine)

5b. Pelage and weight various, but not in above combination 6

6a. Pelage buff, with black legs, feet, and tip of tail; mask across the eyes.. *Mustela nigripes* (Black-footed Ferret)

6b. Pelage not as above; no black mask 7

7a. Foreclaws partially concealed by fur; no median white stripe or spots on forehead (but winter pelage may be almost entirely white) 8

7b. Foreclaws not partially concealed by fur, enlarged for digging; median white mark on forehead; pelage never almost totally white .... *Taxidea taxus* (Badger)

8a. Entire underparts white or yellowish white, or pelage entirely white, sometimes with a black tail tip 9

8b. Entire underparts not white or yellowish white; pelage never all white, with or without a black tail tip 15

9a. Tail without distinct black tip; total length usually less than 200 mm; tail length less than one third the length of the body........................... .... *Mustela nivalis* (Least Weasel)

9b. Tail with distinct black tip; total length usually greater than 200 mm; tail length more than one third the length of the body 10

10a. Sex unknown 11

10b. Sex known 12

11a. Tail under 30% of total length... .... *Mustela erminea* (Ermine, Short-tailed Weasel)

11b. Tail more than 30% of total length .... *Mustela frenata* (Long-tailed Weasel)

12a. Adult male 13

12b. Adult female 14

13a. Total length under 340 mm; tail under 100 mm .... *Mustela erminea* (Ermine)

13b. Total length more than 360; tail over 120 mm.................................... .... *Mustela frenata* (Long-tailed Weasel)

14a. Total length under 295 mm; tail under 85 .. *Mustela erminea* (Ermine)

14b. Total length over 295 mm; tail over 95 mm....................................... .... *Mustela frenata* (Long-tailed Weasel)

15a. Ear brown, with a narrow light buff border.................................... 16

15b. Ear uniform brown, without a buff border 17

16a. Dorsal pelage uniform dark to light brown; ear light brown; belly pale brown; light buff or orange spots or patch on throat and chest; total length under 675 mm; tail length under 225 mm.. *Martes americana* (American Marten)

16b. Dorsal pelage dark brown, occasionally lighter; head and shoulders grizzled; underparts blackish brown; occasionally small white spots on chest and belly; total length more than 675 mm; tail more than 225 mm ....*Martes pennanti* (Fisher)

17a. Toes webbed; total length more than 800 mm; tail more than 250 mm, strongly tapering from a thick base.. *Lontra canadensis* (River Otter)

17b. Toes not webbed; total length under 700 mm; tail under 250 mm, not tapered from a thick base ....*Neovison* (Mustela) vison (Mink)

18a. White spot on forehead; four to six broken longitudinal white stripes or rows of spots on sides ....*Spilogale putorius* (Spotted Skunk)

18b. Median white stripe on forehead and base of nose; top of head white, from which two white stripes diverge over the back; white stripes vary in width, but do not break into rows of spots ....*Mephitis mephitis* (Striped Skunk)

19a. Claws non-retractile, usually blunt; tail usually 200 to 500 mm .... Family Canidae 20

19b. Claws retractile and sharp; tail under 200 mm or over 500 mm .... Family Felidae 24

20a. Tail more than 33% of the total length; hind foot under 185 mm ...21

20b. Tail under 33% of the total length; hind foot more than 185 mm ...23

21a. Tail white-tipped; legs and feet blackish......... *Vulpes vulpes* (Red Fox)

21b. Tail black-tipped; legs and feet rusty, or not markedly different from rest of pelage 22

22a. Weight more than 3 kg; legs and feet rusty; back gray, with a mid-dorsal line extending onto the tail and expanding to form a black tail tip....................................... *Urocyon cinereoargenteus* (Gray Fox)

22b. Weight under 3 kg; upper parts and outer surface of legs buffy; legs creamy white ....*Vulpes velox* (Swift Fox)

23a. Total length under 1370 mm; nose pad under 25 mm wide; hind foot under 220 mm; weight under 20 kg.................*Canis latrans* (Coyote)

23b. Total length more than 1400 kg; nose pad more than 30 mm wide; hind foot more than 225 mm; weight greater than 25 kg... *Canis lupus* (Gray Wolf)

24a. Adults unspotted; upper parts uniform tan or buff; total length more than 1.8 m; tail more than 30% of total length .... *Puma concolor* (Mountain Lion, Cougar)

24b. Adults more or less spotted or grizzled; total length less than 1.25 m; tail under 20% of total length 25

25a. Tip of tail black above and below; tail less than 14% of total length; hind foot greater than 180 mm; ear tufts more than 35 mm .... *Lynx canadensis* (Canada Lynx)

25b. Tip of tail black only on top; tail typically more than 14% of total length; hind foot usually under 200 mm; ear tufts absent or under 35 mm .... *Lynx rufus* (Bobcat)

Order Artiodactyla

1a. Animals horned in both sexes 2

1b. Animals antlered (males only); antlers shed annually .... Family Cervidae 4

2a. Horns with one prong (may be lacking in females), tip recurved, sheath shed annually; striped brown or black markings on neck........... .... Family Antilocapridae, *Antilocapra americana* (Pronghorn)

2b. Horns unbranched, not shed; neck lacking distinctive markings ....Family Bovidae 3

3a. Horns brown, heavily ridged, curving backward, forming massive spiral in adult males; color grayish brown, with rump, backs of legs, and muzzle white .... *Ovis canadensis* (Bighorn Sheep)

3b. Horns black, smooth, curving upward and toward each other; color uniformly dark brown .... *Bison bison* (Bison)

4a. Antlers palmate; snout pendulous; dewlap present; dark brown above and below; about the size of a large horse..... *Alces americanus* (Moose)

4b. Antlers not palmate; snout not pendulous; no dewlap; pelage not uniformly dark brown, may have creamy rump patch; size distinctly smaller 5

5a. Underparts brownish, similar color to back; creamy buff rump patch; total length of females more than 1.75 m, of males more than 2.2 m..... .... *Cervus elaphus* (Elk)

5b. Underparts paler than back; rump patch small or absent; total length of females 2 m or less, of males 2.3 m or less 6

6a. Antlers with single main beam giving rise to a series of tines; hind quarters mostly white; tail large and white below, held high when running... .... *Odocoileus virginianus* (White-tailed Deer)

6b. Antlers with main beam forked, forming two secondary branches, each bearing tines; hind quarters mostly grayish; tail smaller, tipped with black, held down when running....... *Odocoileus hemionus* (Mule Deer)

Keys to Skulls

Key to Orders

1a. Thirteen teeth on upper jaw, 12 teeth on the lower jaw; dental formula 5/4, 1/1, 3/3, 4/4.. ....Didelphimorphia, *Didelphis virginiana* (Virginia Opossum)

1b. Never more than 11 teeth on upper and lower jaws; maximum dental formula 3/3, 1/1, 4/4, 3/3 2

2a. Orbit enclosed by a bony ring formed by the junction of the postorbital process of the frontal bone with the zygomatic arch ....Artiodactyla (Even-toed Ungulates)

2b. Orbit not enclosed by a bony ring formed by the union of the postorbital process of the frontal bone with the zygomatic arch 3

3a. Prominent sharp-pointed canines in both upper and lower jaws 4

3b. Canines absent, or if present not noticeably different from the adjoining teeth 5

4a. Premaxillaries separated in the front by a distinct gap (diastema); upper incisors minute, not forming a continuous row between the canines, never more than two on each side; skull length under 25 mm... ....Chiroptera (Bats)

4b. Premaxillaries fused (sutured) in front; upper incisors form a continuous row between the canines, always three on each side; skull length more than 25 mm ....Carnivora (Carnivores)

5a. Total number of teeth never less than nine on each side of the upper jaw; no great space between incisors and cheek teeth; cheek teeth with sharply pointed cusps; zygomatic arch absent....................................... .... Soricomorpha (Insectivores)

5b. Total number of cheek teeth never more than eight on each side of the upper jaw; prominent space between incisors and cheek teeth; cheek teeth with flat grinding surfaces or rounded cusps; zygomatic arch always present 6

6a. Two incisors above; total teeth never more than 22; bones on side of rostrum not porous or forming lattice-like network.............................. .... Rodentia (Rodents)

6b. Four incisors above (two small ones directly behind large ones); total teeth 28; bones on side of rostrum porous, forming a delicate network .... Lagomorpha (Rabbits and Hares)

Keys to Genera and Species

Order Eulipotyphla (Soricomorpha)

1a. 1a. Condylobasal length more than 22 mm; width more than 11 mm; skull heavy, angular .... *Blarina brevicauda* (Short-tailed Shrew)

1b. Condylobasal length under 22 mm; width under 11 mm; skull delicate, rounded 2

2a. Three unicusps plainly visible in upper jaw when viewed from side........ ....*Sorex hoyi* (Pygmy Shrew)

2b. Four to five unicusps visible from the side 3

3a. Condylobasal length more than 17.5 mm...4

4a. Skull total length longer than 21 mm; mandible longer than 13.3 mm ... .*Sorex palustris* (Northern Water Shrew)

4b. Skull length less than 21 mm; mandible under 13.3 mm........................ ....*Sorex arcticus* (Arctic Shrew)

3b. Condylobasal length under 17.5 mm..5

5a. Maxillary breadth greater than 4.6 mm; first upper incisors lacking medial accessory cusps.....................*Sorex merriami* (Merriam's Shrew)

5b. Maxillary breadth under 4.6 mm; first upper incisors having medial accessory cusps .... *Sorex cinereus* (Masked Shrew) or *Sorex haydeni* (Hayden's Shrew)

*Identification of skulls of these 2 species is very difficult and should be referred to a specialist.

Order Chiroptera

1a. Incisors 1/3 2

1b. Incisors 2/3 3

2a. Greatest length of skull more than 17 mm .... *Lasiurus cinereus* (Hoary Bat)

2b. Greatest length of skull under 14.5 mm...... *Lasiurus borealis* (Red Bat)

3a. Cheek teeth 5/6 4

3b. Cheek teeth 4/5 or 6/6 5

4a. Auditory bullae not much enlarged; rostrum broad, concave on each side; forehead nearly flat...*Lasionycteris noctivagans* (Silver-haired Bat)

4b.Auditory bullae much enlarged; evenly convex above; forehead conspicuously elevated .... *Corynorhinus townsendii* (Townsend's Big-eared Bat)

5b. Cheek teeth 6/6 .... *Myotis* (Mouse-eared bats)
*Skulls of these bats require careful measurement, and may need to be referred to a specialist 6

6a. Mastoid breadth under 7.5 mm; interorbital breadth under 3.5 mm..... .... *Myotis ciliolabrum* (Western Small-footed Myotis)

6b. Mastoid breadth more than 7.5 mm; interorbital breadth more than 3.5 mm 7

7a. Greatest length of skull more than 15.7 mm; breadth across upper molars more than 6.0 mm 8

7b. Greatest length of skull under 15.7 mm; breadth across upper molars under 6.0 mm 9

8a. Greatest length of skull 16.4 mm.; Zygomatic breadth 9.4 mm; Sagittal crest present but not conspicuous ....*Myotis evotis* (Long-eared Myotis)

8b. Greatest length of skull 16.2 mm; Zygomatic breadth 10.0 mm; Sagittal crest well developed .... *Myotis thysanodes* (Fringed Myotis)

9a. Length of maxillary tooth row more than 5.5 mm; mandibular tooth row more than 6.9 mm .... *Myotis septentrionalis* (Northern Long-eared Bat)

9b. Length of maxillary tooth row under 5.5 mm; mandibular tooth row under 6.9 mm 10

10a. Brain case arises abruptly from rostrum .... *Myotis volans* (Long-legged Myotis)

10b. Brain case arises gradually from rostrum .... *Myotis lucifugus* (Little Brown Myotis)

Order Lagomorpha

1a. Postorbital process flares out from skull, and posterior tip not fused to the frontal bone; interparietal bone not distinct in adult....... Hares 2

1b. Postorbital process fused to brain case or running parallel to it, not diverging from it; interparietal bone distinct in adult, surrounded by a clear suture line.. Cottontails 3

2a. Supraorbital process broad, anterior end more than 2 mm higher than adjacent frontal bone; skull longer than 90 mm..*Lepus townsendii* (White-tailed Jackrabbit)

2b. Supraorbital process small, only slightly higher than the adjacent frontal bone; skull length under 90 mm..*Lepus americanus* (Snowshoe Hare)

3a. Auditory bulla relatively large and inflated, anterior-posterior diameter about 13.7 mm ....*Sylvilagus audubonii* (Desert Cottontail)

3b. Auditory bulla smaller, less than 11 mm 4

4a. Diameter of external auditory meatus greater than the crown length of the last three cheek teeth...*Sylvilagus nuttallii* (Mountain Cottontail)

4b. Diameter of external auditory meatus less than the crown length of the last three cheek teeth........*Sylvilagus floridanus* (Eastern Cottontail)

Order Rodentia

1a. Infraorbital foramen larger than the foramen magnum; lower jaw Y-shaped in ventral view; adult skull length at least 80 mm .... Family Erethizontidae, *Erethizon dorsata* (North American Porcupine)

1b. Infraorbital foramen smaller than the foramen magnum; lower jaw V-shaped in ventral view; skull length variable 2

2a. Infraorbital foramen on the side of the rostrum, anterior to the zygomatic plate, opening visible when viewed from the side 3

2b. Infraorbital foramen variable, but if on side of the rostrum the opening is not visible in side view 8

3a. Skull mouselike; zygomata delicate; auditory bullae inflated and exposed on posterodorsal part of the skull .... Family Heteromyidae 4

3b. Skull massive; zygomata sturdy; auditory bullae not inflated ....Family Geomyidae 7

4a. Auditory bullae greatly enlarged; greatest length of skull more than 35 mm; cheek teeth high crowned, with single lake of dentine surrounded by enamel .... *Dipodomys ordii* (Ord's Kangaroo Rat)

4b. Auditory bullae not greatly enlarged; greatest length of skull under 35 mm; cheek teeth low crowned, with two rows of cusps 5

5a. Greatest length of skull more than 30 mm .... *Chaetodipus hispidus* (Hispid Pocket Mouse)

5b. Greatest length of skull under 25 mm 6

6a. Auditory bullae generally not meeting anteriorly at midline; lower premolar nearly equal to last molar .... *Perognathus fasciatus* (Olive-backed Pocket Mouse)

6b. Auditory bullae generally meeting anteriorly at midline; lower premolar distinctly smaller than last molar .... *Perognathus flavescens* (Plains Pocket Mouse)

7a. Upper incisors with two longitudinal grooves (one may be faint) .... *Geomys bursarius* (Plains Pocket Gopher)

7b. Upper incisors with a single faint medial groove .... *Thomomys talpoides* (Northern Pocket Gopher)

8a. Postorbital processes prominent ....Family Sciuridae 9

8b. Postorbital processes absent or rudimentary 19

9a. Postorbital process projects at right angle to the long axis of the skull; anterior surface of incisors white; greatest length of skull more than 70 mm .... *Marmota monax* (Woodchuck)

9b. Postorbital process projects downward or backward; anterior surface of incisors usually orange; greatest length of skull under 70 mm 10

10a. No infraorbital canal, infraorbital foramen pierces the zygomatic plate 11

10b. Infraorbital canal present...12

11a. Cheek teeth 4/4; skull length more than 39 mm.................................. ....*Tamias striatus* (Eastern Chipmunk)

11b. Cheek teeth 5/4, the first premolar reduced to a peg tooth; skull length under 39 mm ....*Neotamias (Tamias) minimus* (Least Chipmunk)

12a. Narrow interorbital region, V-shaped, often with a distinct U-shaped notch.......... *Glaucomys sabrinus* (Northern Flying Squirrel)

12b. Interorbital region not V-shaped 13

13a. Zygomata not parallel, but converging anteriorly; anterior portions twisted toward horizontal plane 14

13b. Zygomata nearly parallel and vertical throughout; not twisted........17

14a. Upper molar rows convergent posteriorly... ....*Cynomys ludovicianus* (Black-tailed Prairie Dog)

14b. Upper molar rows nearly parallel 15

15a. Greatest width of skull more than 60% of the skull length; supraorbital ridges higher than the roof of the skull between them...... ....*Urocitellus (Spermophilus) richardsonii* (Richardson's Ground Squirrel)

15b. Greatest width of skull less than 60% of the skull length; supraorbital ridges lower than the roof of the skull between them...16

16a. Greatest length of skull more than 50 mm; interorbital constriction more than 10 mm.. ....*Poliocitellus (Spermophilus) franklinii* (Franklin's Ground Squirrel)

16b. Greatest length of skull under 45 mm; interorbital constriction under 9 mm .... *Ictidomys (Spermophilus) tridecemlineatus* (Thirteen-lined Ground Squirrel)

17a. Greatest length of skull under 50 mm; zygomatic plate curving abruptly inward to a short rostrum; nasal bones under 30% of the skull length............................*Tamiasciurus hudsonicus* (Red Squirrel)

17b. Greatest length of skull more than 55 mm; zygomatic plate curves gradually into the rostrum; nasal bones more than 30% of the skull length...18

18a. Cheek teeth 5/4, the first cheek tooth a simple peg tooth ....*Sciurus carolinensis* (Gray Squirrel)

18b. Cheek teeth 4/4 .... *Sciurus niger* (Fox Squirrel)

19a. Infraorbital foramen a long oval; cheek teeth 4/4 .... Family Dipodidae 20

19b. Infraorbital foramen a vertical slit; cheek teeth 3/3 21

20a. Incisive foramen under 4.6 mm; palatal breadth at last molar under 4.2 mm; condylobasal length under 20 mm; maxillary tooth row under 3.7 mm .... *Zapus hudsonius* (Meadow Jumping Mouse)*

20b. Incisive foramen more than 4.7 mm; palatal breadth at last molar more than 4.4 mm; condylobasal length more than 21 mm; maxillary tooth row more than 3.8 mm .... *Zapus princeps* (Western Jumping Mouse)*

*Distinguishing between the skulls of these two species is difficult and should be referred to a specialist.

21a. Upper molars with three longitudinal rows of cusps .... Family Muridae 22

21b. Upper molars with two longitudinal rows of cusps, or consisting of numerous triangles and transverse enamel folds .... Family Cricetidae 23

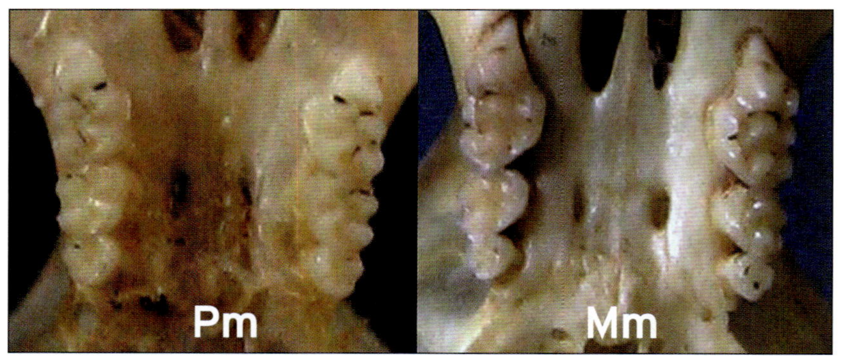

Photo by Robert Seabloom

Upper cheek tooth rows of *Peromyscus maniculatus* (left) and *Mus musculus* (right) showing two longitudinal rows of cusps in *Peromyscus* and three longitudinal rows in *Mus*.

22a. Greatest length of skull over 25 mm .... *Rattus norvegicus* (Norway Rat)

22b. Greatest length of skull under 25 mm .... *Mus musculus* (House Mouse)

23a. Cheek teeth with cusps (may be flattened into sigmoid-shaped lophs or may be ground flat, in which case the tooth will be outlined with an even rim of enamel) 24

23b. Cheek teeth without cusps; flat occlusal area composed of tracts of dentine surrounded by enamel and separated by reentrant angles resulting in more or less prismatic patterns 28

24a. Upper incisors grooved 25

24b. Upper incisors not grooved 26

25a. Breadth of braincase under 9.6 mm; greatest length of skull under 20 mm; depth of braincase under 7.4 mm ....*Reithrodontomys montanus* (Plains Harvest Mouse)*

25b. Breadth of braincase over 9.5 mm; greatest length of skull over 20 mm; depth of braincase over 7.7 mm... ....*Reithrodontomys megalotis* (Western Harvest Mouse)*

*Distinguishing between the skulls of these two species is difficult and should be referred to a specialist.

26a. Coronoid process of the mandible strongly developed, projecting more than 2 mm back from the front end of the notch beneath it; skull longer than 26.5 mm; zygomatic breadth more than 14.5 mm.... .... *Onychomys leucogaster* (Northern Grasshopper Mouse)

26b. Coronoid process of the mandible weakly developed, no more than 1 mm high; skull shorter than 26.5 mm; zygomatic breadth under 14.5 mm 27

27a. Anterior border of the zygomatic plate covers the infraorbital foramen when viewed from the side (a very narrow part of the foramen may show along the anterior border of the plate) .... *Peromyscus leucopus* (White-footed Mouse)*

27b. Anterior border of the zygomatic plate does not cover the infraorbital foramen when viewed from the side (a space about 0.5 mm or more shows in front of the plate)... *Peromyscus maniculatus* (Deer Mouse)*

*Identification of the skulls of these species is difficult, and should be referred to a specialist; notes on habitat may also be useful

28a. Cranium not constricted abruptly in front of the brain case; first and second molars with middle loop undivided, reaching completely across the full width of the tooth ....*Neotoma cinerea* (Bushy-tailed Wood Rat)

28b. Cranium constricted abruptly in front of the brain case; molar patterns consisting of sharp-edged prisms, only terminal loops extending across the full width of the tooth...................................29

29a. Greatest length of skull more than 50 mm; lower molar row longer than 14 mm .... *Ondatra zibethicus* (Muskrat)

29b. Greatest length of skull under 50 mm; lower molar row under 14 mm 30

30a. Posterior border of the palate a simple shelf, lacking a median projection; last upper molar (M3) with four re-entrant angles on either side ....*Myodes (Clethrionomys) gapperi* (Southern Red-Backed Vole)

30b. Posterior border of the palate with a median projection; less than three re-entrant angles on the last upper molar (M3) 31

31a. Third lower molar with two transverse loops and two median triangles; auditory bullae large, extending beyond the occipital condyles and visible in dorsal view ....*Lemmiscus curtatus* (Sagebrush Vole)

31b. Third lower molar with three closed loops and no closed triangles; auditory bullae not extending beyond the occipital condyles and not visible in dorsal view 32

32a. Auditory bullae large, rounded, dorsoventral diameter usually more than 6 mm; second upper molar with five closed loops; last upper molar with three re-entrant angles on the outside .... *Microtus pennsylvanicus* (Meadow Vole)

32b. Auditory bullae small, somewhat flattened, usually under 6 mm; second upper molar with four closed loops; last upper molar with two re-entrant angles on the outside .... *Microtus ochrogaster* (Prairie Vole)

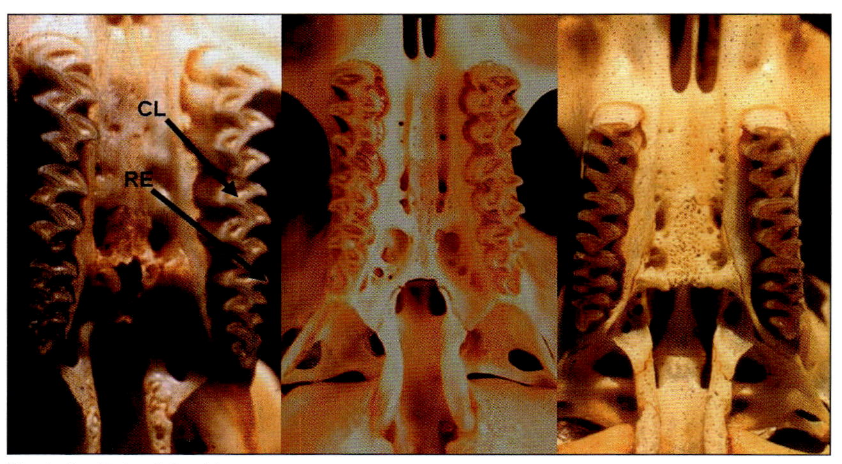

Photo by Robert Seabloom

Upper cheek teeth and bony palates of *Microtus pennsylvanicus* (left), *Microtus ochrogaster* (center), and *Myodes gapperi* (right). Note the closed loops (CL) on cheek tooth 2, and re-entrant angles (RE) on cheek tooth 3, and structure of the bony palate.

Order Carnivora

1a. Total teeth 34 or more, eight or more on the upper jaw, nine or more on the lower jaw; last upper molar sometimes reduced but always more than twice the size of the outer incisor....................................2

1b. Total teeth 30 or less, eight or less on the upper jaw, seven on the lower jaw; last upper molar equal to or larger than the outer incisor.. Family Felidae 19

2a. Dentition somewhat bunodont, with flat-crowned teeth; carnassials (shearing teeth) not well developed...3

2b. Dentition not bunodont; carnassials well-developed 4

3a. Greatest length of skull under 175 mm; total teeth 40 ....Family Procyonidae, *Procyon lotor* (Raccoon)

3b. Greatest length of skull over 225 mm; total teeth 42 ....Family Ursidae, *Ursus americanus* (Black Bear)

4a. Total teeth 38 or less; never more than nine in the upper jaw............5

4b. Total teeth 42, 10 in the upper jaw, 11 in the lower jaw...................... .... Family Canidae 15

5a. Bony palate terminates at or slightly anterior to the plane of the posterior borders of the last upper molars; last upper molar squarish, distinctly larger than the tooth in front of it .... Family Mephitidae 6

5b. Bony palate extends posteriorly beyond the plane of the posterior borders of the last upper molars; last upper molar squarish, about the same size or smaller than the preceeding........Family Mustelidae 7

6a. Greatest length of skull more than 60 mm; skull highly arched, deepest over the orbital region........*Mephitis mephitis* (Striped Skunk)

6b. Greatest length of skull under 60 mm; skull flattened, deepest over the occipital region ....*Spilogale putorius* (Spotted Skunk)

7a. Five cheek teeth in the upper jaw behind the canine...........................8

7b. Four cheek teeth in the upper jaw behind the canine 11

8a. Greatest length of skull more than 130 mm; tooth formula 3/3, 1/1, 4/4, 1/2 ....*Gulo gulo* (Wolverine)

8b. Greatest length of skull less than 130 mm; tooth formula same as above or 3/3, 1/1, 4/3, 1/2..9

9a. Infraorbital foramen greatest diameter more than 6 mm; greatest width of the brain case across the mastoids more than 58 mm; auditory bulla flattened; rostrum broader than long .... *Lontra canadensis* (River Otter)

9b. Infraorbital foramen under 6 mm; greatest width of the brain case under 58 mm; auditory bulla rounded; rostrum longer than broad..10

10a. Greatest length of skull under 85 mm........ *Martes americana* (Marten)

10b. Greatest length of skull over 95 mm.. .... *Pekania (Martes) pennanti* (Fisher)

11a. Greatest length of skull more than 113 mm; skull wedge-shaped, broad posteriorly; last upper molar triangular-shaped .... *Taxidea taxus* (Badger)

11b. Greatest length of skull under 80 mm; skull not wedge-shaped or broad posteriorly; last upper molar dumbbell-shaped......................13

12a. Greatest length of skull more than 55 mm..12

12b. Greatest length of skull under 55 mm..14

13a. Distance between the canines greater than the distance between the medial margins of the auditory bullae... .... *Mustela nigripes* (Black-footed Ferret)

13b. Distance between the canines less than the distance between the medial margins of the auditory bullae... .... *Neovison (Mustela) vison* (Mink)

14a. Greatest length of the skull under 36 mm, usually under 34 mm .... *Mustela nivalis* (Least Weasel)*

14b. Greatest length of the skull more than 32 mm, usually more than 34 mm 15

15a. Greatest length of the skull usually more than 44 mm .... *Mustela frenata* (Long-tailed Weasel)*

15b. Greatest length of the skull usually less than 44 mm........................... .... *Mustela erminea* (Ermine)*

*Skulls of weasels are difficult to distinguish, especially if you do not know the sex. For example, a male ermine may overlap with a female long-tailed weasel.

16a. Cranial ridges lyrate, forming prominent ridges on the outer dorsal surfaces of the skull; depression in the frontal bone just posterior to the postorbital process; upper incisors not lobed .... *Urocyon cinereoargenteus* (Gray Fox)

16b. Cranial ridges not lyrate, but form a sagittal crest posteriorly; no frontal bone depression; upper incisors lobed.................................17

17a. Greatest length of skull under 160 mm; postorbital processes not inflated, slightly concave above; length of lower cheek tooth row behind the canine under 70 mm..18

17b. Greatest length of skull over 160 mm; postorbital processes inflated, slightly convex above; length of lower cheek tooth row usually more than 70 mm 19

18a. Adult greatest length of skull over 120 mm ... *Vulpes vulpes* (Red Fox)

18b. Adult greatest length of skull under 120 mm .... *Vulpes velox* (Swift Fox)

19a. Adult greatest length of skull under 225 mm; length of lower first molar under 25 mm .... *Canis latrans* (Coyote)

19b. Adult greatest length of skull over 230 mm; length of lower first molar over 25 mm .... *Canis lupus* (Gray Wolf)

20a. Teeth 30; length of skull more than 175 mm .... *Puma concolor* (Mountain Lion)

20b. Teeth 28; length of skull under 150 mm 21

21a. Length of carnassial (fourth upper premolar) over 16.6 mm; lower carnassial (first lower molar) more than 13.5 mm; anterior condyloid and posterior lacerate foramina open separately in the basioccipital region; greatest width of the presphenoid more than 5 mm .... *Lynx canadensis* (Canada Lynx)

21b. Length of carnassial under 16.6 mm; lower carnassial under 13.5 mm; anterior condyloid and posterior lacerate foramina open in a common depression; greatest width of the presphenoid usually under 5 mm ....Lynx rufus (Bobcat)

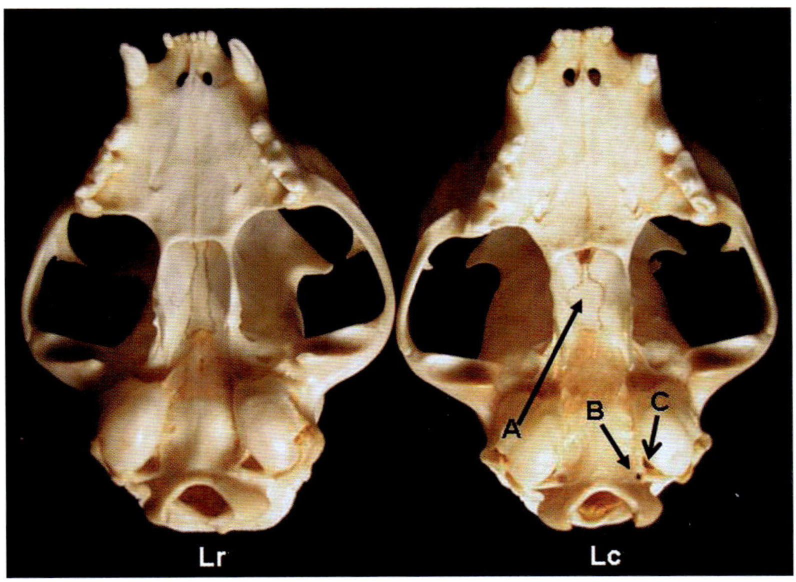

Photo by Robert Seabloom

Ventral views of skulls of *Lynx rufus* (left) and *Lynx Canadensis* (right). Note the relative breadth of the presphenoid bone (A) and the degree of separation of the anterior condyloid foramen (B) from the posterior lacerate foramen (C)

Order Artiodactyla

1a. Conspicuous gap between the lacrimal and nasal bones; antlers present in males, or deciduous horns in both sexes 2

1b. No gap separating the lacrimal and nasal bones; permanent horns present in both sexes..Family Bovidae 6

2a. Deciduous horns present directly above the orbits in both sexes, the sheath shed annually; horns with one prong (prong may be lacking in females); tip of the horn recurved; no conspicuous depression (lacrimal pit) in the lacrimal bone just anterior to the orbit .... Family Antilocapridae, *Antilocapra americana* (Pronghorn)

2b. Antlers present in males (shed annually), the bony base arising well behind the orbits; a conspicuous depression (lacrimal pit) in the lacrimal bone just anterior to the orbit.................................. Family Cervidae 3

3a. Antlers palmate; nasal bones short, less than half as long as the space between their anterior tips and the anterior tips of the premaxillae; skull length more than 530 mm ....*Alces americanus* (Moose)

3b. Antlers not palmate; nasal bones more than half as long as the space between the anterior tips and the anterior tips of the premaxillae; skull length under 530 mm..4

4a. Internal nares not separated by the vomer; upper canine teeth present; skull length over 430 mm .... *Cervus elaphus* (Elk)

4b. Internal nares separated by the vomer; upper canine teeth absent; skull length under 400 mm 5

5a. Antlers with a single main beam giving rise to a series of tines; lacrimal pit shallow, concealing less than 4 mm of a probe viewed from above... ....*Odocoileus virginianus* (White-tailed Deer)

5b. Antlers with main beam forked, forming two secondary branches, each bearing tines; lacrimal pit deep, concealing at least 5 mm of probe viewed from above................ *Odocoileus hemionus* (Mule Deer)

6a. Horns brown, heavily ridged, curved backward, forming massive spiral in adult males .... *Ovis canadensis* (Bighorn)

6b. Horns black, completely smooth, curving upward .. *Bison bison* (Bison)

Glossary

Agonistic - Behavior during conflict with a member of the same species, including aggression, threats, and retreat.

Allogrooming - Care of the body surface performed by one individual upon various parts of the body of a second individual.

Allopatric - Species or populations having non-overlapping distributions.

Altricial - Young born in a relatively undeveloped condition, with their eyes closed, minimal fur, poor temperature control, and may require extensive parental care.

Anterior - Pertaining to the front of an animal.

Antler - Usually branched bony outgrowth from the frontal bone of deer, which is shed annually. During growth, it is covered by skin and fine hair (velvet).

Arboreal - Adapted for living and moving about in trees.

Beam - Basal trunk or segment of an antler.

Biome - The largest recognizable land community, named for the life form of its dominant vegetation, e. g., "prairie," "deciduous forest."

Blastocyst - A very early embryo; the 32- to 64-cell stage which implants in the uterus.

Boreal - Referring to the northern, primarily coniferous, forests bordering the subpolar regions.

Browse - Leaves, stems, twigs, etc., of woody plants consumed by animals.

Bunodont - Low-crowned teeth that have blunt cusps for crushing.

Cache - A large store of food, such as a beaver cache.

Calcar - A cartilaginous process extending medially from the ankle of bats which helps to support the interfemoral membrane or uropatagium.

Canine - A tooth with one cusp, located between the incisors and premolars; if present, there is never more than one in a quadrant.

Carnassial - Shearing cheek teeth found in most carnivores, consisting of the last upper premolar and first lower molar.

Carnivore - An animal that eats meat; a member of the order Carnivora.

Carrying Capacity - The number of individuals of a population the resources of a given area can support.

Cheek Teeth - The teeth posterior to the canines; the premolars and molars.

Circumboreal - Throughout the Northern Hemisphere.

Class - A classification category including one or more orders, e. g., Mammalia.

Climax Community - The stable, end community of succession. It continues in this state indefinitely unless there is a change in environmental conditions.

Colony - A social group of animals remaining in one place, such as a beaver colony or prairie dog town.

Commensal - Different species living in close association with each other, as in house mice living in association with humans.

Community - An interacting group of plants and animals occurring in a given area.

Competitive Exclusion - The hypothesis that two or more species depending on the same resource cannot coexist; one will displace the other.

Coprophagy - Ingestion of feces; seen in many shrews, lagomorphs, and rodents.

Copulation - The act of sexual intercourse.

Corpora (Corpora lutea) – Structures formed from the remnents of the ovarian follicle following ovulation. Corpora counts are useful in estimation of potential litter size.

Cortex - The outer layer of a structure, as in the cortex of a hair.

Coterie - A close social grouping of animals of the same species; the basic social unit in a prairie dog town.

Crepuscular - Activity during the twilight hours around sunrise and/or sunset.

Delayed Fertilization - A reproductive adaptation in certain hibernating bats in which mating takes place in the fall prior to hibernation, and sperm is stored in the female reproductive tract. Ovulation and fertilization then takes place upon arousal in the spring.

Delayed Implantation - A reproductive adaptation in which mating and fertilization take place, the embryo develops to an early stage (blastocyst), but then becomes dormant before implanting in the uterine wall for a period of days to months; common in the family Mustelidae.

Dental Formula - A shorthand description of the number and position of teeth in a species, given as upper and lower incisors, canines, premolars, and molars; e.g., 3/3, 1/1, 4/4, 2/3.

Diastema - A gap between adjacent teeth, as that between the incisors and cheek teeth of rodents.

Digits - Fingers or toes.

Diurnal - Activity during the hours of daylight.

Dorsal - Pertains to the back, or upper surface.

Drift (Glacial drift) - Sands, gravels, and rocks deposited by a glacier.

Echolocation - High frequency sound pulses emitted by bats and some shrews to gain information about their surroundings by the returning echoes.

Ecological Niche - The functional role of species in its community, resulting from its adaptations to exploit resources and occupation of habitat.

Ecosystem - The living and non-living components of an environment occurring together in a given area.

Endothermy - Maintenance of body temperature by an internal source of heat (metabolism).

Epiphysial Closure - A cartilaginous zone between the bony cap and the shaft of a long bone, which gradually ossifies with age; a useful aging criterion in certain mammals.

Epizootic - A widespread animal disease.

Erectile - Capable of being erected, or set upright, as in erectile hairs.

Estrous Cycle - A series of reproductive events in a female mammal which prepares the uterus for pregnancy, followed by ovulation.

Estrus - The stage of the estrous cycle during which the female permits copulation and ovulation usually occurs.

Eutheria - An infraclass of mammals which includes the placentals, those bearing living young.

Family - A classification category within an order, and including one or more genera, e. g., Felidae.

Fenestration - A portion of the skull having a light, lattice-like bone structure.

Foramen - An opening in a bone through which a nerve or blood vessel passes.

Forb - An herbaceous plant, other than grasses and sedges.

Fossorial - A life style and adaptation involving digging underground for foraging or shelter.

Genus - A classification category within a family, and including one or more species, e.g., *Peromyscus*.

Gestation - The period of time between fertilization and birth.

Graminoid - Grass-like plants; grasses and sedges.

Guard Hair - Coarse hairs forming a protective outer coat in most mammals.

Herbivore - An animal that feeds on plants.

Hibernaculum - A site where hibernation occurs.

Hibernation - A state of dormancy in which the animal's metabolism is minimal, but maintains a body temperature above freezing.

Home Range - The area in which an animal spends most of its time in its normal activities; home ranges can expressed as daily, seasonal, or annual.

Homeothermy - The ability to regulate body temperature by controlling the rate of heat loss through behavioral or physiological mechanisms.

Horn - Bony process extending from the frontal bone of bovids, covered by a sheath of keratinized material derived from the epidermis.

Incisive Foramen – A foramen (of which there are two) in the anterior portion of the bony palate.

Incisor - Single-cusped (usually) teeth, anterior to the canines, used for cutting and gnawing.

Infraorbital Foramen – An opening through the maxillary bone (zygomatic plate) from the orbit to the face.

Insectivore - An animal having a diet of insects; a member of the order Soricomorpha.

Interfemoral Membrane – The membrane between the hind legs of bats, enclosing the tail. It is also referred to as the uropatagium.

Interorbital Region - Space on the dorsal surface of the skull between the two orbits.

Interparietal – A bone in the most posterior part of the skull between the two parietals.

Keystone Species - Species which enrich ecosystem function in a unique and significant manner, and the effect is disproportionate to their numerical abundance.

Lactation - Production of milk by mammary glands.

Mammary Gland - Milk-producing glands, unique to mammals.

Marsupial - A member of the infraclass Metatheria; collectively, the orders Didelphimorphia, Paucituberculata, Microbiotheria, Dasyuromorphia, Peramelemorphia, Diprotodontia, Notoryctermorphia.

Marsupium - A pouch formed from skin on the abdominal wall, containing teats to which the young of marsupials remain attached for an extended period.

Medulla - The inner or core area of a structure, such as a hair.

Metatheria - The infraclass containing the marsupials.

Microtine - A member of the rodent subfamily Microtinae, the voles, lemmings, and muskrats.

Migration - A periodic departure from and return to an area.

Molars - Permanent cheek teeth with multiple cusps, behind the premolars.

Monestrous - Having only one period of estrus per year.

Monogamous - A pair bond involving only one member of the opposite sex at a time.

Moraine - A mass of debris deposited by a glacier.

Nocturnal - Activity during the hours of darkness.

Omnivore - An animal that consumes both plant and animal food, such as a raccoon or bear.

Order - A category of classification below class, containing one or more families.

Palmate - Flattened, or web-like.

Parturition - The process of giving birth.

Patagium - A thin membrane of skin, often forming a gliding surface, as between the fore and hind limbs of flying squirrels or between the hind limbs of bats (uropatagium).

Pelage - All of the hairs on a mammal.

Piloerection – Erection of hair on the skin; "fluffing" of fur in response to cold.

Placenta - A vascularized structure consisting of both maternal and fetal tissue, forming a connection between the fetus and the uterus, and providing for transport of nutrients, gases, and waste.

Placental Scar - A dark, pigmented area on the uterus, indicating the prior attachment of a placenta; often used as an indicator of litter size.

Polyestrous - Having more than one period of estrus, or heat, per year.

Polygamous - A mating system involving more than one mate of the opposite sex.

Polygynous - A mating system involving more than one female per male.

Post-partum Heat - A period of estrus occurring shortly after giving birth.

Posterior - Pertaining to the rear portion of an animal.

Precocial - Young born in a relatively advanced state of development, with the eyes open, fully furred, some degree of temperature control, able to move about, and requiring minimal parental care.

Prehensile - Usually involves a tail capable of grabbing branches or other objects.

Premolars - Cheek teeth in front of the molars and behind the canines.

Promiscuous - A mating system involving no persistent pair bond between the sexes.

Prototheria - Subclass including the order Monotremata, the egg-laying mammals.

Reentrant Angle - Inward-pointing angle of enamel along the margin of a cheek tooth.

Riparian - Along banks of rivers and streams, as in riparian woodland.

Rostrum - The anterior portion of the facial region of the skull

Rumen - The first compartment of the four-part stomach of ungulates.

Rut - The mating season for deer and other ungulates.

Scansorial – An adaptation for climbing.

Scat - Feces or droppings.

Species - Classification unit within a genus. It is the most definitive classification unit, defined as "groups of naturally or potentially interbreeding natural populations which are reproductively isolated from other such groups." The name includes both the genus and species, e. g., *Peromyscus maniculatus*.

Steppe - An extensive area of dry grassland, such as the North American prairie.

Stotting - A stiff-legged, bounding gait used by some ungulates, such as mule deer.

Submaxillary Gland - A salivary gland beneath the lower jaw; in some shrews it produces a toxin which immobilizes prey.

Subspecies - A relatively uniform population of a species living in a given area. For example, *Peromyscus maniculatus bairdii* is a subspecies of *Peromyscus maniculatus*.

Succession - The orderly process of community change to a stable, or climax, state.

Sympatric - Species which have overlapping distributions.

Territory - A portion of the home range which is defended by an animal or group of animals.

Thoracic Vertebra - Vertebra in the chest area.

Tine - A spike or branch off of the beam of an antler.

Tragus - A fleshy projection from the lower margin of the external ear; in bats the tragus aids in echolocation.

Underfur - The short, fine, insulating hair of a mammal, underneath the guard hairs.

Ungulate - Any hoofed mammal.

Unicusp – A tooth with a single cusp.

Uropatagium – The membrane between the hind legs of bats, enclosing the tail. It is also called the interfemoral membrane.

Uterus - the portion of the female reproductive tract where embryonic development occurs.

Valvular – Ears or nostrils which can be closed when the animal is under water.

Ventral - The underside, or belly of an animal.

Vibrissae - Long, stiff hairs which mainly serve as touch receptors.

Zygomatic Plate – A flat region in the zygomatic process of the maxilla.

References

Adams, A. W. 1961. Furbearers of North Dakota. North Dakota Game and Fish Department, Bismarck.

Allardyce, D., and M. A. Sovada. 2003. A review of the ecology, distribution, and status of swift foxes in the United States. Pp. 3–18 *in* The swift fox: ecology and conservation of swift foxes in a changing world (M. A. Sovada and L. Carbyn, eds.). Canadian Plains Research Center, Regina.

Allen, S. H. 1984. Some aspects of reproductive performance in female red fox in North Dakota. Journal of Mammalogy 65: 246–255.

Allen, S. H., J. O. Hastings, and S. C. Kohn. 1987. Composition and stability of coyote families and territories in North Dakota. Prairie Naturalist 19:107–114.

Anderson, A. E., and O. C. Wallmo. 1984. *Odocoileus hemionus.* Mammalian Species 219:1-9.

Anderson, E. M., and M. J. Lovallo. 2003. Bobcat and lynx. Pp. 758–786 *in* Wild mammals of North America: biology, management, and conservation (G. A. Feldhamer, B. C. Thompson, and J. A. Chapman, eds.). Johns Hopkins University Press, Baltimore.

Anderson, R. M. 1965. Methods of collecting and preserving vertebrate animals. National Museum of Canada, Bulletin 69.

Apps, C. D., N. J. Newhouse, and T. A. Kinley. 2002. Habitat associations of American badgers in southeastern British Colombia. Canadian Journal of Zoology 80:1228–1239.

Armitage, K. B. 2003. Marmots. Pp. 188–210 *in* Wild mammals of North America: biology, management, and conservation (G. A. Feldhamer, B. C. Thompson, and J. A. Chapman, eds.). Johns Hopkins University Press, Baltimore.

Armstrong, D. M., J. R. Choate, and J. K. Jones, Jr. 1986. Distributional patterns of mammals in the plains states. Occasional papers, The Museum, Texas Tech University 105:1–27.

Armstrong, D. M., and J. K. Jones, Jr. 1971. *Sorex merriami.* Mammalian Species 2:1–2.

Arnett, E. B., W. K. Brown, W. P. Erickson, J. K. Fiedler, B. L. Hamilton, T. H. Henry, A. Jain, G. D. Johnson, J. Kerns, R. R. Koford, C. P. Nicholson, T. J. O'Connell, M. D. Piorkowski, and R. D. Tankersley. 2008. Patterns of bat fatalities at wind energy facilities in North America. Journal of Wildlife Management 72:61–78.

Arnold, T. W., and E. K. Fritzell. 1987. Activity patterns, movements, and home ranges of prairie mink. Prairie Naturalist 19:25–32.

Arnold, T. W., and E. K. Fritzell. 1989. Spring and summer prey remains collected from male mink dens in southwestern Manitoba. Prairie Naturalist 21:189–192.

Arnold, T. W., and E. K. Fritzell. 1990. Habitat use by male mink in relation to wetland characteristics and avian prey abundances. Canadian Journal of Zoology 68:2205–2208.

Aubry, K. B., K. S. McKelvey, and J. P. Copeland. 2007. Distribution and broadscale habitat relations of the wolverine in the contiguous United States. Journal of Wildlife Management 71:2147–2158.

Autenreith, R. E., D.E. Brown, R. M. Lee, R. A. Ockenfels, B. W. O'Gara, T. M Pojar, and J. D. Yoakum, eds. 2006. Pronghorn management guide, Fourth edition. Pronghorn Workshop and North Dakota Game and Fish Department, Bismarck, ND.

Baerwald, E. F., G. H. D'Amours, B. J. Klug, and R. M. R. Barclay. 2008. Barotrauma is a significant cause of bat fatalities at wind turbines. Current Biology 18:R695-R696.

Bailey, V. 1926. A biological survey of North Dakota: I. Physiography and life zones; II. The mammals. United States Department of Agriculture, Bureau of Biological Survey, North American Fauna 49.

Baker, B. W., and E. P. Hill. 2003. Beaver. Pp. 288–310 *in* Wild Mammals of North America: Biology, Management, and Conservation (G. A. Feldhamer, B. C. Thompson, and J. A. Chapman, eds.). Johns Hopkins University Press, Baltimore.

Baker, R. J, R. D. Bradley, and L. R. McAililey, Jr. 2003. Pocket gophers. Pp. 276–287 *in* Wild Mammals of North America: Biology, Management, and Conservation (G. A. Feldhamer, B. C. Thompson, and J. A. Chapman, eds.). Johns Hopkins University Press, Baltimore.

Barnhart, P. R., and E. H. Gilam. 016. Understanding peripheral bat populations using maximum suitability modeling. PLoS ONE 11(12):e0152508.doi:10.1371/journal.pone.0152508.

Barnhart, P. R., and E. H Gilam. 2017. Documentation of overwintering bat species presence and hibernacula use in the badlands of North Dakota. Northwestern Naturalist 98:48–56.

Beer, J. R. 1955. Movements of tagged beaver. Journal of Wildlife Management 19:492–493.

Beer, J. R. 1962. Emergence of thirteen-lined ground squirrels from hibernation. Journal of Mammalogy 43:109.

Bekoff, M. 1977. *Canis latrans*. Mammalian Species 79:1–9.

Bekoff, M., and E. M. Gese. 2003. Coyote. Pp. 467–481 *in* Wild mammals of North America: biology, management, and conservation (G. A. Feldhamer, B. C. Thompson, and J. A. Chapman, eds.). Johns Hopkins University Press, Baltimore.

Beneski, J. T., and D. W. Stinson. 1987. Sorex palustris. Mammalian Species 296:1–6.

Berg, W. E. 1982. Reintroduction of fisher, pine marten, and river otter. Pp. 159–173 *in* Midwest furbearer management (G. C. Sanderson, ed.). Proceedings of a symposium at the 43rd Midwest Fish and Wildlife Conference, Wichita, Kansas.

Bergeron, D. J. 1977. Habitat selection and reproduction in two species of cottontails (*Sylvilagus audubonii* and *Sylvilagus floridanus*) in southwestern North Dakota. M. S. thesis, University of North Dakota, Grand Forks.

Bergeron, D. J., and R. W. Seabloom. 1981. Habitat partitioning by eastern and desert cottontails in southwestern North Dakota. Prairie Naturalist 13:105–110.

Berry, R. J. 1981. Population dynamics of the house mouse. Pp. 395–425 *in* Biology of the house mouse (R. J. Berry, ed.). Symposia of the Zoological Society of London 47:1–715.

Biggins, D., B. Miller, B. Oakleaf, A. Farmer, R. Crete, and A. Dood. 1989. A system for evaluating black-footed ferret habitat. Report prepared for the interstate coordinating committee by the reintroduction site group. U. S. Fish and Wildlife Service, National Ecology Research Center, 2627 Redwing Road, Creekside One, Fort Collins, CO.

Bihrle, C. 1996. EHD in North Dakota: Profile of a whitetail killer. North Dakota Outdoors 59(8):2–8.
Birney, E. C., and R. P. Lampe. 1972. Sagebrush vole *(Lagurus curtatus)* in South Dakota. American Midland Naturalist 88:466.
Bishop, N. G., and J. L. Culbertson. 1976. Decline of prairie dog towns in southwestern North Dakota. Journal of Range Management 29:217–220.
Bjorge, R. R., J. R. Gunson, and W. M. Samuel. 1981. Population characteristics and movements of striped skunks (Mephitis mephitis) in central Alberta. Canadian Field-Naturalist 95:149–155.
Blackwell, J. M. 1981. The role of the house mouse in disease and zoonoses. Pp. 591–615 *in* Biology of the house mouse (R. J. Berry, ed.). Symposia of the Zoological Society of London 47:1–715.
Bleich, V. C., G. A. Sargeant, and B. P. Wiedmann. 2018. Ecotypic variation in population dynamics of reintroduced bighorn sheep. Journal of Wildlife Management 82:8–18.
Blood, D. A., and A. L. Lovaas. 1966. Measurements and weight relationships in Manitoba elk. Journal of Wildlife Management 30:135–140.
Bowyer, R. T., V. Van Ballenberghe, and J. G. Kie. 2003. Moose. Pp. 931–964 *in* Wild mammals of North America: biology, management, and conservation (G. A. Feldhamer, B. C. Thompson, and J. A. Chapman, eds.). Johns Hopkins University Press, Baltimore.
Brillhart, D. E., and D. W. Kaufman. 1994. Temporal variation in coyote prey in tallgrass prairie of eastern Kansas. Prairie Naturalist 26:93–105.
Bronson, F. H. 1984. The adaptability of the house mouse. Scientific American 250(3):116–125.
Brown, E. B. 1971. Some aspects of the ecology of the small winter-active mammals of a field and adjacent woods in Itasca State Park, Minnesota. PhD dissertation, University of Minnesota, Minneapolis.
Brown J. H., and B. A. Harney. 1993. Population ecology of heteromyid rodents in temperate habitats. Pp. 618–651 in Biology of the Heteromyidae (Hugh H. Genoways and James H. Brown, eds.). American Society of Mammalogists, Special Publication 10.
Bryant, L. D., and C. Maser. 1982. Classification and distribution. Pp. 1–59 *in* Elk of North America: ecology and management (J. W. Thomas and D. E. Toweill, eds.). Stackpole, Harrisburg.
Bubenik, A. B. 1982. Physiology. Pp. 125–179 in Elk of North America: ecology and management (J. W. Thomas and D. E. Toweill, eds.). Stackpole, Harrisburg.
Buckner, C. H. 1957. Population studies on small mammals of southeastern Manitoba. Journal of Mammalogy 38:87–97.
Buskirk, S. W., and W. J. Zielinski. 1997. American marten *(Martes americana)* ecology and conservation. Pp. 17–22 *in* Mesocarnivores of northern California: biology, management, and survey techniques, workshop manual (J. E. Harris and C. V. Ogan, eds.). August 12–15, 1997, Humboldt State University, Arcata, CA. The Wildlife Society, California North Coast Chapter, Arcata.
Butsch, R. S. 1954. The life history and ecology of the red-backed vole, *Clethrionomys gapperi gapperi* Vigors, in Minnesota. PhD dissertation, University of Michigan, Ann Arbor.
Byers, J. A. 1997. American pronghorn: social adaptations and the ghosts of predators past. University of Chicago Press, Chicago.

Byers, J. A. 2003. Pronghorn. Pp. 998–1008 *in* Wild mammals of North America: biology, management, and conservation (G. A. Feldhamer, B. C. Thompson, and J. A. Chapman, eds.). Johns Hopkins University Press, Baltimore.

Caceres, M. C., and R. M. R. Barclay. 2000. *Myotis septentrionalis.* Mammalian Species 634:1–4.

Cadieux, C. L. 1986. Pronghorn: North America's unique antelope. The practical guide for hunters. Stackpole, Harrisburg.

Calhoun, J. B. 1963. The ecology and sociology of the Norway rat. United States Department of Health, Education and Welfare, Public Health Service Publication 1008:1–288.

Carbyn, L. N. 1987. Gray wolf and red wolf. Pp. 359–376 *in* Wild furbearer management and conservation in North America (M. Novak, J. A. Baker, M. E. Obbard, and B. Malloch, eds.). Ontario Trappers Association and Ministry of Natural Resources, Toronto.

Carroll, L. E., and H. H. Genoways. 1980. *Lagurus curtatus.* Mammalian Species 124:1–6.

Chamberlain, M. J., and B. D. Leopold. 2000. Spatial use patterns, seasonal habitat selection, and interactions among adult gray foxes in Mississippi. Journal of Wildlife Management 64:742–751.

Chapman, J. A. 1975. *Sylvilagus nuttallii.* Mammalian Species 56:1–3.

Chapman, J. A., J. G. Hockman, and M. M. Ojeda C. 1980. *Sylvilagus floridanus.* Mammalian Species 136:1–8.

Chapman, J. A., and J. A. Litvaitis. 2003. Eastern cottontail. Pp. 101–125 *in* Wild mammals of North America: biology, management, and conservation (G. A. Feldhamer, B. C. Thompson, and J. A. Chapman, eds.). Johns Hopkins University Press, Baltimore.

Chapman, J. A., and G. R. Willner. 1978. *Sylvilagus audubonii.* Mammalian Species 106:1–4.

Choromanski-Norris, J., E. K. Fritzell, and A. B. Sargeant. 1986. Seasonal activity cycle and weight changes of the Franklin's ground squirrel. American Midland Naturalist 116:101–107.

Choromanski-Norris, J., E. K. Fritzell, and A. B. Sargeant. 1989. Movements and habitat use of Franklin's ground squirrels in duck-nesting habitat. Journal of Wildlife Management 53:324–331.

Christie, K. S. 2017. Pronghorn resource selection and habitat fragmentation in North Dakota. Journal of Wildlife Management 81:154–162.

Ciuti, S., et al. 2015. Predicting mule deer recruitment from climate oscillations for harvest management on the northern Great Plains. Journal of Wildlife Management 79:1226–1238.

Clambey, G. K. 1986. Elm-ash-basswood forest. North Dakota Outdoors 49(1):10–11.

Clark, T. W., E. Anderson, C. Douglas, and M. Strickland. 1987. *Martes americana.* Mammalian Species 289:1–8.

Clark, T. W., J. Grensten, M. Gorges, R. Crete, and J. Gill. 1987. Analysis of black-footed ferret translocation sites in Montana. Prairie Naturalist 19:43–56.

Clark, W. R., J. J. Hasbrouck, J. M. Kienzler, and T. F. Glueck. 1989. Vital statistics and harvest of an Iowa raccoon population. Journal of Wildlife Management 53:982–990.

Cockrum, E. L., and S. P. Cross. 1964. Time of bat activity over water holes. Journal of Mammalogy 45:635–636.

Collins, A. R., J. P. Workman, and D. W. Uresk. 1984. An economic analysis of black-tailed prairie dog *(Cynomys ludovicianus)* control. Journal of Range Management 37:358–361.

Copeland, J. P., and J. S. Whitman. Wolverine. Pp. 672-682 in Wild mammals of North America: biology, management, and conservation (G. A. Feldhamer, B. C. Thompson, and J. A. Chapman, eds.). Johns Hopkins University Press, Baltimore.

Coues, E. 1897. The manuscript journals of Alexander Henry, fur trader of the Northwest Company, and of David Thompson, official geographer of the same company, 1799–1814. Ross and Haines, Minneapolis, 2 vol.

Cowan, W. F. 1973. Ecology and life history of the raccoon *(Procyon lotor)* in the northern part of its range. PhD dissertation, University of North Dakota, Grand Forks.

Criddle, S. 1939. The thirteen-striped ground squirrel in Manitoba. Canadian Field Naturalist 53:1–6.

Cryan, P. M. 2003. Seasonal distribution of migratory tree bats (*Lasiurus* and *Lasionycteris*) in North America. Journal of Mammalogy 84:579–593.

Cryan, P. M., M. A. Bogan, and G. M. Yanega. 2001. Roosting habits of four bat species in the Black Hills of South Dakota. Acta-Chiroptologica 3:43–52.

Currier, M. J. P. 1983. *Felis concolor.* Mammalian Species 200:1–7.

Cypher, B.L. 2003. Foxes. Pages 511–546 *in* Wild Mammals of North America: Biology, Management, and Conservation (G. A. Feldhamer, B. C. Thompson, and J A. Chapman eds.). John Hopkins University Press, Baltimore.

Davis, D. E. 1953. The characteristics of rat populations. Quarterly Review of Biology 28:373–401.

Dice, L. R. 1940. Intergradation between two subspecies of deer mouse *(Peromyscus maniculatus)* across North Dakota. Laboratory of Vertebrate Biology, University of Michigan, Contributions 13:1–14.

Dieter, C. D. 1987. Habitat use by beaver along the Big Sioux River. MS thesis, South Dakota State University, Brookings.

Dieter, C. D., and T. R. McCabe. 1989. Factors influencing beaver lodge site selection on a prairie stream. American Midland Naturalist 122:408–411.

Douglas, C. W., and M. A. Strickland. 1987. Fisher. Pp. 511-529 *in* Wild furbearer management and conservation in North America (M. Novak, J. A. Baker, M. E. Obbard, and B. Malloch, eds.). Ontario Ministry of Natural Resources, Toronto.

Dusek, G. L., R. J. Mackie, J. D. Herriges, Jr., and B. B. Compton. 1989. Population ecology of white-tailed deer along the lower Yellowstone River, USA. Wildlife Monographs 104:1–68.

Dusek, G. L., A. K. Wood, and R. J. Mackie. 1988. Habitat use by white-tailed deer in prairie-agricultural habitat in Montana. Prairie Naturalist 20:135–142.

Dyke, S., S. Hagen, and P. Isakson. 2004. North Dakota's 100 species of conservation priority. North Dakota Outdoors 67(1):2–21.

Eagle, T. C. 1989. Movement patterns of mink in the Prairie Pothole Region of North Dakota. Dissertation. University of Minnesota, St. Paul.

Eagle, T. C., and J. S. Whitman. 1987. Mink. Pp. 615–624, *in* Wild furbearer management and conservation in North America (M. Novak, J. A. Baker, M. E. Obbard, and B. Malloch, eds.). Ontario Trappers Association and Ontario Ministry of Natural Resources, Toronto.

Eberhardt, L., and A. B. Sargeant. 1977. Mink predation on prairie marshes during the waterfowl breeding season. Pages 33–43 *in* R. L. Philips and C. Jonkel, editors. Proceedings of the 1975 Predator Symposium. Montana Forest and Conservation Experiment Station, University of Montana, Missoula.

Edwards, J., M. Ford, and D. Guynn. 2003. Fox and gray squirrels. Pp. 248–267 *in* Wild Mammals of North America: Biology, Management, and Conservation (G. A. Feldhamer, B. C. Thompson, and J. A. Chapman, eds.). Johns Hopkins University Press, Baltimore.

Egoscue, H. J. 1979. *Vulpes velox*. Mammalian Species 122:1–5.

Eisenberg, J. F. 1968. Behavior patterns. Pp. 451–495 *in* Biology of *Peromyscus* (Rodentia) (J. A. King, ed.). American Society of Mammalogists Special Publication 2:1-593.

Eisenberg, J. F. 1981. The mammalian radiations. University of Chicago Press, Chicago.

Erb, J., and H. R. Perry, Jr. 2003. Muskrats. Pp. 311–348 *in* Wild Mammals of North America: Biology, Management, and Conservation (G. A. Feldhamer, B. C. Thompson, and J. A. Chapman, eds.). Johns Hopkins University Press, Baltimore.

Erlien, D. A., and J. R. Tester. 1984. Population ecology of sciurids in northwestern Minnesota. Canadian Field-Naturalist 98:1–6.

Errington, P. L. 1937. Summer food habits of the badger in north-western Iowa. Journal of Mammalogy 18:213–216.

Errington, P. L. 1963. Muskrat populations. Iowa State University Press, Ames.

Fairaizl, S. D. 1978. Bighorn sheep in North Dakota: population estimates, food habits and their biogeochemistry. MS thesis, University of North Dakota, Grand Forks.

Falls, J. B. 1968. Activity. Pp. 543–570, *in* Biology of *Peromyscus* (Rodentia) (J. A. King, ed.). American Society of Mammalogists Special Publication 2:1–593.

Feist, J. J. 1997. Bighorn sheep *(Ovis canadensis)* ecology and demography in the North Dakota badlands. MS thesis, University of North Dakota, Grand Forks.

Feldhammer, G. A., L. C. Drickamer, S. H. Vessey, J. F. Merritt, and C.Krajewski. 2007. Mammalogy: adaptation, diversity, and ecology. Johns Hopkins University Press, Baltimore.

Fenton, M. B., and R. M. R. Barclay. 1980. *Myotis lucifugus*. Mammalian Species 142:1–8.

Forbes, R. B. 1966. Studies of the biology of Minnesota chipmunks. American Midland Naturalist 76:290–308.

Forde, J. D., N. F. Sloan, and D. A. Shown. 1984. Grassland habitat management using prescribed burning in Wind Cave National Park, South Dakota. Prairie Naturalist 16:97–110.

Foresman, K. R. 2001. The wild mammals of Montana. American Society of Mammalogists, Special Publication 12.

Foresman, K. R., and R. E. Jensen, Jr. 1992. A morphometric comparison of the masked shrew *(Sorex cinereus)* and Hayden's shrew *(Sorex haydeni)* in Montana and bordering states. Northwest Naturalist 73:15–21.

Foresman, K. R., and D. E. Pearson. 1999. Activity patterns of American martens, *Martes americana,* snowshoe hares, *Lepus americanus,* and red squirrels, *Tamiasciurus hudsonicus,* in westcentral Montana. Canadian Field Naturalist 113:386–389.

Fox, R. A. 1989. Mule deer *(Odocoileus hemionus)* home range and habitat use in an energy-impacted area of the North Dakota badlands. MS thesis, University of North Dakota.

Franzmann, A. W. 1978. Moose. Pp. 67–81 *in* Big game of North America: ecology and management (J. L. Schmidt and D. L. Gilbert, eds.). Stackpole, Harrisburg.

Franzmann, A. W. 1981. *Alces alces*. Mammalian Species 154:1–7.

Franzmann, A. W., and C. C. Schwartz, eds. 1998. Ecology and management of North American moose. Smithsonian Institution Press, Washington.

French, A. R., and S. Forand. 2000. Role of soil temperature in timing of emergence from hibernation in the jumping mouse, *Zapus hudsonius*. Pp. 111–118 *in* Life in the Cold: Eleventh International Hibernation Symposium (G. Heldmaier and M. Klingespor, eds. Springer Verlag.
Fritzell, E. K. 1977. Dissolution of raccoon sibling bonds. Journal of Mammalogy 58:427–428.
Fritzell, E. K. 1978a. Aspects of raccoon *(Procyon lotor)* social organization. Canadian Journal of Zoology 56:260–271.
Fritzell, E. K. 1978b. Habitat use by prairie raccoons during the waterfowl breeding season. Journal of Wildlife Management 42:118–127.
Fritzell, E. K. 1978c. Reproduction of raccoons *(Procyon lotor)* in North Dakota. American Midland Naturalist 100:253–256.
Fritzell, E. K. 1987. Gray and island gray fox. Pp. 409–420 *in* Wildfurbearer management and conservation in North America (M. Novak, J. A. Baker, M. E. Obbard, B. Malloch, eds.). Ontario Trappers Association and Ontario Ministry of Natural Resources, Toronto.
Fritzell, E. K. 1989. Mammals in prairie wetlands. Pp. 268–301 *in* Northern prairie wetlands (A. van der Valk, ed.). Iowa State University Press, Ames.
Fritzell, E. K., and R. J. Greenwood. 1984. Mortality of raccoons in North Dakota. Prairie Naturalist 16:14.
Fritzell, E. K., and K. J. Haroldson. 1982. *Urocyon cinereoargenteus*. Mammalian Species 189:1–8.
Fritzell, E. K., and J. W. Matthews. 1975. A large raccoon litter. Prairie Naturalist 7:87–88.
Fuller, T. K., W. E. Berg, and D. W. Kuehn. 1985. Bobcat home range size and daytime cover-type use in northcentral Minnesota. Journal of Mammalogy 66:568–571.
Garrison, T. E., and T. L. Best. 1990. *Dipodomys ordii*. Mammalian Species 353:1–10.
Gehrt, S. D. 2003. Raccoon. Pp. 611–634 *in* Wild Mammals of North America: Biology, Management, and Conservation (G. A. Feldhammer, B. C. Thompson, and J. A. Chapman, eds.). Johns Hopkins University Press, Baltimore.
Geist, V. 1971. Mountain sheep: a study in behavior and evolution. University of Chicago Press, Chicago.
Genoways, H. H. 1967. Second report of *Myotis volans* from North Dakota. Transactions of the Kansas Academy of Science 69:355.
Genoways, H. H., and J. K. Jones, Jr. 1972. Mammals from southwestern North Dakota. Occasional Papers, The Museum, Texas Tech University 6:1–36.
George, S. B., J. R. Choate, and H. H. Genoways. 1986. *Blarina brevicauda*. Mammalian Species 261:1–9.
Gerads, J. 2002. Chronic wasting disease: It's rare. It's a killer. North Dakota Outdoors 65(10):18–21.
Giddings, B. J., G. L. Risdahl, and L. R. Irby. 1990. Bobcat habitat use in southeastern Montana during periods of high and low lagomorph abundance. Prairie Naturalist 22:249–258.
Goehring, H. H. 1972. Twenty-year study of *Eptesicus fuscus* in Minnesota. Journal of Mammalogy 53:201–207.
Greenwood, R. J., and A. B. Sargeant. 1994. Age-related reproduction in striped skunks (Mephitis mephitis) in the upper Midwest. Journal of Mammalogy 73:657–662.
Greenwood, R. J., A. B. Sargeant, J. L. Piehl, D. A. Buhl, and B. A. Hanson. 1999. Foods and foraging of prairie striped skunks during the avian nesting season. Wildlife Society Bulletin 27:823–832.

Greenwood, R. J., W. E. Newton, G. L. Pearson, and G. J. Schamber. 1997. Population and movement characteristics of radio-collared striped skunks in North Dakota during an epizootic of rabies. Journal of Wildlife Diseases 33:226–241.

Grondahl, C. R. 1973. Status of the black-tailed prairie dog and the black-footed ferret in North Dakota. Pp. 51–60 *in* Proceedings of the black-footed ferret and prairie dog workshop (R. L. Linder and C. N. Hillman, eds.). South Dakota State University, Brookings.

Gunderson, H. L. 1950. A study of some small mammal populations at Cedar Creek Forest, Anoka County, Minnesota. Minnesota Museum of Natural History, Occasional Papers 4:1–49.

Gunderson, H. L. 1962. An eight and one-half year study of the red-backed vole (*Clethrionomys gapperi* Vigors) at Cedar Creek Forest, Anoka and Isanti Counties, Minnesota. PhD dissertation, University of Michigan, Ann Arbor.

Gunderson, H. L. 1978. A mid-continent irruption of Canada lynx, 1962–63. Prairie Naturalist 10:71–80.

Hagmeier, E. M. 1966. A numerical analysis of the distributional patterns of North American mammals. II. Re-evaluation of the provinces. Systematic Zoology 15:279–299.

Hagmeier, E. M., and C. D. Stults. 1964. A numerical analysis of the distributional patterns of North American mammals. Systematic Zoology 13:125–155

Hall, E. R. 1981. The mammals of North America. Second edition. John Wiley and Sons, New York. 2 vols.

Halls, L. K., ed. 1984. White-tailed deer: ecology and management. Stackpole, Harrisburg.

Hammond, M. C. 1943. Beaver on the Lower Souris Refuge. Journal of Wildlife Management. 7:316–321.

Hanson, J. R. 1984. Bison ecology in the northern plains and a reconstruction of bison patterns for the North Dakota region. Plains Anthropologist 29:93–113.

Harmoning, A. K. 1976. White-tailed deer dispersion and habitat utilization in central North Dakota. MS thesis, North Dakota State University, Fargo.

Hart, E. B., M. C. Belk, E. Jordan, and M. W. Gonzalez. 2004. *Zapus princeps*. Mammalian Species 749:1–7.

Hazard, E. B. 1982. The mammals of Minnesota. University of Minnesota Press, Minneapolis.

Helgen, K. M., F. R. Cole, L. E. Helgen, and D. E. Wilson. 2009. Generic revision of the holarctic ground squirrel genus *Spermophilus*. Journal of Mammalogy 90:270–305.

Hendricks, P., and H. F. Allard. 1988. Winter food habits of prairie porcupines in Montana. Prairie Naturalist 20:1–6.

Hibbard, E. A. 1956. Range and spread of the gray and fox squirrels in North Dakota. Journal of Mammalogy 37:525–531.

Hibbard, E. A. 1958. Movements of beaver transplanted in North Dakota. Journal of Wildlife Management 22:209–211.

Hibbard, E. A. 1963. Another hoary bat found hanging on a fence. Journal of Mammalogy 44:265.

Hibbard, E. A. 1972. Vertebrate ecology and zoogeography of the Missouri River valley in North Dakota. PhD thesis, North Dakota State University, Fargo.

Hibbard, E. A., and J. R. Beer. 1960. The plains pocket mouse in Minnesota. Flicker 32:89–94.

Higgins, K. F., E. D. Stukel, J. M. Goulet, and D. C. Backlund. 2000. Wild mammals of South Dakota. South Dakota Department of Game, Fish and Parks, Pierre.

Hillman, C. N., and T. W. Clark. 1980. *Mustela nigripes*. Mammalian Species 126:1–3.
Hillman, C. N., R. L. Linder, and R. B. Dahlgren. 1979. Prairie dog distribution in areas inhabited by black-footed ferrets. American Midland Naturalist 102–185–187.
Hillman, C. N., and J. C. Sharps. 1978. Return of the swift fox to the northern Great Plains. Proceedings of the South Dakota Academy of Science 57:154–162.
Hines, T. D., and R. M. Case. 1991. Diet, home range, movements, and activity periods of swift fox in Nebraska. Prairie Naturalist 23:131–138.
Hoganson, J. W. 2006. Dinosaurs, sharks, and wooly mammoths: glimpses of life in North Dakota's prehistoric past. North Dakota Geological Survey Educational Series 31:1–60.
Hoganson, J. W., and E. C. Murphy. 2003. Geology of the Lewis and Clark trail in North Dakota. Mountain Press Publishing Company, Missoula, Montana.
Hohn, B. M., and W. H. Marshall. 1966. Annual and seasonal weight changes in a thirteen-lined ground squirrel population, Itasca State Park, Minnesota. Journal of the Minnesota Academy of Science 33:102–106.
Holloway, G. L., and R. M. R. Barclay. 2001. *Myotis ciliolabrum*. Mammalian Species 670:1–5.
Hoofer, S. R., J. R. Choate, and N. E. Mandrak. 1999. Mensural discrimination between *Reithrodontomys megalotis* and *R. montanus* using cranial characters. Journal of Mammalogy 80:91–101.
Hoogland, J. L. 1995. The black-tailed prairie dog: social life of a burrowing mammal. University of Chicago Press, Chicago.
Hoogland, J. L. 1996. *Cynomys ludovicianus*. Mammalian Species 535:1–10.
Hoogland, J. L. 2003. Black-tailed prairie dog. Pp. 232–247 *in* Wild Mammals of North America: Biology, Management, and Conservation (G. A. Feldhammer, B. C. Thompson, and J. A. Chapman, eds.). Johns Hopkins University Press, Baltimore.
Hooper, E. T. 1968. Classification. Pp. 27–74 *in* Biology of *Peromyscus* (Rodentia) (J. A. King, ed.). American Society of Mammalogists Special Publication 2:1–593.
Hopkins, R. B. 1983. Small mammal component of habitat types in Theodore Roosevelt National Park. Prairie Naturalist 15:45–46.
Iverson, S. L., R. W. Seabloom, and J. M. Hnatiuk. 1967. Small mammal distributions across the prairie-forest transition of Minnesota and North Dakota. American Midland Naturalist 78:188–197.
Iverson, S. L., and B. N. Turner. 1972a. Natural history of a Manitoba population of Franklin's ground squirrels. Canadian Field-Naturalist 86:145–149.
Iverson, S. L., and B. N. Turner. 1972b. Winter coexistence of *Clethrionomys gapperi* and *Microtus pennsylvanicus* in a grassland habitat. American Midland Naturalist 88:440–445.
Jackson, S. D. 1990. Ecology of mule deer on a sagebrush-grassland habitat in northeastern Montana. MS thesis, Montana State University, Bozeman.
Jacques, C. N., J. A. Jenks, and R. W. Klaver. 2009. Seasonal movements and home range use by female pronghorns in sagebrush-steppe communities of western South Dakota. Journal of Mammalogy 90:433–441.
James, T. R. 1967. The ecology and life history of the white-tailed jack rabbit *(Lepus townsendii campanius)* in North Dakota. PhD Dissertation, University of North Dakota, Grand Forks.
James, T. R., and R. W. Seabloom. 1969a. Reproductive biology of the white-tailed jack rabbit in North Dakota. Journal of Wildlife Management 33:558–568.
James, T. R., and R. W. Seabloom. 1969b. Aspects of growth in the white-tailed jack rabbit. Proceedings North Dakota Academy of Science 23:7–14.

Jatnieks-Straumanis, S. A. 1983. A study of pronghorn range in southwestern North Dakota. MS thesis, University of North Dakota, Grand Forks.

Jenkins, S. H., and P. E. Busher. 1979. *Castor canadensis*. Mammalian Species 120:1–8.

Jensen, W. F. 1988. Summer and fall ecology of mule deer in the North Dakota badlands. PhD dissertation, University of North Dakota, Grand Forks.

Jensen, W. F. 1992. Mule deer habitat use in the North Dakota badlands. Prairie Naturalist 24:97–108.

Jensen, W. F. 2000. Estimating big game weights. North Dakota Outdoors. September-October. 26–27.

Jensen, W. F. 2017. Estimating North Dakota moose reproductive parameters during a population increase. Alces. In Press.

Jensen, W. F., et al. 2013. Mass, morphology, and growth rates of moose in North Dakota. Alces 49:1–15.

Jensen, W. F., B. M. Hosek, and W. L. Rudd. 2004. Mapping continental range distribution of pronghorn using geographic information systems technology. Pronghorn Workshop Proceedings 21:18–36.

Jensen, W. F., and R. W. Seabloom. 1993. Pronghorn skull found along edge of historic eastern distribution in North Dakota. Prairie Naturalist 25:289–290.

Johnson, D. H., and A. B. Sargeant. 1977. Impact of red fox predation on the sex ratio of prairie mallards. U.S. Fish and Wildlife Service Wildlife Research Report 6. Washington.

Johnson, D. H., A. B. Sargeant, and R. J. Greenwood. 1989. Importance of individual species of predators on nesting success of ducks in the Canadian Prairie Pothole Region. Canadian Journal of Zoology 67:291–297.

Jones, J. K., Jr., D. M. Armstrong, R. S. Hoffmann, and C. Jones. 1983. Mammals of the northern great plains. University of Nebraska Press, Lincoln.

Jones, J. K., Jr., and H. H. Genoways. 1966. Records of bats from western North Dakota. Transactions of the Kansas Academy of Science 69:88–90.

Jones, J. K., Jr., R. P. Lampe, C. A. Spenrath, and T. H. Kunz. 1973. Notes on the distribution and natural history of bats in southeastern Montana. Occasional Papers, The Museum, Texas Tech University 15:1–12.

Jones, J. K., Jr., and W. C. Stanley. 1962. *Myotis subulatus* in North Dakota. Journal of Mammalogy 43:263.

Jones, W. T. 1993. The social systems of heteromyid rodents. Pp. 575–595 *in* Biology of the Heteromyidae (Hugh H. Genoways and James H. Brown, eds.). American Society of Mammalogists, Special Publication 10.

Jonkel, C. 1978. Black, brown (grizzly), and polar bears. Pp. 227–248 *in* Big game of North America: ecology and management (J. L. Schmidt and D. L. Gilbert, eds.). Stackpole, Harrisburg.

Kalin, O. T. 1964. Soils and other factors affecting the distribution of two species of pocket gophers (*Geomys bursarius* and *Thomomys talpoides*) in northeastern North Dakota, and a taxonomic study of two *Geomys* populations on opposite sides of the Red River. MS thesis, University of North Dakota, Grand Forks.

Karch, T., and G. Roloff. 1986. Native forests. North Dakota Outdoors 49(1):6–8.

Keller, B. L. 1985. Reproductive patterns. Pp. 725–778 *in* Biology of New World *Microtus* (R. H. Tamarin, ed.). American Society of Mammalogists, Special Publication 8:1–893.

Kelt, D. A., M. S. Hafner, and the American Society of Mammalogists' ad hoc Committee for Guidelines. 2010. Updated guidelines for protection of mammalogists and wildlife researchers from Hantavirus Pulmonary Syndrome (HPS). Journal of Mammalogy 91:1524–1527.

Kernohan, B. J., J. A. Jenks, and D. E. Naugle. 1994. Movement patterns of white-tailed deer at Sand Lake National Wildlife Refuge, South Dakota. Prairie Naturalist 26:293–300.

Kernohan, B. J., J. A. Jenks, and D. E. Naugle. 2002. Localized movements and site fidelity of white-tailed deer in the northern Great Plains. Prairie Naturalist 34:1–12.

Kilgore, D. L., Jr. 1969. An ecological study of the swift fox *(Vulpes velox)* in the Oklahoma panhandle. American Midland Naturalist 81:512–534.

King, C. M. 1983. *Mustela erminea.* Mammalian Species 195:1–8.

Kirkland, G. L., and D. F. Schmidt. 1996. *Sorex arcticus.* Mammalian Species 524:1–5.

Kistner, T. P. 1982. Diseases and parasites. Pp. 181–217 *in* Elk of North America: ecology and management (J. W. Thomas and D. E. Toweill, eds.). Stackpole, Harrisburg.

Knapp, S. J. 1986. New range records for the fox squirrel in the Yellowstone River drainage, Montana. Prairie Naturalist 18:128.

Knight, R. R. 1970. The Sun River elk herd. Wildlife Monographs 23:1–66.

Knowles, C. J. 1985. Observations on prairie dog dispersal in Montana. Prairie Naturalist 17:33–40.

Knowles, C. J. 1987. Reproductive ecology of black-tailed prairie dogs in Montana. Great Basin Naturalist 47:202–206.

Knowlton, F. F. 1972. Preliminary interpretation of coyote population mechanics with some management implications. Journal of Wildlife Management 36:369–382.

Knudsen, G. J., and J. B. Hale. 1968. Food habits of otters in the Great Lakes region. Journal of Wildlife Management 32:89–93.

Knue, J. 1991. Big game in North Dakota: a short history. North Dakota Game and Fish Department, Bismarck.

Koford, C. B. 1958. Prairie dogs, whitefaces, and blue grama. Wildlife Monographs 3:1–78.

Kolar, J. L. 2009. Pronghorn migration and resource selection in southwestern North Dakota. MS thesis, University of Missouri, Columbia. Columbia.

Koprowski, J. L. 1994a. *Sciurus carolinensis.* Mammalian Species 480:1–9.

Koprowski, J. L. 1994b. *Sciurus niger.* Mammalian Species 479:1–9.

Koucky, R. W. 1983. The buffalo disaster of 1882. North Dakota History 50(1):23–30.

Krausman, P.R., and R. T. Bowyer. 2003. Mountain sheep. Pp. 1095–1115 *in* Wild mammals of North America: biology, management, and conservation (G. A. Feldhamer, B. C. Thompson, and J. A. Chapman, eds.). Johns Hopkins University Press, Baltimore.

Kunz, T. H. 1982. *Lasionycteris noctivagans.* Mammalian Species 172:1-5.

Kunz, T. H., and R. E. Martin. 1982. *Plecotus townsendii.* Mammalian Species 175:1–6.

Kurta, A., and R. H. Baker. 1990. *Eptesicus fuscus.* Mammalian Species 356:1–10.

Kwiecinski, G. G. 1998. *Marmota monax.* Mammalian Species 591:1–8.

Lackey, J. A., D. G. Huckaby, and B. G. Ormiston. 1985. *Peromyscus leucopus.* Mammalian Species 247:1–10.

Lampe, R. P. 1982. Food habits of badgers in east central Minnesota. Journal of Wildlife Management 46:790–795.

Lampe, R. P., J. K. Jones, Jr., R. S. Hoffmann, and E. C. Birney. 1974. The mammals of Carter County, southeastern Montana. Occasional Papers, Museum of Natural History, University of Kansas 25:1–39.

Lampe, R. P., and M. A. Sovada. 1981.Seasonal variation in home range of a female badger *(Taxidea taxus)*. Prairie Naturalist 13:55–58.

Lariviere, S. 1999. *Mustela vison*. Mammalian Species 608: 1–9.

Lariviere, S. 2001. *Ursus americanus*. Mammalian Species 647:1–11.

Lariviere, S. 2003. Mink. Pp. 662–671 *in* Wild mammals of North America: biology, management, and conservation (G. A. Feldhamer, B. C. Thompson, and J. A. Chapman, eds.). Johns Hopkins University Press, Baltimore.

Lariviere, S., and M. Pasitschniak-Arts. 1996. *Vulpes vulpes*. Mammalian Species 537:1–11.

Lariviere, S., and L. R. Walton. 1997. *Lynx rufus*. Mammalian Species 563:1–8.

Lariviere, S., and L. R. Walton. 1998. *Lontra canadensis*. Mammalian Species 587:1–8.

Lausen, C. L., and R. M. R. Barclay. 2006. Winter bat activity on the Canadian prairies. Canadian Journal of Zoology 84:1079–1086.

Layne, J. N. 1968. Ontogeny. Pp. 148–253 *in* Biology of *Peromyscus* (Rodentia) (J. A. King, ed.). American Society of Mammalogists Special Publication 2:1–593.

Leonard, R. D. 1986. Aspects of reproduction of the fisher *(Martes pennanti)* in Manitoba. Canadian Field Naturalist 100:32–44.

Lewis, T. L., W. F. Jensen, K. A. Keehr, and R. W. Seabloom. 1994. Summer and fall food habits of coyotes in southwestern North Dakota. Prairie Naturalist 26:287–292.

Lim, B. K. 1987. *Lepus townsendii*. Mammalian Species 288:1–6.

Lindzey, F. 1987. Mountain Lion. Pp. 657–668 *in* Wild furbearer management and conservation in North America (M. Novak, J. A. Baker, M. E. Obbard, and B. Malloch, eds.). Ontario Ministry of Natural Resources, Toronto.

Lindzey, F. 2003. Badger. Pp. 683–691 *in* Wild mammals of North America: biology, management, and conservation (G. A. Feldhamer, B. C. Thompson, and J. A. Chapman, eds.). Johns Hopkins University Press, Baltimore.

Lokemoen, J. T., and K. F. Higgins. 1972. Population irruption of the least weasel *(Mustela nivalis)* in east central North Dakota. Prairie Naturalist 4:96.

Long, C. A. 1974. *Microsorex hoyi* and *Microsorex thompsoni*. Mammalian Species 33:1–4.

Long, C. A. 1973. *Taxidea taxus* in Wild mammals of North America: biology, management, and conservation (G. A. Feldhamer, B. C. Thompson, and J. A. Chapman, eds.). Johns Hopkins University Press, Baltimore.

Long, C. A., and C. A. Killingley. 1983. The North American badger. Pages 84–155 *in* Badgers of the World. Charles C. Thomas Publisher, Springfield.

Longley, W. H., and J. B. Moyle. 1963. The beaver in Minnesota. Minnesota Department of Conservation, Division of Game and Fish, Technical Bulletin 6:1–87.

Lysne, L. 1991. Small mammal demographics in North Dakota conservation reserve program plantings. MS thesis, University of North Dakota, Grand Forks.

MacCracken, J. G., and D. W. Uresk. 1984. Big game habitat use in southeastern Montana. Prairie Naturalist 16:135–139.

MacCracken, J. G., D. W. Uresk, and R. M. Hansen. 1983. Rodent-vegetation relationships in southeastern Montana. Northwest Science 57:272–278.

Mackie, R. J. 1970. Range ecology and relations of mule deer, elk, and cattle in the Missouri River Breaks, Montana. Wildlife Monographs 20:1–79

Mackie, R. J., J. G. Kie, D. F. Pac, and K. L. Hamlin. 2003. Mule Deer. Pp. 889–905 *in* Wild mammals of North America: biology, management, and conservation (G. A. Feldhamer, B. C. Thompson, and J. A. Chapman, eds.). Johns Hopkins University Press, Baltimore.

Mackintosh, J. F. 1981. Behavior of the house mouse. Pp. 337–365 *in* Biology of the house mouse (R. J. Berry, ed.). Symposia of the Zoological Society of London 47:1–715.

Madison, D. M. 1985. Activity rhythms and spacing. Pp. 373–419 *in* Biology of New World *Microtus* (R. H. Tamarin, ed.). American Society of Mammalogists, special publicantion 8:1–893.

Manning, R. W., and J. K. Jones, Jr. 1989. *Myotis evotis*. Mammalian Species 329:1–5.

Manning, R. W., and J. K. Jones, Jr. 1988. *Perognathus fasciatus*. Mammalian Species 303:1–4.

Marshall, W. H. 1935. A study of the winter activities of the mink. Journal of Mammalogy 17:382–392.

Marshall, W. H., G. W. Gullion, and R. G. Schwab. 1962. Early summer activities of porcupines as determined by radio-positioning techniques. Journal of Wildlife Management 26:75–79.

Martin, A. C., H. S. Zim, and A. L. Nelson. 1951. American wildlife and plants. McGraw-Hill, New York.

Martin, R. E., R. H. Pine, and A. F. DeBlase. 2001. A manual of mammalogy with keys to families of the world. McGraw-Hill, New York.

Maskey, J. J., and J. R. Smith. 2017. Population ecology of moose in northwestern North Dakota. North Dakota Game and Fish Department Report No. A-250. 52 p.

Maskey, J. J., Jr. 2008. Movements, resource selection, and risk analyses for parasitic diseases in an expanding moose population in the northern Great Plains. PhD dissertation, University of North Dakota, Grand Forks.

Matthews, W. L., and J. E. Swenson. 1982. The mammals of east-central Montana. Proceedings, Montana Academy of Science 41:1–13.

Mattson, T. A., S. W. Buskirk, and N. L. Stanton. 1996. Roost sites of the silver-haired bat *(Lasionycteris noctivagans)* in the black hills of South Dakota. Great Basin Naturalist 56:247–253.

McCabe, R. E., B. W. O'Gara, H. M. Reeves, and D. P. Metz. 2004. Prairie ghost. Pronghorn and human interaction in early America. University of Colorado Press, Boulder.

McCarty, R. 1978. *Onychomys leucogaster.* Mammalian Species 87:1–6.

McManus, J. J. 1974. *Didelphis virginiana*. Mammalian Species 40:1–6.

Meagher, M. 1978. Bison. Pp. 123-133 *in* Big game of North America: ecology and management (J. L. Schmidt and D. L. Gilbert, eds.). Stackpole, Harrisburg.

Meagher, M. 1986. *Bison bison.* Mammalian Species 266:1–8.

Mech, L. D. 1970. The wolf. Natural History Press (Doubleday), New York.

Mech, L. D. 1974. *Canis lupus.* Mammalian Species 37:1–6.

Mech, L. D. 1980. Age, sex, reproduction, and spatial organization of lynxes colonizing northeastern Minnesota. Journal of Mammalogy 61:261–267.

Mech, L. D., and L. D. Frenzel. Jr. (ed.). 1971. Ecological studies of the timber wolf in northeastern Minnesota (Research Paper NC-52), USDA Forest Service, North Central forest Experiment Station, St. Paul.

Mech, L. D., and F. J. Turkowski. 1966. Twenty-three raccoons in one winter den. Journal of Mammalogy 47:529–530.
Melquist, W. E., and A. E. Dronkert. 1987. River otter. Pp. 625–641 *in* Wild furbearer management and conservation in North America (M. Novak, J. A., Baker, M. E. Obbard, and B. Malloch, eds). Ontario Trappers Association and Ministry of Natural Resources, Toronto.
Melquist, W. E., P. J. Polechia, Jr., and D. Toweill. 2003. River otter. Pp. 709–734 *in* Wild mammals of North America: biology, management, and conservation (G. A. Feldhamer, B. C. Thompson, and J. A. Chapman, eds.). Johns Hopkins University Press, Baltimore.
Merritt, J. F. 1981. *Clethrionomys gapperi.* Mammalian Species 146:1–9.
Messick, J. P. 1987. North American badger. Pages 586–597 in Wild furbearer management and conservation in North America (M. Novak, J. A. Baker, M. E. Obbard, and B. Mallock, eds.). Ontario Trappers Association and Ministry of Natural Resources, Toronto.
Messick, J. P., and M. G. Hornocker. 1981. Ecology of the badger in southwestern Idaho. Wildlife Monographs 76.
Michel, E. S., et al. 2018. Weather and landscape factors affect white-tailed deer neonate survival at ecologically important life stages in the northern Great Plains. PLoS ONE 13(4):e0195247.
Michener, G. R. 1984. Age, sex, and species differences in the annual cycles of ground-dwelling sciurids: implications for sociality. Pp. 81–107 *in* The biology of ground-dwelling squirrels (J. O. Murie and G. R. Michener, eds.). University of Nebraska Pres, Lincoln.
Michener, G. R. 2000. Caching of Richardson's ground squirrels by North American badgers. Journal of Mammalogy 81:1106–1117.
Michener, G. R., and J. W. Koeppel. 1985. *Spermophilus richardsonii.* Mammalian Species 243:1–8.
Miller, K. V., L. I. Muller, and S. Demarais. 2003. White-tailed deer. Pp. 906–930 *in* Wild mammals of North America: biology, management, and conservation (G. A. Feldhamer, B. C. Thompson, and J. A. Chapman, eds.). Johns Hopkins University Press, Baltimore.
Milne, S. A. 2004. Population ecology and expansion dynamics of black-tailed prairie dogs in western North Dakota. MS thesis, University of North Dakota, Grand Forks.
Monk, R. R., and J. K. Jones, Jr. 1996. *Perognathus flavescens.* Mammalian Species 525:1–4.
Monty, A. M., and R. E. Emerson. 2003. Eastern woodrat (*Neotoma floridana* and allies). Pp. 381–393, in Wild mammals of North America: biology, management, and conservation (G. A. Feldhamer, B. C. Thompson, and J. A. Chapman, eds.). Johns Hopkins University Press, Baltimore.
Moore, T. D., L. E. Spence, and C. E. Dugnolle. 1974. Identification of the dorsal guard hairs of some mammals of Wyoming. Bulletin 14, Wyoming Game and Fish Department, Cheyenne.
Mullican, T. R. 1994. First record of Merriam's shrew from South Dakota. Prairie Naturalist 26:173.
Mullican, T. R. 1999. Earliest seasonal record of reproduction in the hoary bat on the northern great plains. Prairie Naturalist 31:249–250.
Murie, O. J. 1954. A field guide to animal tracks. Houghton Mifflin, Boston.

Murray, D. L. 2003. Snowshoe hare and other hares. Pp. 147–175 *in* Wild mammals of North America: biology, management, and conservation (G. A. Feldhamer, B. C. Thompson, and J. A. Chapman, eds.). Johns Hopkins University Press, Baltimore.

Naylor, V. J. 1987. Utilization of grazed and ungrazed woody draws by small mammals in the North Dakota badlands. Institute for Ecological Studies, University of North Dakota, Grand Forks. Contribution 7:1–46.

Nellermoe, R. L. 1983. A comparison of small mammal populations in available habitats of an agricultural area of the Red River of the North. Unpublished Doctor of Arts report, University of North Dakota, Grand Forks.

Nelson, E. W. 1925. Status of pronghorned antelope, 1922–24. U. S. Department of Agriculture Bulletin 1346.

Nelson, J. J., P. R. Barnhart, and E. H. Gillam. 2015. Distribution and occurrence of bat species in North Dakota. Prairie Naturalist 47:84-93.

Nelson, J. P. 1981. Seasonal activity and time budgets of a North Dakota fox squirrel population. MS thesis, University of North Dakota, Grand Forks.

Nero, R. W., and R. E. Wrigley. 1977. Status and habits of the cougar in Manitoba. Canadian Field Naturalist 91:28-40.

Nowak, R. M. 1999. Walker's Mammals of the World, 6th ed. Johns Hopkins University Press, Baltimore. 2 vols.

O'Gara, B. W. 1978. *Antilocapra americana.* Mammalian Species 90:1–7.

O'Gara, B. W., and J. D. Yoakum. 2004. Pronghorn: Ecology and management. University of Colorado Press, Boulder.

Osborn, R. G., J. A. Jenks, and W. F. Jensen. 1997. Diet of North Dakota elk determined from rumen and fecal analyses. Prairie Naturalist 29:237–247.

Ostroff, A. C., and E. J. Finck. 2003. *Spermophilus franklinii.* Mammalian Species 724:1–5.

Paquet, P. C., and L. N. Carbyn. 2003. Gray wolf. Pp. 482–510 *in* Wild mammals of North America. (G. A. Feldhamer, B. C. Thompson, and J. A. Chapman, eds.). Johns Hopkins University Press, Baltimore.

Pasitschiniak-Arts, M., and L. Lariviere. 1995. Gulo gulo. Mammalian Species 499: 1–10.

Paulson, D. D. 1988. *Chaetodipus hispidus.* Mammalian Species 320:1–4.

Peek, J. M. 2003. Wapiti. Pp. 877–888 *in* Wild mammals of North America: biology, management, and conservation (G. A. Feldhamer, B. C. Thompson, and J. A. Chapman, eds.). Johns Hopkins University Press, Baltimore.

Peek, J. M., D. L. Urich, and R. J. Mackie. 1976. Moose habitat selection and relationships to forest management in northeastern Minnesota. Wildlife Monographs 48:1-65.

Pefaur, J. E., and R. S. Hoffmann. 1974. Notes on the biology of the olive-backed pocket mouse *(Perognathus fasciatus)* on the northern great plains. Prairie Naturalist 6:7–15.

Pelton, M. R. 2003. Black Bear. Pp. 547–555 *in* Wild mammals of North America: biology, management, and conservation (G. A. Feldhamer, B. C. Thompson, and J. A. Chapman, eds.). Johns Hopkins University Press, Baltimore.

Peterson, L. E. 1984. Northern plains. Pp. 441–448 *in* White-tailed deer: ecology and management (L. K. Halls, ed.). Wildlife Management Institute. Stackpole Books, Harrisburg.

Peterson, R. O., J. D. Woolington, and T. N. Bailey. 1984. Wolves of the Kenai Peninsula, Alaska. Wildlife Monographs 88:1–52.

Phillips, M. L., W. R. Clark, M. A. Sovada, D. J. Horn, R. R. Koford, and R. J. Greenwood. 2003. Predator selection of prairie landscapes features and its relation to duck nest success. Journal of Wildlife Management 67:104–114.

Phillips, R. L., W. E. Berg, and D. B. Siniff. 1973. Movement patterns and range use of moose in northwestern Minnesota. Journal of Wildlife Management 37:266–278.

Pierce, B. M., and V. C. Bleich. 2003. Mountain Lion. Pp. 744–757 *in* Wild mammals of North America: biology, management, and conservation (G. A. Feldhamer, B. C. Thompson, and J. A. Chapman, eds.). Johns Hopkins University Press, Baltimore.

Pietz, P. J., and J. R. Tester. 1983. Habitat selection by snowshoe hares in north central Minnesota. Journal of Wildlife Management 47:686–696.

Pigage, J. C. 1975. Reproduction of *Spermophilus richardsonii richardsonii* at the eastern edge of its range. MS thesis, University of North Dakota, Grand Forks.

Powell, R. A. 1981. *Martes pennanti.* Mammalian Species 156:1–6.

Powell, R. A., S. W. Buskirk, and W. J. Zielinski. 2003. Fisher and marten. Pp. 635–649 *in* Wild mammals of North America: biology, management, and conservation (G. A. Feldhamer, B. C. Thompson, and J. A. Chapman, eds.). Johns Hopkins University Press, Baltimore.

Pruitt, W. O., Jr. 1960. Animals in the snow. Scientific American 202(1):60–68.

Pugh, S. R., S. Johnson, and R. H. Tamarin. 2003. Voles. Pp. 349-370 *in* Wild mammals of North America: biology, management, and conservation (G. A. Feldhamer, B. C. Thompson, and J. A. Chapman, eds.). Johns Hopkins University Press, Baltimore.

Quanstrom, W. R. 1966. Flood tolerance in Richardson's ground squirrel. Journal of Mammalogy 47:323.

Quanstrom, W. R. 1968. Some aspects of the ethoecology of Richardson's ground squirrel in eastern North Dakota. PhD dissertation, University of Oklahoma, Norman.

Quimby, D. C. 1951. The life history and ecology of the jumping mouse, *Zapus hudsonius.* Ecological Monographs 21:61–95.

Quinn, N. W. S., and G. Parker. 1987. Lynx. Pp. 685–694 *in* Wild furbearer management and conservation in North America (M. Novak, J. A. Baker, M. E. Obbard, and B. Malloch, eds.). Ontario Ministry of Natural Resources, Toronto.

Raine, R. M. 1983. Ranges of juvenile fisher, *Martes pennanti,* and marten, *Martes americana,* in southeastern Manitoba. Canadian Field Naturalist 96:431–438.

Raine, R. M. 1983. Winter habitat use and response to snow cover of fisher, *Martes pennanti,* and marten, *Martes americana,* in southeastern Manitoba. Canadian Journal of Zoology 61:25–34.

Reading, R. P., and R. Matchett. 1997. Attributes of black-tailed prairie dog colonies in northcentral Montana. Journal of Wildlife Management 61:664–673.

Reich, L. M. 1981. *Microtus pennsylvanicus.* Mammalian Species 159:1–8.

Reid, N. J. 1954. The distribution of the black-tailed prairie dog in the badlands of southwestern North Dakota. M.S. thesis, Iowa State University, Ames.

Reid, R., and C. L. Gannon. 1927. Birds and mammals observed by Lewis and Clark in North Dakota. North Dakota Historical Quarterly 1(4):14–36.

Reynolds, H. W., C. C. Gates, and R. D. Glaholt. 2003. Bison. Pp. 1009–1060 *in* Wild mammals of North America: biology, management, and conservation (G. A. Feldhamer, B. C. Thompson, and J. A. Chapman, eds.). Johns Hopkins University Press, Baltimore.

Roe, K. A., and C. M. Roe. 2003. Habitat selection guidelines for black-tailed prairie dog relocations. Wildlife Society Bulletin 31:1246–1253.

Rogers, L. L. 1987. Effects of food supply and kinship on social behavior, movements, and population growth of black bears in northeastern Minnesota. Wildlife Monographs 97:1–72.

Rolley, R. E. 1987. Bobcat. Pp. 671–681 *in* Wild furbearer management and conservation in North America (M. Novak, J. A. Baker, M. E. Obbard, and B. Malloch, eds.). Ontario Ministry of Natural Resources, Toronto.

Rollings, C. T. 1945. Habits, foods, and parasites of the bobcat in Minnesota. Journal of Wildlife Management 9:131–145.

Rosatte, R. C. 1987. Striped, spotted, hooded, and hog-nosed skunk. Pp. 599–613 *in* Wild furbearer management and conservation in North America (M. Novak, J. A. Baker, M. E. Obbard, and B. Mallock, eds.). Ontario Trappers Association and Ministry of Natural Resources, Toronto.

Rosatte, R. C., and S. Lariviere. 2003. Skunks. Pp. 692–707 *in* Wild Mammals of North America: Biology, Management, and Conservation (G. A. Feldhammer, B. C. Thompson, and J. A. Chapman, eds.). Johns Hopkins University Press, Baltimore.

Roze, U., and L. M. Ilse. 2003. Porcupine. Pp. 371–380 *in* Wild Mammals of North America (G. A. Feldhamer, B. C. Thompson, and J. A. Chapman, eds.). Johns Hopkins University Press, Baltimore.

Rubbelke, D., and S. G. Saupe. 1984. Water shrews in Minnesota. Project Summary, Minnesota of Natural Resources. Unpublished manuscript.

Ruggiero, L. F., et al. 2007. Wolverine conservation and management. Journal of Wildlife Management 71:2145–2146.

Sargeant, A. B. 1972. Red fox spatial characteristics in relation to waterfowl predation. Journal of Wildlife Management 36:225–236.

Sargeant, A. B. 1982. A case history of a dynamic resource–the red fox. Pages 122–137 *in* Midwest furbearer management (G. C. Sanderson, ed.) Proceedings of the 43rd Midwest Fish and Wildlife Conference, Wichita.

Sargeant, A.B., S. H. Allen, and R. T. Eberhardt. 1984. Red fox predation on breeding ducks in mid-continent North America. Wildlife Monographs 89.

Sargeant, A. B., S. H. Allen, and J. P. Fleskes. 1986. Commercial sunflowers: food for red foxes in North Dakota. Prairie Naturalist 18:91–94.

Sargeant, A. B., S. H. Allen, and J. O. Hastings. 1987. Spatial relations between sympatric coyotes and red foxes in North Dakota. Journal of Wildlife Management 51:285–293.

Sargeant, A. B., S. H. Allen, and J. O. Hastings. 1987a. Spatial relations between sympatric coyotes and red foxes in North Dakota. Journal of Wildlife Management 51:285–293.

Sargeant, A. B., M. A. Sovada, and R. J. Greenwood. 1987b. Responses of three prairie ground squirrel species, *Spermophilus franklinii, S. richardsonii,* and *S. tridecemlineatus,* to duck eggs. Canadian Field-Naturalist 101:95–97.

Sargeant, A. B., R. J. Greenwood, M. A. Sovada, and T. L. Shaffer. 1993. Distribution and abundance of predators that affect duck production—Prairie Pothole Region. U.S. Fish and Wildlife Service, Resource Publication 194, Washington.

Sargeant, A. B., and D. W. Warner. 1972. Movements and denning habits of a badger. Journal of Mammalogy 53:207–210.

Sargeant, G. A., and M. W. Oehler. 2007. Dynamics of newly established elk populations. Journal of Wildlife Management. 71:1141–1148.

Sasmal, I, S. Datta, and G. Soiseth. 2018. Intensive survey of small mammal and herptofauna in the Turtle Mountains of North Dakota. Interim report to North Dakota Game and Fish Department.

Sather, J. H. 1958. Biology of the great plains muskrat in Nebraska. Wildlife Monographs 2:1–35.
Sayre, R. W. 1996. Ecology of bighorn sheep in relation to habitat and oil development in the North Dakota badlands. PhD dissertation, University of North Dakota, Grand Forks.
Schantz, V. S. 1950. The gray fox in North Dakota. Journal of Mammalogy 31:197–198.
Schmidt, C. A. 2003. Conservation assessment for the Townsend's big-eared bat in the Black Hills National Forest South Dakota and Wyoming. http://www.fs.fed.us/planning/assessments/big_ear_bat.pdf.
Scott-Brown, J. M., S. Herrero, and J. Reynolds. 1987. Swift fox. Pp. 433–441 *in* Wild furbearer management and conservation in North America (M. Novak, J. A. Baker, M. E. Obbard, and B. Malloch, eds.). Ontario Trappers Association and Ministry of Natural Resources, Toronto.
Seabloom, R. W. 1958. A study of production, movements, and survival of the Great Plains muskrat *(Ondatra zibethicus cinnamominus)* in North Dakota. MS thesis, University of Minnesota, St. Paul.
Seabloom, R. W. 2002. Additional records of the hispid pocket mouse in North Dakota. Prairie Naturalist 34:61–62.
Seabloom, R. W., and J. R. Beer. 1963. Observations of a muskrat population decline in North Dakota. Proceedings, North Dakota Academy of Science 17:66–70.
Seabloom, R. W., R. D. Crawford, and M. G. McKenna. 1978. Vertebrates of southwestern North Dakota. Institute for Ecological Studies, University of North Dakota Research Report 24 (REAP 78-4).
Seabloom, R. W., J. J. Feist, and S. L. McDonough. 1999. Hantavirus infection in North Dakota small mammals: 1994,1995. Prairie Naturalist 31:129–139.
Seabloom, R. W., M. G. McKenna, and R. D. Crawford. 1980. Recent records of mammals from southwestern North Dakota. Prairie Naturalist 12:119–123.
Seabloom, R. W., and T. L. Shaffer. 2005. Plains harvest mouse in North Dakota. Prairie Naturalist 37:247–250.
Seabloom, R. W., and P. W. Theisen. 1990. Breeding biology of the black-tailed prairie dog in North Dakota. Prairie Naturalist 22:65–74.
Selting, J. P., and L. R. Irby. 1997. Agricultural land use patterns of native ungulates in southeastern Montana. Journal of Range Management 50:338–345.
Seton, E. T. 1909. Life histories of northern animals. Charles Scribner's Sons, New York. 2 vols.
Shackleton, D. M. 1985. *Ovis canadensis*. Mammalian Species 230:1–9.
Shaughnessy, M. 2016. Survey of "fringe" mammals in western North Dakota. North Dakota Game and Fish Department Final Report T-39-R-1. 23 p.
Shaughnessy, M., and N. Woodman. 2015. New records of Merriam's Shrew (*Sorex merriami*) from western North Dakota. Check List 11(3):1623.
Sheets, R. G., and R. L. Linder. 1969. Food habits of the black-footed ferret (*Mustela nigripes*) in South Dakota. South Dakota Academy of Science Proceedings 48:58–61.
Sheffield, S. R., and C. M. King. 1994. *Mustela nivalis*. Mammalian Species 454:1–10.
Sheffield, S. R., and H. H. Thomas. 1997. *Mustela frenata*. Mammalian Species 570:1–9.
Sherwin, R. E., W. L. Gannon, and J. S. Altenbach. 2003. Managing complex systems simply: understanding inherent variation in the use of roosts by Townsend's big-eared bat. Wildlife Society Bulletin 31:62–72.

Sherwin, R. E., D. Stricklan, and D. S. Rogers. 2000. Roosting affinities of Townsend's big-eared bat *(Corynorhinus townsendii)* in northern Utah. Journal of Mammalogy 81:939–947.

Shump, K. A., and A. U. Shump. 1982a. *Lasiurus borealis*. Mammalian Species 183:1–6.

Shump, K. A., and A. U. Shump. 1982b. *Lasiurus cinereus*. Mammalian Species 185:1–5.

Sidle, J. G., D. H. Johnson, and B. R. Euliss. 2001. Estimated aerial extent of colonies of black-tailed prairie dogs in the northern great plains. Journal of Mammalogy 84:928–936.

Skaley, J. E. 1967. A study of two adjacent microtine populations. M.S. thesis, University of North Dakota, Grand Forks.

Skovlin, J. M. 1982. Habitat requirements and evaluations. Pp. 369–413 *in* Elk of North America: ecology and management (J. W. Thomas and D. E. Toweill, eds.). Stackpole, Harrisburg.

Smith, F. A. 1997. *Neotoma cinerea*. Mammalian Species 564:1–8

Smith, J. R. 2004. Population ecology of white-tailed deer in the drift prairie-couteau of North Dakota. MS thesis. Department of Biology, University of North Dakota, Grand Forks.

Smith, W. P. 1991. *Odocoileus virginianus*. Mammalian Species 388:1–13.

Snyder, D. P. 1982. *Tamias striatus*. Mammalian Species 168:1–8.

Snead, E., and G. O. Hendrickson. 1942. Food habits of the badger in Iowa. Journal of Mammalogy 23:380–391.

Sovada, M. A., Roaldson, J. M., and Sargeant, A. B. 1999. Foods of American badgers in west-central Minnesota and southwestern North Dakota during the duck nesting season. American Midland Naturalist 142:410–414.

Sovada, M. A., C. C. Roy, J. B. Bright, and J. R. Gillis. 1998. Causes and rates of mortality of swift foxes in western Kansas. Journal of Wildlife Management 62:1300–1306.

Sparrowe, R. D., and P. F. Springer. 1970. Seasonal activity patterns of white-tailed deer in eastern South Dakota. Journal of Wildlife Management 34:420–431.

Stalling, D. T. 1990. *Microtus ochrogaster.* Mammalian Species 355:1-9.

Steele, M. A. 1998. *Tamiasciurus hudsonicus*. Mammalian Species 586:1–9.

Stevens, O. A. 1943. Audubon's journey up the Missouri River, 1843. North Dakota Historical Quarterly 10(2):63–82.

Stickel, L. F. 1968. Home range and travels. Pp. 373–411 *in* Biology of *Peromyscus* (Rodentia) (J. A. King, ed.). American Society of Mammalogists Special Publication 2:1–593.

Stockrahm, D. M. R. B. 1979. Comparison of population structures of black-tailed prairie dog, *Cynomys l. ludovicianus* (Ord) towns in western North Dakota. M.S. thesis, University of North Dakota, Grand Forks.

Stockrahm, D. M., and R. W. Seabloom. 1988. Comparative reproductive performance of black-tailed prairie dog populations in North Dakota. Journal of Mammalogy 69:160–164.

Strassler, A. 1996. Ecology of the northern badlands elk herd. M.S. thesis, University of North Dakota, Grand Forks.

Streubel, D. P., and J. P. Fitzgerald. 1978. *Spermophilus tridecemlineatus*. Mammalian Species 103:1–5.

Strickland, M. A., and C. W. Douglas. 1987. Marten. Pp. 531–546 *in* Wild furbearer management and conservation in North America (M. Novak, J. A. Baker, M. E. Obbard, and B. Malloch, eds.). Ontario Ministry of Natural Resources, Toronto.

Sullivan, M. G. 1988. Distribution, habitat use, and food habits of reintroduced elk in Theodore Roosevelt National Park, North Dakota. MS thesis, Montana State University, Bozeman.

Svendsen, G. G. 2003. Weasels and black-footed ferret. Pp. 650–661 *in* Wild mammals of North America: biology, management, and conservation (G. A. Feldhamer, B. C. Thompson, and J. A. Chapman, eds.). Johns Hopkins University Press, Baltimore.

Svihovec, L. K. 1967. A comparison study of the ecological distribution of small mammals in southwestern North Dakota. MS thesis, University of North Dakota, Grand Forks.

Swenson, J. E. 1981. Distribution of Richardson's ground squirrel in eastern Montana. Prairie Naturalist 13:27–30.

Swenson, J. E., and S. J. Knapp. 1980. Composition of beaver caches on the Tongue River in Montana. Prairie Naturalist 12:33–36

Swenson, J. E., S. J. Knapp, and H. J. Wentland. 1983. Winter distribution and habitat use by muse deer and white-tailed deer in southeastern Montana. Prairie Naturalist 15:97–112.

Taber, R. D., K. Raedeke, and D. A. McCaughran. 1982. Population characteristics. Pp. 279–298 *in* Elk of North America: ecology and management (J. W. Thomas and D. E. Toweill, eds.). Stackpole, Harrisburg.

Tait, M. J., and C. J. Krebs. 1985. Population dynamics and cycles. Pp. 567–620 *in* Biology of New World *Microtus* (R. H. Tamarin, ed.). American Society of Mammalogists, special publication 8:1–893.

Tenneson, C., and L. W. Oring. 1985. Winter food preferences of porcupines. Journal of Wildlife Management 49:28–33.

Terman, C. R. 1968. Population dynamics. Pp. 412–450 *in* Biology of *Peromyscus* (Rodentia) (J. A. King, ed.). American Society of Mammalogists Special Publication 2:1–593.

Tester, J. R. 1965. Effects of a controlled burn on small mammals in a Minnesota oak-savannah. American Midland Naturalist 74:240–243.

Tester, J. R., S. Malchow, C. McLain, and J. B. Lehrer. 1993. Movements and habitat use by meadow jumping mice in northwestern Minnesota. Prairie Naturalist 25:33–37.

Tester, J. R., and W. H. Marshall. 1961. A study of certain plant and animal interrelations on a prairie in northwestern Minnesota. Occasional Papers, Minnesota Museum of Natural History, University of Minnesota, Minneapolis 8:1–51.

Theisen, P. W. 1981. Age-specific reproduction in the black-tailed prairie dog *(Cynomys l. ludovicianus)*. MS thesis, University of North Dakota, Grand Forks.

Toweill, D. E., and J. W. Thomas, eds. 2002. North American elk: ecology and management. Smithsonian Institute Press. Washington.

Trevor, J. T., R. W. Seabloom, and S. H. Allen. 1989. Food habits in relation to sex and age of bobcats from southwestern North Dakota. Prairie Naturalist 21:163–168.

Tumlison, R. 1987. *Felis lynx*. Mammalian Species 269:1–8.

Turner, B. N., S. L. Iverson, and K. L. Severson. 1976. Postnatal growth and development of captive Franklin's ground squirrels *(Spermophilis franklinii)*. American Midland Naturalist 95:93–102.

Turner, R. W. 1974. Mammals of the Black Hills of South Dakota and Wyoming. Miscellaneous Publications, University of Kansas Museum of Natural History 60:1–178.

Turner, R. W., and J. B. Bowles. 1967. Comments on reproduction and food habits of the olive-backed pocket mouse in western North Dakota. Transactions of the Kansas Academy of Science 70:266–267.

United States Fish and Wildlife Service. 2013. Black-footed ferret draft recovery plan. U.S. Fish and Wildlife Service, Denver. 130 p.

United States Fish and Wildlife Service. 1988. Black-footed ferret recovery plan. U. S. Fish and Wildlife Service, Denver.

United States Fish and Wildlife Service. 2000. Endangered and threatened wildlife and plants: 12 month finding for a petition to list the black-tailed prairie dog as threatened. Federal Register 65(24):5476–5488.

Uresk, D. W. 1984. Black-tailed prairie dog food habits and forage relationships in western South Dakota. Journal of Range Management 37:325–329.

Van Ballenberghe, V, A. W. Erickson, and D. Byman. 1975. Ecology of the timber wolf in northeastern Minnesota. Wildlife Monographs 43:1–43.

van Zyll de Jong, C. G. 1980. Systematic relationships of woodland and prairie forms of the prairie shrew, *Sorex cinereus cinereus* Kerr) and *S. c. haydeni* Baird, in the Canadian prairie provinces. Journal of Mammalogy 61:66–75.

Vermeire, L. T. et al. 2004. The prairie dog story: do we have it right? Bioscience 54:589–695.

Verts, B. J. 1967. The biology of the striped skunk. University of Illinois Press, Urbana

Verts, B. J., and L. N. Carraway. 1999. *Thomomys talpoides*. Mammalian Species 618:1–11.

Verts, B. J., and L. N. Carraway. 2001. *Tamias minimus*. Mammalian Species 653:1–10.

Voigt, D. R., and W. E. Berg 1987. Coyote. Pages 345–356 *in* Wild furbearer management and conservation in North America. M. Novak, J. A. Baker, M. E. Obbard, and B. Mallock, eds. Ontario Trappers Association and Ministry of Natural Resources, Toronto.

Vosburgh, T. C., and L. R. Irby. 1998. Effects of recreational shooting on prairie dog colonies. Journal of Wildlife Management 62:363–372.

Wade-Smith, R. A., and B. J. Verts. 1982. *Mephitis mephitis*. Mammalian Species 173:1–7.

Walley, W. J. 1970. Habitat preferences of *Microtus pennsylvanicus* (Ord) on a burned native prairie. MS thesis, University of North Dakota, Grand Forks.

Wallmo, O. C., ed. 1981. Mule and black-tailed deer of North America. University of Nebraska Press, Lincoln.

Walsh, L. L, R. Seabloom, and C. W. Thompson. 2017. Range extension of the Virginia Opossum (*Didelphis virginina*) in North Dakota. Prairie Naturalist 49:72–75.

Warner, R. M., and N. J. Czaplewski. 1984. *Myotis volans*. Mammalian Species 224:1–4.

Warren, E. R. 1927. The beaver: its works and its ways. American Society of Mammalogists Monograph 2. Williams and Wilkins, Baltimore.

Webster, W. D., and J. K. Jones, Jr. 1982. *Reithrodontomys megalotis*. Mammalian Species 167:1–5.

Wells-Gosling, N., and L. R. Heaney. 1982. *Glaucomys sabrinus*. Mammalian Species 229:1–8.

Westfall, J. A. 1989. The ecology of reintroduced elk in Theodore Roosevelt National Park, North Dakota. MS thesis, Montana State University, Bozeman.

Wheatley, M. 1997a. Beaver *(Castor canadensis)* home range size and patterns of use in the taiga of southeastern Manitoba. 1. Seasonal variation. Canadian Field Naturalist 111:204–210.

Wheatley, M. 1997b. Beaver, *Castor canadensis*, home range size and patterns of use in the taiga of southeastern Manitoba. 2. Sex, age, and family status. Canadian Field Naturalist 111:211–216.

Wheatley, M. 1997c. Beaver, *Castor canadensis*, home range size and patterns of use in the taiga of southeastern Manitoba. 3. Habitat variation. Canadian Field Naturalist 111:217–222.

Whitaker, J. O. 1972. *Zapus hudsonius*. Mammalian Species 11:1–7.

Whitaker, J. O. 2004. *Sorex cinereus*. Mammalian Species 743:1–9.

Wiedmann, B. P., and V. C. Bleich. 2014. Responses of bighorn sheep to recreational activities: a trial of a trail. Wildlife Society Bulletin 38:773–782.

Wiedmann, B. P., and B. Hosek. 2013. North Dakota bighorn sheep management plan (2013–2023). Wildlife Division Report A-213. North Dakota Game and Fish Department, Bismarck, USA.

Wiedmann, B. P., and G. A. Sargeant. 2014. Ecotypic variation in recruitment of reintroduced bighorn sheep: implications for translocation. Journal of Wildlife Management 78:397–401.

Wilkins, K. T. 1986. *Reithrodontomys montanus*. Mammalian Species 257:1–5.

Williams, H. H. Genoways, and J. K. Braun. 1993. Taxonomy and systematics. Pp. 38–196 *in* Biology of the Heteromyidae (Hugh H. Genoways and James H. Brown, eds.). American Society of Mammalogists, Special Publication 10.

Willner, G. R., G. A. Feldhamer, E. E. Zucker, and J. A. Chapman. 1980. *Ondatra zibethicus*. Mammalian Species 141:1–8.

Wilson, D. E., and D. M. Reeder, eds. 2005. Mammal species of the world: a taxonomic and geographic reference. Johns Hopkins University Press, Baltimore, MD. 2 vol.

Wilson, J. W., III. 1974. Analytical zoogeography of North American mammals. Evolution 28:124–140.

Wobeser, G. A. 1985. Handbook of diseases of Saskatchewan wildlife. Saskatchewan Parks and Renewable Resources, Saskatoon.

Wolff, J. O. 1985. Behavior. Pp. 340–372 *in* Biology of New World *Microtus* (R. H. Tamarin, ed.). American Society of Mammalogists, special publication 8:1–893.

Wood, A. K. 1986. Population ecology and habitat relationships of mule deer in prairie-agricultural habitat. PhD thesis, Montana State University, Bozeman.

Wood, A. K. 1988. Use of shelter by mule deer during winter. Prairie Naturalist 20:15–22.

Woods, C. A. 1973. *Erethizon dorsatum*. Mammalian Species 29:1–6.

Wrigley, R. E. 1974. Mammals of the sandhills of southwestern Manitoba. Canadian Field Naturalist 88:21–39

Wrigley, R. E., and J. E. Dubois. 1973. Distribution of the pocket gophers *Geomys bursarius* and *Thomomys talpoides* in Manitoba. Canadian Field-Naturalist 87:167-169.

Wrigley, R. E., J. E. DuBois, and H. W. R. Copland. 1979. Habitat, abundance, and distribution of six species of shrews in Manitoba. Journal of Mammalogy 60:505 – 520.

Yahner, R. H. 2003. Pine squirrels. Pp. 268-275 *in* Wild mammals of North America: biology, management, and conservation (G. A. Feldhamer, B. C. Thompson, and J. A. Chapman, eds.). Johns Hopkins University Press, Baltimore.

Yeager, L. E. 1937. Naturally sustained yield in a farm fur crop in Mississippi. Journal of Mammalogy 35:406-415.

Yoakum, J. D. 1978. Pronghorn. Pp. 103-121 *in* Big game of North America: ecology and management (J. L. Schmidt and D. L. Gilbert, eds.). Stackpole, Harrisburg.

Index to Common and Scientific Names

A

Alces americanus 51, 362, 423, 436
American badger 139, 189, 190, 444–45, 457, 462
American black bear 48, 146
American marten 48, 159–60, 422, 446, 449
American mink 48, 185
American porcupine 50, 338, 416, 428
Antilocapra americana 51, 386, 423, 436, 458,
Antilocapridae 51, 361, 386–87, 389, 423, 436
Arctic shrew 28, 31, 33, 35–36, 47, 61–62, 413, 425
Artiodactyla 6, 5, 360, 412, 423–24, 436

B

Badger, American 139, 189, 190
Bat
    Big brown 44, 47, 74, 78–80, 414
    Fringed myotis, 47, 101–104, 414, 426
    Hoary 47, 78, 82, 87–88, 415, 426
    Little brown bat 74, 80, 91, 94–95, 97–99, 104
    Little brown myotis 47, 95–98, 415, 426
    Long-eared myotis 41, 47, 92–94, 99, 102, 415, 426
    Long-legged myotis 31, 33, 41, 47, 95, 104–106, 415, 426
    Northern myotis 47, 95, 99, 100
    Red 47, 82, 84–85, 86, 87, 88, 415, 426,
    Silver-haired 47, 81–83, 87, 415, 426
    Townsend's big-eared 47, 75–77, 414, 426
    Western small-footed myotis 33, 35, 41, 44, 47, 90–91, 415, 426
Bear
    Black 48, 107, 146–47, 150, 420, 433
    Grizzly 42, 48
Beaver, North American 49, 256, 416
Big brown bat 44, 47, 74, 78–80, 414
Bighorn sheep 8, 36, 40, 42, 43, 51, 395, 397–98, 399, 400, 401, 423
Bison 7, 8, 20, 21, 34, 42, 51, 131, 213, 361, 390, 391–396, 397, 423, 436
Bison bison 20, 391, 423, 436
Black-footed ferret 32, 42, 43, 48, 176–80, 213, 421, 434
Black-tailed prairie dog 32, 41, 43, 44, 49, 177, 208–214, 417, 429
Blarina brevicauda 47, 58, 413, 425
Bobcat 47, 108, 113–17, 225, 254, 341, 423, 435
Bovidae 51, 361, 391, 393, 395, 397, 399, 401, 423, 436
Bushy-tailed wood rat 40, 43, 50, 419, 431

C

Canada lynx 47, 108–12, 423, 435
Canidae 48, 107, 124, 125, 127, 129, 131, 133, 135, 137, 139, 141, 143, 145, 422, 433
Canis latrans 48, 124, 126, 422, 435
Canis lupus 48, 129, 422, 435
Caribou 405
Carnivora 6, 47, 107, 412, 420, 426, 433, 437
Castor canadensis 22, 49, 256, 453,
Castoridae 49, 207, 256–57, 259, 261, 416
Cervidae 51, 361, 362, 363, 365, 367, 369, 371, 373, 375, 377, 379, 381, 383, 385, 423, 436
Cervus elaphus 51, 368, 423, 436
Chaetodipus hispidus 50, 271, 418, 428

Chipmunk
  Eastern 45, 49, 215, 244, 247–49, 250, 417, 429
  Least 33, 45, 49, 244, 245, 248, 250, 417, 429
Chiroptera 47, 73, 412, 414, 424, 426
Condylura cristata 404
Corynorhinus (Plecotus) townsendii 47, 75, 414, 426
Cottontail
  Desert (Audubon's) 40, 44, 351–52, 355, 358, 416, 427
  Eastern 51, 352, 353, 354, 355, 356, 357, 358, 359, 416, 427
  Mountain 355, 358, 359, 416, 427
Coyote 31, 32, 44, 48, 124, 125, 126, 127, 128, 139, 140, 144, 349, 384, 389, 422, 435
Cricetidae 50, 205, 288, 289, 291, 293, 295, 297, 299, 301, 303, 305, 307, 309, 311, 311, 313, 315, 317, 319, 321, 323, 323, 327, 329, 419, 430
Cynomys ludovicianus 49, 208, 417, 429

D

Deer
  Mule 32, 36, 40, 51, 373, 374, 375, 376, 377, 378, 379, 381, 395, 423, 436, 442
  White-tailed 8, 31, 32, 33, 35, 40, 51, 366, 375, 380, 381, 382, 383, 384, 385, 395, 423, 436
Deer mouse 28, 44, 50, 316, 319, 320, 321, 323, 407, 420, 431
Desert (Audubon's) cottontail 40, 44, 351, 352, 355, 358, 416, 427
Didelphidae 52, 53, 55
Didelphimorphia 5, 47, 52, 412, 424, 441
Didelphis virginiana 47, 53, 412, 424, 456
Dipodidae 50, 207, 282, 283, 285, 287, 419, 430
Dipodomys ordii 50, 274, 418, 428

E

Eastern chipmunk 45, 49, 215, 244, 247, 248, 249, 250, 417, 429
Eastern cottontail 51, 352, 353, 354, 355, 356, 357, 358, 359, 416, 427
Eastern gray squirrel 49, 222
Eastern mole 403, 404
Eastern spotted skunk 28, 49, 198
Elk 7, 8, 23, 33, 34, 35, 36, 40, 42, 51, 131, 153, 361, 363, 368, 369, 370, 371, 372, 373, 378, 384, 395, 423, 436,
Eptesicus fuscus 47, 78, 414, 450
Erethizon dorsata 50, 338, 416, 428
Erethizontidae 50, 207, 338, 339, 341, 416, 428
Ermine 36, 48, 167, 168, 169, 170, 171, 172, 173, 174, 181, 184, 421, 434
Eulipotyphla 6, 47, 57, 412, 413, 425

F

Felidae 47, 107, 108, 109, 111, 113, 115, 117, 119, 121, 123, 422, 433, 439
Ferret, black-footed 32, 42, 43, 48, 176–80, 213, 421, 434,
Fisher 33, 42, 48, 160–67, 341, 342, 422, 434
Flickertail 49, 235
Fox
  Gray 45, 48, 133–36, 422, 434
  Red 31, 40, 42, 43, 48, 125, 133, 140, 141–45, 349, 422, 434
  Swift 32, 41, 48, 137, 139, 422, 435
Fox squirrel 36, 42, 49, 222, 223, 227–31, 418, 430
Franklin's ground squirrel 31, 49, 232–35, 418, 429
Fringed myotis 47, 101–104, 414, 426

G

Geomyidae 50, 207, 263, 265, 267, 269, 271, 273, 275, 277, 279, 281, 418, 428
Geomys bursarius 50, 263, 418, 428
Glaucomys sabrinus 49, 215, 416, 429
Glaucomys volans 404
Gray fox 45, 48, 133, 134–36, 422, 434
Gray wolf 33, 48, 125, 129–32, 142, 422, 435
Grizzly bear 42, 48
Gulo gulo 48, 151, 421, 433

H

Hare, snowshoe 33, 43, 44, 45, 51, 109, 110, 112, 115, 344–47, 348, 349, 350, 416, 427
Hayden's shrew 47, 63, 64, 65, 413, 425
Heteromyidae 50, 207, 271, 418, 428
Hispid pocket mouse 32, 50, 271–73, 418, 428

Hoary bat 47, 78, 82, 87, 88, 415, 426
House mouse 42, 50, 330, 333, 334, 335, 420, 430

I

Ictidomys tridecemlineatus 49, 240, 418, 429

J

Jackrabbit, white-tailed 40, 43, 44, 51, 345, 347, 348, 349, 350, 416, 427

K

Kangaroo rat, Ord's 40, 43, 44, 50, 274, 275, 276, 418, 428

L

Lagomorpha 6, 51, 343, 412, 416, 424, 427
Lasionycteris noctivagans 47, 73, 414, 415, 426
Lasiurus borealis 47, 84, 415, 426
Lasiurus cinereus 47, 87, 414, 415, 426
Least chipmunk 33, 45, 49, 244, 247, 248, 249, 250, 417, 429
Least weasel 36, 48, 107, 168, 181, 183, 184, 421, 434
Lemming, southern bog 404
Lemmiscus curtatus 50, 69, 419, 432
Leporidae 51, 343, 344, 345, 347, 349, 351, 353, 355, 357, 359
Lepus americanus 51, 110, 344, 416, 427
Lepus townsendii 51, 347, 416, 427
Little brown bat 74, 80, 91, 94–95, 97–99, 104
Little brown myotis 47, 95–98, 415, 426
Long-eared myotis 41, 47, 92–94, 99, 102, 415, 426
Long-legged myotis 31, 33, 41, 47, 95, 104–106, 415, 426
Long-tailed weasel 40, 48, 168, 172, 173, 175, 177, 184, 421, 434
Lontra canadensis 48, 155, 422, 433,
Lynx canadensis 47, 108, 423, 435
Lynx rufus 47, 113, 423, 435

M

Marmota monax 49, 219, 416, 428
Marten, American 48, 159, 160, 422,
Martes americana 48, 159, 422, 434
Martes pennanti 48, 422,
Masked shrew 47, 63, 64, 65, 413, 425
Meadow jumping mouse 50, 282, 284, 285, 286, 287, 419, 430
Meadow vole 28, 50, 292, 296, 298, 299, 300, 404, 405, 419, 432
Mephitidae 49, 107, 193, 420, 433
Mephitis mephitis 49, 193, 422, 433
Merriam's shrew 47, 68, 69, 413, 425
Microtus ochrogaster 50, 291, 419, 432
Microtus pennsylvanicus 50, 296, 419, 432
Mink, American 48, 185–89
Mole
  Eastern 403, 404
  Star-nosed 404
Moose 33, 35, 42, 43, 45, 51, 110, 131, 153, 360, 361, 362, 363, 365, 366, 367, 368, 384, 395, 423, 436
Mountain cottontail 35, 358, 359, 416, 427
Mountain lion 40, 47, 115, 118–123, 135, 372, 378, 405, 422, 435
Mouse
  Deer 28, 44, 50, 316, 319, 320–23, 407, 420, 431
  House 43, 50, 330–33, 334–35, 420, 430
  Meadow jumping 50, 282–84, 285, 286, 287, 419, 430
  Northern grasshopper 50, 312–15, 419, 431
  Plains harvest 50, 325, 327–29, 420, 430
  Western harvest 50, 324–27, 328, 329, 420, 431
  Western jumping 50, 282, 285–87, 419, 430
  White-footed 50, 312, 316–19, 321, 322, 330, 420, 431
Mule deer 32, 36, 40, 51, 373, 374–79, 381, 395, 423, 436, 442
Muridae 50, 207, 330–37, 419, 430
Muskrat 2, 19, 22, 31, 40, 50, 71, 110, 157, 173, 186, 187, 194, 207, 308–11, 419, 431, 441
Mus musculus 50, 330–33, 420, 430
Mustela erminea 48, 167–71, 421, 434
Mustela frenata 172–76, 421, 434
Mustela nigripes 48, 176–80, 421, 434
Mustela nivalis 48, 181-84, 421, 434
Mustela vison 48, 185–89, 422, 434
Mustelidae 48, 107, 151–201, 420, 433, 438
Myodes gapperi 50, 432
Myotis ciliolabrum 90–94, 415, 426

Myotis evotis 47, 92–94, 415
Myotis lucifugus 47, 95–98, 415, 426
Myotis septentrionalis 47, 99–101, 415, 426
Myotis thysanodes 47, 99–101, 415, 426
Myotis volans 47, 104–106, 415, 426

N

Neotamias minimus 49, 244–47, 417, 429
Neotoma cinerea 50, 305–307, 334, 419, 431
Neovison vison 48, 166, 177, 185–189, 422, 434
North American beaver 49, 256–62, 416
North American porcupine 50, 338–42, 416, 428
Northern flying squirrel 10, 49, 207, 215–18
Northern grasshopper mouse 50, 312–15, 419, 431
Northern myotis 47, 95, 99–101
Northern pocket gopher 50, 263, 264, 265, 267–70, 418, 428
Northern short-tailed shrew 47, 58–60, 413
Northern water shrew 47, 70–72, 413, 425
Norway rat 42, 50, 305, 334–37, 419, 430

O

Odocoileus hemionus 51, 374–79, 423, 436
Odocoileus virginianus 51, 380-85, 423, 436
Olive-backed pocket mouse 50, 271, 277–79, 280, 418, 428
Ondatra zibethicus 2, 22, 50, 308–12, 419, 431
Onychomys leucogaster 50, 312–15, 419, 431
Opossum, Virginia 53–55, 412, 424
Ord's kangaroo rat 40, 43, 44, 50, 274–76, 418, 428
Otter 28, 33, 35, 36, 48, 155–59, 422, 433
Ovis canadensis 51, 397–402, 423, 436

P

Perognathus fasciatus 50, 277–79, 418, 428, 456, 458
Perognathus flavescens 50, 279–81, 418, 428, 457
Peromyscus leucopus 50, 312, 316, 322, 420, 431, 440
Peromyscus maniculatus 50, 312, 320, 322, 420, 430, 431, 440, 442
Plains harvest mouse 50, 325, 327–29, 420, 431
Plains pocket gopher 50, 263–66, 268, 269, 271
Plains pocket mouse 28, 31, 50, 279–81, 418, 428
Pocket gopher
  Northern 50, 263, 264
  Plains 50, 263–66, 268, 269, 418, 428
Pocket mouse
  Hispid 32, 41, 44, 50, 271–73, 277, 418, 428
  Olive-backed 50, 271–73, 277–79, 280, 418, 428
  Plains 28, 31, 50, 271, 279–81, 418, 428
Poliocitellus franklinii 49, 232–35, 417, 429
Porcupine, North American 3, 50, 153, 164, 162, 207, 338–42, 416, 428
Prairie dog, black-tailed 32, 41, 43, 44, 49, 177, 208–14, 417, 429
Prairie vole 44, 50, 291–95, 296, 297, 298, 299, 419, 432, 470
Procyonidae 49, 107, 202
Procyon lotor 49, 202–206, 420, 433
Pronghorn 8, 32, 40, 44, 51, 360, 361, 386–90, 395, 423, 436
Puma concolor 47, 118–23, 422, 435
Pygmy shrew 28, 31, 33,33,5, 36, 47, 66–68, 413, 425

R

Raccoon 9, 36, 40, 49, 107, 202–206, 310, 420, 433, 441
Rangifer tarandus 405
Rattus norvegicus 50, 334–37, 419, 430
Red bat 47, 82, 84–86, 87, 88, 415, 426
Red fox 9, 31, 40, 42, 43, 48, 125, 126, 133, 134, 137, 141–45, 194, 196, 254, 349, 422, 434
Red squirrel 33, 42, 43, 44, 45, 49, 161, 174, 223, 228, 249, 251–55, 303, 417, 429
Reithrodontomys megalotis 50, 324–27, 420, 431
Reithrodontomys montanus 50, 327–29, 420, 431

Richardson's ground squirrel 28, 31, 43, 44, 49, 110, 120, 143, 153, 174, 186, 191, 208, 211, 212, 223, 235–39, 418, 429
River otter 28, 33, 35, 36, 48, 155–59, 422, 433
Rodentia 6, 49, 207, 412, 416, 424, 428

S

Sagebrush vole 32, 41, 45, 50, 69, 288–91, 419, 432
Scalopus aquaticus 403
Sciuridae 49, 207, 208, 416, 428
Sciurus carolinensis 49, 222–26, 418, 429
Sciurus niger 49, 227–31, 418, 430
Sheep, bighorn viii, 8, 36, 37, 40, 42, 43, 51, 361, 395, 397-402, 423, 436
Short-tailed weasel 48, 167–71, 421
Shrew
  Arctic 31, 33, 35, 36, 45, 47, 61–62, 413, 425
  Dwarf 403
  Hayden's 47, 61, 63–66, 413, 425
  Masked 47, 61, 63–66, 413, 425
  Merriam's 47, 68–69, 413, 425
  Northern short-tailed 47, 58–60, 413
  Northern water 47, 70–72, 413, 425
  Preble's 284, 403
  Pygmy 28, 31, 33, 35, 36, 66–68, 207, 413, 425
Silver-haired bat 47, 81–83, 87, 415, 426
Skunk
  Eastern spotted 28, 49, 198–201
  Striped 43, 49, 193–97, 199, 200, 422, 433
Snowshoe hare 32, 43, 44, 45, 51, 109, 110, 112, 161, 343, 344–47, 348, 349, 350, 412, 416, 424, 427
Sorex arcticus 47, 61–62, 413, 425
Sorex cinereus 47, 63, 413, 425
Sorex haydeni 47, 63, 413, 425
Sorex hoyi 47, 66, 413, 425
Sorex merriami 47, 68, 413, 425
Sorex nanus 403
Sorex preblei 403
Sorex palustris 47, 70–72, 413, 425
Soricidae 47, 57, 58
Soricomorpha 424
Southern bog lemming 404
Southern flying squirrel 215, 404
Southern red-backed vole 50, 301, 304, 419, 432
Spermophilus franklinii 43, 49, 232–35, 417, 429
Spermophilus richardsonii 46, 49, 235–39, 418, 429
Spermophilus tridecemlineatus 46, 49, 240–43, 418, 429
Spilogale putorius 49, 198, 422, 433
Squirrel
  Eastern gray 36, 45, 49, 222–26, 227, 227, 228, 229, 230, 251, 418, 429
  Fox 36, 42, 49, 223, 227–31, 252, 254, 418, 430
  Northern flying 49, 215–18, 404, 416, 429
  Red 33, 42, 43, 44, 45, 49, 161, 174, 223, 228, 249, 251–55, 303, 417, 429
  Southern flying 215, 404
Star-nosed mole 404
Striped skunk 43, 49, 193–97, 198, 199, 200, 422
Swift fox 32, 41, 48, 134, 137–40, 422, 435
Sylvilagus audubonii 51, 351–53, 416, 427
Sylvilagus floridanus 51, 354–57, 416, 427
Sylvilagus nuttallii 51, 357–59, 416, 427
Synaptomys cooperi 404

T

Tamiasciurus hudsonicus 49, 228, 251–55, 417, 429
Tamias striatus 49, 247–50, 417, 429
Taxidea taxus 49, 189–92, 421, 434
Thirteen-lined ground squirrel 44, 49, 240–243, 418, 429
Thomomys talpoides 50, 267–70, 418, 428
Townsend's big-eared bat 47, 75–77, 103, 414, 426

U

Urocyon cinereoargenteus 48, 133–36, 422, 434
Urocitellus richardsonii 49, 235–39, 418, 429
Ursidae 48, 107, 146
Ursus americanus 48, 146–50, 420, 433
Ursus arctos 48

V

Vespertilionidae 47, 75
Virginia opossum 5, 53–56

Vole
Meadow 28, 40, 296–300, 303, 404
Prairie 291–95
Sagebrush 288–91
Southern red-backed 301–304
Vulpes velox 48, 137–40, 422, 435
Vulpes vulpes 48, 141–45, 422, 434

W

Weasel
Least 107, 168, 176, 181–84
Long-tailed 168, 172–76, 177, 184
Short-tailed 167–71, 176
Western harvest mouse 50, 324–27, 328, 329, 420, 431
Western jumping mouse 3, 207, 282, 285–87, 419, 430
Western small-footed myotis 33, 35, 41, 44, 47, 90–92, 415, 426
White-footed mouse 312, 316–20, 321, 322, 330, 420, 431
White-tailed deer 8, 31, 32, 33, 35, 40, 51, 366, 375, 380–85, 395, 423, 436
White-tailed jackrabbit 8, 40, 43, 44, 45, 51, 345, 347–50, 416, 427
Wolf, gray 33, 42, 48, 125, 129–32, 142, 144, 422, 435
Wolverine 48, 151–54, 341, 421, 433
Woodchuck 49, 219–22, 416, 428
Wood rat, bushy tailed 40, 43, 50, 334, 419, 431

Z

Zapus hudsonius 50, 282–85, 419, 430
Zapus princeps 50, 285–87, 419, 430